CHARACTER LIMIT

Character Limit

How Elon Musk Destroyed Twitter

**Kate Conger
Ryan Mac**

Cornerstone Press

1 3 5 7 9 10 8 6 4 2

Cornerstone Press
20 Vauxhall Bridge Road
London SW1V 2SA

Cornerstone Press is part of the Penguin Random House group of companies whose addresses can be found at global.penguinrandomhouse.com.

Copyright © Kate Conger and Ryan Mac 2024

Image credits appear on page 453

While the authors have made every effort to provide accurate internet addresses at the time of publication, neither the publisher nor the authors assume any responsibility for errors or for changes that occur after publication. Further, the publisher does not have any control over and does not assume any responsibility for author or third-party websites or their content.

Kate Conger and Ryan Mac have asserted their right to be identified as the authors of this Work in accordance with the Copyright, Designs and Patents Act 1988.

First published in the US by Penguin Press, an imprint of Penguin Random House LLC, New York, in 2024
First published in the UK by Cornerstone Press in 2024

www.penguin.co.uk

A CIP catalogue record for this book is available from the British Library.

ISBN 9781529914696 (hardback)
ISBN 9781529914702 (trade paperback)

Designed by Amanda Dewey and Christina Nguyen

Printed and bound in Great Britain by Clays Ltd, Elcograf S.p.A.

The authorised representative in the EEA is Penguin Random House Ireland, Morrison Chambers, 32 Nassau Street, Dublin D02 YH68

www.greenpenguin.co.uk

Penguin Random House is committed to a sustainable future for our business, our readers and our planet. This book is made from Forest Stewardship Council® certified paper.

for Kim Mac and Marie Aro

The Golden Sir
@screaminbutcalm

Me sowing: Haha fuck yeah!!! Yes!!
Me reaping: Well this fucking sucks. What the fuck.

2:14 PM Mar 12, 2019

220K　48K　139K

CONTENTS

Introduction 1

Act I >

1 \ back at twttr   11
2 \ #StayWoke   18
3 \ "This is actually me"   23
4 \ OneTeam   40
5 \ An Invasion   49
6 \ A Polynesian Vacation   59
7 \ Resource Plan   76
8 \ Parag   80
9 \ Bluesky   89
10 \ Twitter in Trouble   94
11 \ Musk's Shopping Spree   99
12 \ An Offer   113
13 \ Poison Pill   122

Act II >>

14 \ "Bring the cattle"   133
15 \ Parag's Last Stand   146
16 \ Just Say Yes   155
17 \ Golden, Golden   168
18 \ 💩   184
19 \ Bots and Horses   193
20 \ Sun Valley   203
21 \ Court of Chancery   215
22 \ "Sorry I didn't get your special"   219
23 \ Mudge   223
24 \ An Accelerant to X   228
25 \ "Not owned by a fucking moron"   231
26 \ Let That Sink In   238
27 \ Trick or Tweet   247
28 \ "The bird is freed"   260

Act III >>>

29 \ Code Reviews   271
30 \ Lords and Peasants   283
31 \ "Educate me"   289
32 \ A Blue Heart   293
33 \ The Snap   301
34 \ The Aftermath   307
35 \ Verified or Not   310
36 \ Elections   320
37 \ Zombie Attack   328
38 \ Fired for Shitposting   343
39 \ Zero-Based Budgeting   357
40 \ "I'm Rich, Bitch!"   372
41 \ Self-Doubt   380
42 \ Red Pilled   389
43 \ Endorsements   400
44 \ Linda   410
45 \ Tell It to Earth   423

Epilogue   431

Acknowledgments   437
Note on Reporting   441
Notes   443
Image Credits   453
Index   455

CHARACTER LIMIT

Introduction ❯ *November 11, 2022*

After waiting for nearly four hours, the senior data scientist was growing restless. He hadn't planned to be in the Twitter office. It was Veterans Day and most of his colleagues were logging out, but he was loitering outside a tenth-floor conference room at the company's San Francisco headquarters, waiting to be summoned by Twitter's new owner.

Elon Musk was in his element. The fifty-one-year-old billionaire thrived on tests of his endurance, thrilled by his own strength when he slept on a conference room couch at Tesla's factory, or stayed up all night putting together the final flourishes for a rocket launch at SpaceX. Now he was pitting himself against Twitter. In the aftermath of his $44 billion takeover, he was straining the bounds of the social network. How quickly could he bend it to his will?

The data scientist, a lanky man with tousled reddish-brown hair and cutting blue eyes, had started working at Twitter only a year earlier. His colleagues had quickly come to see him as a deep thinker who was fascinated by the good—and bad—that came from connecting communities of humans online. He had honed his talent for distilling the vast landscape of social media into digestible sound bites during his five years at Facebook, where he had delved into reams of user data to explore thorny topics like hate speech and misinformation that contributed to people rioting at the U.S. Capitol on January 6, 2021. An academic at heart, the data scientist often critiqued his employer's faults with a level of candor that executives rarely heard from other employees.

When Musk had offered to buy Twitter that April, the data scientist was optimistic. Musk was a man who had revolutionized two industries, mainstreaming electric cars and privatizing space exploration. Perhaps he was the visionary who could give the social media company a much-needed shot in the arm.

But over the past two weeks, Musk had fired half of the data scientist's colleagues with no plan and little explanation of his vision. He had alienated advertisers, undermining the foundation of Twitter's business. And he had fallen for a blatant conspiracy theory, tweeting out a fake story about the husband of Nancy Pelosi, the Speaker of the U.S. House of Representatives, suggesting he was enmeshed in a tryst with a disturbed man who had attacked him in his home. It was the kind of absurd fiction that only someone with a warped mind—radicalized by hours spent online every day in their own filter bubble—would believe. The data scientist was horrified. Musk, apparently, was one of the easily misled conspiracists he had studied in his work.

Despite all the changes Musk had already made to Twitter, the billionaire had signaled he wouldn't be resting that Friday. Early that morning, he sent out an email to his employees extending "a note of appreciation to those who were there with me."

"I will be in the office again today," Musk wrote. "Stop by the 10th floor if you'd like to talk about taking Twitter to the next level. The priority is near-term actions."

The data scientist decided to take the plunge. He trekked through the fog and gloom of the San Francisco winter, making his way up Market Street to Twitter's looming Art Deco headquarters. Just after 10:00 a.m., he set up camp at a bank of desks outside Caracara, a conference room with sweeping views of downtown and the gleaming dome of city hall. Its glass walls allowed people passing by to gaze in at executives as though they were lions at the zoo. Musk was the main attraction, and the workers waiting outside to meet with him whispered quietly about what they hoped to tell their new boss. As the data scientist clacked away on his laptop, working on a pair of memos he planned to share with Musk, he eavesdropped. Some employees murmured concerns about how few people had signed up for Twitter's new subscription product. Others were trading tips on how to best communicate with their new leader.

Behnam Rezaei, a warm man who wore round, tortoiseshell glasses and had led engineering teams at Twitter for more than five years, approached the data scientist to offer advice. Rezaei had climbed the ladder into Musk's good graces, avoiding the firings that took out his fellow managers and rising to a vice president role. Rezaei thought highly of the data scientist and had cashed in some of his newfound political capital to get him face time with Musk.

"Elon only wants to hear positive things," Rezaei instructed him. "Don't tell him about what we can't do or try to justify the status quo."

"Elon just wants to do what benefits humanity."

Rezaei wasn't aware, however, that the data scientist had already decided to quit. When he saw Musk's email early Friday morning, he delayed his departure by one day so that he could speak directly to the new owner. He still believed in Twitter and the power of large social networks, and he hoped Musk would listen. Maybe Musk's cohort of yes-men, who had arrived at the company following his takeover, hadn't dared to tell him how badly he was screwing up.

As he waited, the data scientist finalized two documents he had prepared for his encounter. The first was a list of ideas for running Twitter more effectively. The second, a bolder statement, outlined why Musk's plans to earn significant revenue from subscriptions and alter content-moderation policies wouldn't work, and how his paranoia and instability were damaging the company.

As the hours ticked by, he scrounged for whatever snacks remained in one of the nearby kitchens. His heart beat heavily as he rehearsed in his head what he would say to Musk. Finally, just after 2:00 p.m., Musk's assistant approached him. Musk was busy, she said. He would have only five minutes.

The data scientist strode into the conference room. Musk sat on one side of a large oak table, his sagging, six-foot-two frame scrunched into a Herman Miller office chair. The data scientist quickly introduced himself before launching into his presentation. Musk listened intently as he explained his ideas for growth, verification of users, and motivating employees. He then sketched out a vision for content moderation that placed decision-making power in the hands of an organization outside of its owner's direct control.

"Newspapers and magazines have editorial independence, meaning owners don't get to make final judgments about what stays and what goes," the data scientist explained. "Social media companies should have the same structure."

Musk wasn't impressed. "Or not," he muttered.

Musk's assistant peeked back into the room and said he had another meeting. "Do you have any final thoughts?" she asked.

"Yes, I want to say one thing," the data scientist said. He took a deep breath and turned to Musk.

"I'm resigning today. I was feeling excited about the takeover, but I was really disappointed by your Paul Pelosi tweet. It's really such obvious partisan misinformation and it makes me worry about you and what kind of friends you're getting information from. It's only really like the tenth percentile of the adult population who'd be gullible enough to fall for this."

The color drained from Musk's already pale face. He leaned forward in his chair. No one spoke to him like this. And no one, least of all someone who worked for him, would dare to question his intellect or his tweets. His darting eyes focused for a second directly on the data scientist.

"Fuck you!" Musk growled.

The data scientist grew bolder. He was not prone to conflict or insults, but Musk's reaction reinforced his belief that the billionaire was nowhere near fit to run a company crucial to the world's online discourse. He remained collected, but uttered something he had not planned to say.

"I hope you'll declare bankruptcy and let someone else run the company."

"Well, resignation accepted," Musk snapped.

The data scientist made his way for the exit.

"I'll take your laptop," Musk's assistant said meekly. He handed the device to her and left.

As the data scientist walked back to the desk where he had left his belongings, he could hear the patter of two of Musk's security guards jogging to catch up with him. He wondered if they would try to hassle him or perhaps even rough him up, but they simply watched over him as he packed his things before escorting him to the elevator bank. As they all got into the elevator car and rode down to the first floor, one of the guards turned to smile at him.

"What did you say to him?" he asked.

"I told him some things he didn't like," the data scientist responded.

"It must have felt good."

"Yep," he said, before stepping out of the elevator shaft, handing over his ID badge, and leaving Twitter's headquarters for the last time. "To be honest, what I said is what everyone is saying behind his back. But nobody's saying it to his face."

>>> ELON MUSK ARRIVED AT TWITTER a conquering hero in the eyes of many—not least of all his own. Surrounded by a cadre of loyalists who enabled him to make many reckless maneuvers—like buying the company in the first place—he was cheered on by millions of online supporters who liked and retweeted his every move. He dragged friendly investors along for the ride as he seized control of one of the world's preeminent online spaces for political and cultural discourse and attempted to transform it to suit his whims. With his successful and eye-wateringly expensive takeover, Musk had seemingly surpassed all other tech executives in wealth, influence, and fame. He was untouchable.

But incidents like the data scientist's confrontational exit scarred him. Whether he fully realized it at the time or not, Musk had gambled his reputation and billions of dollars on the haphazard acquisition of his favorite toy.

His takeover had not been welcomed by the company's leaders nor many of its users, but Twitter had placed itself on the auction block through years of mismanagement. Its embittered founder, Jack Dorsey, had neglected the service in the twilight of his tenure as chief executive. Over time, Dorsey came to believe the company he once prized shouldn't be a business at all— but as he ignored Twitter's profit margins, investors swooped in to squeeze it for cash. When he stepped down in 2021, the company embarked on a hasty cleanup job to assuage Wall Street.

But no one was prepared for Musk's hyper-aggressive campaign, and no one could stop him. He saw Twitter not only as a business but as an ideological tool, a weapon, that was being wielded by San Francisco liberals who suppressed views he enjoyed. Policies at Twitter set the tone for other social media companies as they debated sensible online speech, and Musk wanted to drive a new set of mores into the conversation.

That he had enough money at his disposal was truly extraordinary, an aberrant feature of twenty-first-century capitalism. By the start of April 2022, Musk had a net worth nearing $270 billion. With the main source of his wealth, Tesla's shares, reaching new heights and giving him unfathomable purchasing power, he set his sights on his one true passion. While most tech billionaires might have spent the money on mega yachts, sports teams, media publications, or faraway islands, Musk coveted a megaphone, a website where his voice could be broadcast directly to hundreds of millions of people. He wanted Twitter.

Musk's blitzkrieg takeover had no cultural or social precedent. This simply wasn't the type of transaction a single human pulled off. Corporations or private equity firms were supposed to buy companies that size, not individuals. But Musk had reached a pinnacle of wealth that only a handful of titans had ever come close to, and the rules of traditional business no longer applied.

Musk's love for Twitter was simple, relatable, and even humanizing. He had spent hours every day scrolling the site—reading posts, laughing at memes, and firing off stream-of-consciousness thoughts—like any regular user. He became intoxicated by the engagement offered to him, and like with so many other hardcore tweeters, the platform became his addiction. The difference between him and the other diehards chasing the constant dopamine rush of Twitter, however, was that he had the means to capture his addiction and the desire to remake it in his own image.

>>> ON THE MORNING OF APRIL 14, 2022, we—Kate Conger and Ryan Mac, reporters at *The New York Times*—woke up to a four-word tweet, the unbelievable but inevitable culmination of two storylines we had pursued for a decade as journalists in Silicon Valley: "I made an offer."

We threw ourselves into reporting a monumental story. Silicon Valley's most consequential entrepreneur was buying one of its most iconic companies. How would this end?

We had written extensively about social media's content moderation wars, Twitter's shortcomings as a business, and Dorsey. We had also covered Musk's enterprises and his constant boundary pushing. Suddenly, those stories became one—and the whole was much, much greater than the sum of its parts.

Musk's decision to buy Twitter seemed to have been a snap judgment. He had assumed Twitter was a knot of technical issues that a great engineering mind like himself could easily untangle, enabling the growth of free speech in the digital town square. But at its core, Twitter was plagued by social and political dilemmas, not merely technological ones. Its leaders constantly grappled with questions about what people should be allowed to say, and they made enemies of governments, activists, celebrities, and even their own employees. The questions confronting Twitter were not simple. They have been debated across the internet since its advent. They may not have static answers at all. It is not for nothing that Twitter's most loyal users referred to it as a "hellsite," a corner of the internet where something—or someone—was always burning. People walked away from a session scrolling through their timelines feeling angry, frustrated, disgusted—and yet they couldn't wait to log back on. The company needed a leader who deeply understood psychology, politics, and history, and the messy ways people connect instantaneously and constantly online. Instead, it got someone whose offer for the company—$54.20 a share—included a weed joke.

Musk found his quixotic ambitions for the platform thwarted as he attempted to govern Twitter, and he became increasingly convinced that employees were rebelling against him. They should be thanking him, he thought. The way he saw it, he alone was brave enough to gamble $44 billion on saving the social media platform he loved. Didn't they realize he was saving humanity? He tasked some of his lackeys with ferreting out other dissenting voices so he could fire them. He instituted code freezes that prevented anyone from making changes to Twitter's apps or website, in case an employee sabotaged them. His personal bodyguards began following him to the office restrooms, ensuring that few employees could encroach upon him during his sacred time.

As his takeover unfolded, Musk's paranoia intensified while those close to him worried about his increasingly fragile state of mind. The chaotic nature of the platform and the repercussions of his actions would expose his limits. The more he tried to impose his will on Twitter, the further it seemed to spin out of his control, and the deeper his obsession became. Holes began to show in the capabilities of an entrepreneur considered by many to be one of mankind's most successful business leaders.

—

>>> AT THE TIME OF WRITING, the story of Musk's conquest is not over. It may yet end with a bang or a whimper—or an improbable success. But what is clear already is that Musk has destroyed the platform. What he owns is no longer Twitter—not in name, but also neither in substance nor in spirit. Gone are the people who built it idealistically, at a time when Silicon Valley's utopian promises seemed much easier to believe, and gone is its company culture of debate, equality, and idealism. What that means for a world in which the news media is in a state of constant existential peril, and democracy itself is in danger, remains to be seen. But the early signs are not good.

Out of Twitter's rubble, Musk is building X, a harsher and much more cynical social media company. With it, he is ushering in a new era of anything-goes online speech that is governed by his own whims. There have been plenty of promises. Musk has said that X can be the world-dominating "everything app," where people will not only be able to post their thoughts but also pay for goods, make phone calls, or watch movies. X will have it all, he asserts, and deep-pocketed investors have bet billions of dollars that he will be right.

But those lofty promises are so far just that. To those users who have hung around, one of the most important modes of global communication has become practically unrecognizable and now serves the interests of one man. What was once called the digital town square is becoming Musk's mirror.

ACT I >

1 > back at twttr

Jack Dorsey's time had finally come. Standing in the cafeteria of Twitter's headquarters on San Francisco's Market Street on June 11, 2015, he watched eagerly as Dick Costolo, a former entrepreneur who had been hired four years earlier to lead Twitter out of turmoil, delivered an unexpected message to hundreds of Twitter's workers during their regular all-hands meeting. Costolo was resigning as CEO. He had been quietly confiding in friends since the start of the year that he had grown anguished with his role at Twitter and was ready to leave. The person to replace him, at least on an interim basis, was the man standing next to him who had helped start it all: @Jack.

Dorsey had strategized for years to complete his Steve Jobs character arc. He was unceremoniously fired as Twitter's chief executive in 2008 and had plotted a route back to the top job ever since, allying with board members and engineering a narrative in the media that framed him as the sole visionary behind the social media service. His comeback would be triumphant—the driving creative force behind Twitter's previous successes, back at last—just as it had been when Jobs returned to Apple.

To Twitter employees, Dorsey was a man returning from a winter in the wilderness. He stood in front of them with a new executive title and a thick brown beard that descended past the collar of his shirt and framed his lean, angular face and icy eyes. Although he had served as the chairman of the company for years and was renowned as a cofounder, few employees had ever seen him in person. They sized up the former model and meditation enthusiast before them and couldn't help but view him with trepidation as he spoke in his distinctive monotone about changing the company.

Twitter needed that change fast, because it was in the dumps. There was little product innovation, no user growth—the key metric by which any meaningful social internet company is judged on—and a snowballing sense that the company couldn't succeed. The history of dysfunction and backstabbing among the company's founders, and the game of musical chairs in leadership, contributed to the feeling that Twitter was going nowhere. Dorsey had instigated some of the early missteps, and chaos was becoming an accepted part of Twitter's culture. With good reason, the workers wondered: *Was Jack the solution? Could he turn this company around? Or would he flame out once again?*

—

>>> BORN IN 1976, Dorsey grew up in St. Louis, raised by a liberal mother and a conservative father, alongside two younger brothers. As a teenager, he became interested in the inner workings of dispatch services—an early sign that he would gravitate to building complex systems for transporting information. He attended college in Missouri and New York, but dropped out before finishing his degree. Dorsey moved to the Bay Area in 1999, on the eve of the dot-com bust.

It was a heady time to be in tech. The visionaries and pioneers of the internet and personal computer had governed themselves with an open, collaborative ethos that relied on loose anarchic consensus combined with unfettered technical expertise. That democratic culture appealed to Dorsey, a punk aficionado who once had blue hair. Upon arriving in the Bay Area, he moved into the Sunshine Biscuit Factory, a warehouse buried deep in the gritty eastern side of Oakland known for housing artists and hosting underground music shows. He tinkered with online programs to dispatch taxis, bike messengers, or even first responders.

Unlike a lot of the scraggly coders who flocked to Silicon Valley in pursuit of million-dollar paychecks, Dorsey was interested in aesthetics—his own and that of the products around him. He toyed with the idea of abandoning tech altogether in favor of becoming a fashion designer. He liked to change his appearance, piercing his nose or using castile soap to turn his hair into dreadlocks. His shape-shifting presentation and interests would stick throughout his life, leaving some of the people around him wondering if he was just looking for a place to fit in.

"I actually got the nose ring because there was just an impulse thing—I thought that might look really cool," Dorsey later said on *60 Minutes*. "There was no real statement that I was trying to make."

Having taken a few freelance coding gigs around the Bay Area, including one building a dispatch service for the ferries to San Francisco's Alcatraz Island, Dorsey came across Odeo, a podcasting start-up that web entrepreneur Ev Williams was building in San Francisco, in 2005. Williams had made a fortune two years earlier by selling his publishing platform Blogger to Google, and Odeo was his next act. With Blogger, Williams had taken pride in delivering publishing to the masses, allowing anyone to post their own content online with a single click. He eschewed content moderation, viewing it as an impossible task, and let most posts remain on his platform.

The twenty-eight-year-old Dorsey sent Williams his résumé and was offered a freelance coding gig at Odeo, where he quickly fell in with the other cyberpunks on staff. But even among that oddball crew, Dorsey stood out. He was exceptionally quiet, preferring online chats to talking in person. He faded into the background when the group worked on projects or went out for drinks. And even though he was working for the renowned Blogger founder, Dorsey did his own journaling on a competing blogging platform, LiveJournal.

For the early days of the social web, Dorsey was a prolific poster. And while his personality came out more in his LiveJournal posts, he felt like he needed something more. On both platforms, a user needed so much intention to post—drafting sentences and paragraphs of a blog or uploading and editing images from their digital cameras—before publishing. There needed to be something quicker, more stream-of-consciousness, where someone could post and share effortlessly and instantly without much thought.

"Real-time, up-to-date, from the road," Dorsey said. His vision would mimic status updates on AOL's instant messaging service, where users posted notes about what they were up to, what they were thinking about, or cryptic song lyrics that revealed their mood.

In July 2000, he had sketched the idea in a legal pad with a blue ballpoint, calling it My.Stat.Us, surrounding the product name with curlicued doodles. In the sketch, Dorsey's status was "reading," but other options included "in bed" and "going to park." At the time, Dorsey frequented South Park in San Francisco, a small oval of green space in the city's South of Market district, nestled among tech offices and apartment buildings.

The idea stayed rooted in the back of his mind as Odeo sputtered along. The start-up was struggling to get users, and when Apple added podcasts to iTunes in 2005, Odeo was dead on arrival. Dorsey saw an opportunity and started pitching his status update concept to Williams and other leaders at Odeo. One of them, Noah Glass, thought the pinging of a status update felt like a twitch. He thumbed through *tw-* words in the dictionary until he landed on *twitter*, the excited chirp of a bird. *Twitter* had a breathless, intriguing air about it. The Odeo leaders shortened it to *Twttr*, in keeping with the trend of vowel-less start-up names that launched in the early 2000s and making it compatible with text message short codes so that its users could send status updates from their cell phones. (The word *tweet* would come in 2007 from enthusiastic third-party developers who were looking to coin a verb for what they were actually doing when they posted.)

In March 2006, an early version of the service was ready to go. "just setting up my twttr," Dorsey wrote. It was the world's first official tweet.

The service had plenty of skeptics. But Dorsey led by example, posting short missives about his travels, the champagne he drank, and the meals he ate. His quiet manner inspired loyalty, and he seemed to listen to and trust his coworkers, rather than seeking to overrule them or order them around.

"I'm happy this idea has taken root; I hope it thrives," Dorsey wrote later, reflecting on Twitter's beginnings sketched in his notebook. "Some things are worth the wait."

Twitter did thrive. In keeping with his terse way of speaking, Twitter users were limited to 140 characters per tweet—a format that also made it possible to send tweets via text message, a necessity in the pre-smartphone era. Dorsey lost the nose ring and was named the company's chief executive, while Williams, who was its largest shareholder after supplying much of the early financing, served as chairman. Williams retained a 70 percent ownership stake and gave 20 percent to Dorsey. They eventually shuttered

Odeo and began focusing on Twitter full time, which exploded after being named best start-up at the 2007 South by Southwest conference in Austin, Texas.

It grew so fast that it often overworked the infrastructure Dorsey, Williams, and a small team of their former Odeo coworkers had pulled together with the digital equivalent of duct tape and prayers. Outages became a regular occurrence, and during downtime, Twitter users were greeted with an illustration of a whale being airlifted by a flock of birds known as the "fail whale." But most of the time, when it was working as intended, the site was governed by a simple principle: the tweets must flow.

Some of them were outright pornographic. Some were threats against other users. Twitter stuck with the philosophy that had worked at Williams's Blogger—there wasn't time for content moderation, and even if there was, no one on the team had the patience to sift through questionable tweets.

Dorsey was supportive of this laissez-faire approach but ducked the obligation of conveying Twitter's stance to the public. He left that job to one of his cofounders, Biz Stone, and other early employees who had worked together at Blogger and believed they could replicate its approach. Besides, Dorsey was overwhelmed with the onslaught of responsibilities he faced as a first-time chief executive, floundering as he struggled to manage employees, balance budgets, and oversee Twitter's brittle infrastructure. He was more than willing to let others tackle the thorny issues of content moderation, as he preferred to work on Twitter's interfaces and technology. He liked convincing people that his pet project could transform their conversations—and, eventually, their lives.

By 2008, the site's constant outages and mounting expenses were unsustainable. Twitter had more than a million users, but it often crashed when people tried to sign up or publish tweets. The problems needed to be fixed—in fact, fixing them was long overdue—and Dorsey wasn't moving fast enough. In a coup that October, Williams and the Twitter board, which consisted of two venture capital investors, fired Dorsey. He was given a consolation seat on the board but none of the voting rights that normally would have come with it. Williams took over as CEO.

With Dorsey on the sidelines, the site he had helped create continued to explode in popularity. In 2009, Iranians flocked to Twitter to protest their country's presidential election, cementing the company's reputation as an online haven for free speech. Twitter became the fastest-growing website that year, skyrocketing from 1.2 million visitors in May 2008 to 18.2 million in May 2009.

Williams, Stone, and Twitter's other leaders took a permissive approach, and the service became fundamental in the Arab Spring as protesters across the Middle East used Twitter and Facebook to protest their governments and organize themselves politically, toppling dictatorships across the region.

The company occasionally took down illegal content, like child exploitation material. But in most cases, Twitter stuck to its maximalist approach to speech. Executives nicknamed their start-up "the free speech wing of the free speech party," a clear and unapologetic middle finger to anyone who criticized the company for refusing to take down tweets.

On the fringes of the company, Dorsey began plotting a comeback. When he wasn't staring frostily at Williams in board meetings, he tinkered with solutions to a problem that small merchants faced when trying to accept credit card payments. Dorsey began work on a digital payments processor, tapping some of his old interests in relaying and dispatching information, and developed an elegant credit card reader, reminiscent of Apple products in its design, that could be plugged into an iPhone's headphone jack. He gave the project a simple name: Square.

After its founding in 2009, Square was quickly adopted by small businesses. But despite his success at Square, Dorsey couldn't take his eye off Twitter. He still was bitter toward Williams and dreamed of going back. While he was able to apply his artistry to Square, it didn't hold the cultural sway of Twitter. Its logo wasn't splashed across the chyrons of national news outlets. Presidential candidates, world-famous actors, and Dorsey's favorite musicians didn't rely on Square. And even though creating Twitter had been a group effort, it was *his* idea, *his* notebook sketch, *his* vision.

The first step to return to the throne was to get rid of Williams. Dorsey waged a whisper campaign, telling board members and senior Twitter employees that Williams wasn't up to the task of running the company. Indeed, Williams was struggling. "We were just hanging on by our fingernails to a rocket ship," he later said.

By 2010, Dorsey's story had taken hold. The board removed Williams as chief executive and replaced him with Dick Costolo. It was payback from Dorsey, who had been ousted by Williams two years earlier. Dorsey eventually succeeded Williams as the board's executive chairman, giving him back some control over the company, while Williams was shunted into a new role overseeing product.

From there, it was just a matter of Dorsey getting himself from the boardroom to the corner office. In 2013, the company went public, valued at more than $18 billion. By then, Dorsey's stake had been diluted to less than 5 percent as the company had taken on additional investors over the years. Williams's ownership had been whittled down to 12 percent.

Twitter had yet to earn a profit. The company boasted 218 million monthly active users and, in the six months leading up to its 2013 public offering, it had run up a loss of almost $70 million.

But Costolo had figured out how to weave some ads into the timeline and analysts thought Twitter showed promise in taking on Facebook. Dorsey, clean-shaven in a crisp white shirt and black suit jacket, beamed as Twitter celebrities including the actor Sir Patrick Stewart rang the bell at the New York Stock Exchange, launching his company onto the public market.

2 > #StayWoke

On the muggy afternoon of August 9, 2014, in Ferguson, Missouri, two teenagers strolled down a winding road that ran through the center of a low-slung Midwest development. Tan apartments with wooden balconies stood on either side, offering little shade. A patrol car rolled up alongside them, and the officer inside told the boys to get out of the road and walk on the sidewalk instead.

Within minutes, one of the boys, Michael Brown, Jr., was dead, shot in the middle of the street by the police officer. Videos and photos of Brown's body began circulating on Twitter. By the next day, protesters were streaming onto the streets in Ferguson just miles from where Dorsey grew up in St. Louis.

A week later, on August 16, 2014, an aggrieved ex-boyfriend posted a lengthy online screed about a video game developer named Zoë Quinn. He falsely accused her of sleeping with a journalist in exchange for positive reviews of a video game she had created. Even though the accusation came from a jilted ex, it spread rapidly through online communities of gamers, who seized on it as an excuse to threaten Quinn with rape or death. When other women stood up for her, they, too, became fair game.

Each event wrought major change and became the subject of significant political debate, and each came with its own hashtag: #Ferguson and #Gamergate.

Dorsey closely followed the protests near his hometown, which began the day Brown was killed and continued throughout the sweltering days of August 2014. Within days of Brown's death, Dorsey was back in Missouri, charging up his cell phone before joining the demonstrators and live-tweeting what he saw. He ditched his business casual attire, dressing in a simple white T-shirt and a St. Louis Cardinals cap, and handed out blooming red roses to demonstrators as they marched down Florissant Road, a thoroughfare within walking distance of where Brown had been shot.

He chafed at the unfairness of the police and the media. "Heartbreaking: witnessed the St. Louis County police run up to a woman, throw her to the ground, & arrest her. For standing still," Dorsey tweeted the night of August 19.

Ferguson became a foundation for the larger Black Lives Matter movement that grew in the following years. But it was also pivotal for Twitter, and for Dorsey himself. His live tweets of the protests spotlighted the demonstrators and emerging citizen journalists who were also tweeting from the ground, and news outlets flocked to write about the white billionaire tech founder who had come back to his home state to engage with issues of Black racial justice.

When he returned to San Francisco, Dorsey kept his protest apparel, wearing casual T-shirts, hoodies, and jeans, and growing out his beard. He stayed in touch with some prominent tweeters from the protests, inviting them to visit the company's headquarters in late 2014. He spoke up about racial justice issues on his personal Twitter account, and had company merch made branded with the hashtag #StayWoke and the Twitter bird logo.

Almost simultaneously, the Gamergate campaign spread like brushfire. If Ferguson highlighted the power of placing the tools of communication directly in the hands of participants, Gamergate illustrated how that power could be abused. Thousands of Twitter users—often using anonymous accounts—launched frenzied attacks against prominent women, posting their private information to contribute to further abuse and death threats. Within a month of the blog post about Quinn, #Gamergate had been shared on Twitter more than a million times.

The protests in Ferguson captured Dorsey's attention. The complaints of men's rights activists fell to Twitter's general counsel—a former corporate deals lawyer named Vijaya Gadde.

Gadde's life in corporate law molded her into a tough, risk-averse thinker with a muted, court-appropriate wardrobe and a polished coif of perfectly wavy hair. She stood out from Twitter's hoodied rank and file, and she didn't think that "the tweets must flow" was a practical policy for online conversation. If everyone on the service screamed at each other, some people would throw up their hands and leave the conversation altogether—and in all likelihood, the people who would leave would be the most vulnerable.

Gadde found a kindred spirit in Del Harvey, a child-safety expert who had been Twitter's twenty-fifth hire and knew the platform's dark side firsthand. The company had recruited Harvey, a woman with a slight stature and malleable voice that had served her well in her previous job: impersonating teens in chat rooms to reel in men for an organization called the Perverted Justice Foundation, which rose to prominence through a partnership with *To Catch a Predator*. At Twitter, Harvey had become the de facto garbage collector, policing the sexual exploitation of children on the platform and battling spammers. She also studied harassment and found that vitriolic threats from a handful of accounts were enough to drive users off the platform, even if their overall experiences were positive—powerful data that she used to convince Twitter executives that good speech wouldn't naturally drown out the bad.

While Harvey could be strident, Gadde was a lawyerly voice of reason. The pair balanced each other and together they crafted a new approach to make sure the loudest voices didn't dominate the conversation on Twitter. They hired dozens of content moderators, added new tools for blocking users and muting conversations, and began building tools that would help them detect abuse before it trended on the platform.

"Freedom of expression means little as our underlying philosophy if we continue to allow voices to be silenced because they are afraid to speak up," Gadde wrote in a 2015 op-ed for *The Washington Post*.

But Gadde would be careful to preserve Twitter's passive stance, she promised. "It is not our role to be any sort of arbiter of global speech. However, we will take a more active role in ensuring that differences of opinion do not cross the line into harassment," she said. Costolo would also acknowledge the company's shifting views on speech. "We suck at dealing with abuse and trolls on the platform and we've sucked at it for years," he wrote in an internal email to Twitter employees following Gamergate.

Ultimately, Costolo would be overwhelmed by the dilemma. At the end of 2014, the social network had stagnated at about 300 million monthly active users and, despite crossing $1 billion in revenue for the first time, lost about $578 million. (For comparison, Facebook in that same period had 1.39 billion monthly active users, generating $12.5 billion in revenue for a $2.94 billion profit.) Stuck in amber with its 140-character limit, Twitter also failed to develop exciting new features. Its early acquisition of Periscope, a live-streaming start-up that hadn't even launched, in 2015, failed to excite. By that summer, he was out, stressed by the hatred and coordinated abuse on Twitter and worn down by its negative impact on user growth.

In July, as Dorsey officially took over for Costolo, uncertainty reigned. Dorsey lost the facial hair—perhaps after pressure from his own mom, who had once tweeted: "Not a fan of the beard. @jack has a great face. Like to see it."—and switched to a uniform that included a gray #StayWoke T-shirt. He was a stabilizing force. After Dorsey pushed a small round of layoffs that October, Twitter began slowly trudging toward its first full year of profit. Still, it finished 2015 with a loss of $521 million, as its valuation fluttered around the $15 billion mark. Dorsey had talked about making Twitter universal, and as easy to use as "looking out your window," but, in reality, it was an addictive app for a subset of influential people and organizations, and hard to use and intimidating for the average person.

Issues with toxic content and misinformation continued. The company had never truly known how to harness its influence over politics nor the ways its platform could be manipulated. Russian intelligence agents set up sock puppet accounts that tweeted divisively about hot-button political issues, including Black Lives Matter, during the 2016 U.S. presidential election. The platform had also been essential to Donald Trump's political career—he leveraged his bombastic Twitter personality to secure constant media attention and outrage, rising from reality TV star to Republican nominee to president.

"Such a beautiful and important evening!" Trump tweeted triumphantly on November 9, 2016, after winning the Oval Office. "The forgotten man and woman will never be forgotten again. We will all come together as never before."

The backlash against Twitter was immediate. Democrats blamed the company for enabling Trump and profiting from his inflammatory statements, while Trump credited Twitter for carrying him into the White House.

As the company dealt with constant controversy, Dorsey became preoccupied with his health, adopting fad diets, yoga, and meditation, seeking balance while juggling the demands of running Twitter and Square. He started each day with what Twitter employees teasingly referred to as "salt juice," a blend of water, lemon, and pink Himalayan salt. In a nod to its kooky founder, the company began serving salt juice in some of its cafeterias around the globe and Dorsey's well-being obsessions began to bleed into his work at Twitter. In the wake of Gamergate and the foreign interference in the 2016 elections, he made it his mission to steer tweeters toward "healthy conversations."

"We're committing Twitter to help increase the collective health, openness, and civility of public conversation, and to hold ourselves publicly accountable towards progress," he wrote in March 2018. "We aren't proud of how people have taken advantage of our service, or our inability to address it fast enough."

Dorsey's personal feelings were an enigma. Sometimes, he could be completely in tune with the problems the company and its employees faced. Other times, he appeared tone-deaf, as when he embarked on a ten-day meditation trip in Myanmar, the site of a recent social media-fueled genocide in November 2018.

When he arrived back in San Francisco that December, his assistant arranged for a birthday surprise. As they traveled around India months earlier, Dorsey had become enamored with the country's monkeys, like the pink-faced rhesus macaques that were common on the grounds of Twitter's Delhi home base. To relive that memory, his assistant had an animal trainer bring a pair of simians into the office.

It provided a brief laugh, before Twitter's chief moved on to meetings, leaving the primates in a conference room for passersby to gawk at. Later, someone filed a complaint with human resources.

Dorsey had gotten control of Twitter, but it was a zoo and he was no zookeeper.

3 > "This is actually me"

On July 15, 2018, Elon Reeve Musk awoke at home in Los Angeles, forty-seven years old and jet-lagged from a trip to Thailand and Shanghai. His thirty-year-old girlfriend, Claire Elise Boucher, an ethereal pop singer who performed as Grimes, was still asleep beside him.

It was early that Sunday morning and, instinctively, Musk did what he always did in a quiet moment—he took out his phone. He would sometimes play mobile strategy games, or check his email, which overflowed with updates from his employees and Google Alerts for his own name, set up tactically to track news about himself. Despite having encouraged coverage of his own antics as an entrepreneur and executive, Musk had thin skin and wanted to know everything about how the public perceived himself and his companies—Tesla Motors, SpaceX, Neuralink, and the Boring Company. That morning, however, he focused on his primary addiction: Twitter.

Musk had amassed more than 22 million followers on the social media service and posted more than 5,000 tweets, opining about his companies' milestones, cracking jokes, and attacking his critics.

After some scrolling, Musk came across a link to a CNN video. He clicked and was confronted with an unfamiliar face. Sitting in a verdant Thai jungle, Vernon Unsworth, a serious-looking Brit in a white tee, wrinkled his forehead as questions were put to him. *This man is talking about me*, Musk realized.

"What are your thoughts on what Elon Musk's idea was?" asked a faint voice off-screen.

Unsworth smiled thinly, as if unsure whether to share his thoughts or not. "He can stick his submarine where it hurts. It just had absolutely no chance of working," he replied.

"Just a PR stunt."

Musk grew furious. He watched the interview again. Then he watched it again. His whirlwind trip to Thailand had revolved around the very thing Unsworth was criticizing. After a made-for-retweets story about a youth soccer team trapped in a cave went viral, he had hustled a coterie of SpaceX engineers and a custom-built mini submarine to the Southeast Asian nation to attempt an emergency rescue. The twelve boys and their adult chaperone had been lost in a partly submerged cave in the northern part of the country for eighteen days. But as Unsworth had pointed out in his interview, there were so many twists and turns in the cave, the metal tube of the submarine would have never made it past the first fifty meters of the dive.

Musk would not let the insult stand. He googled Vernon Unsworth and found news articles about the British expat who apparently lived around Chiang Rai, Thailand. Unsworth, who began caving as a teenager and had assisted in rescues in the United Kingdom, had moved to Thailand to explore its network of caves and had been brought into the effort because of his knowledge of the caverns where the boys were trapped. He lived in the country with his partner, a forty-year-old nail salon owner.

"This guy is creepy," Musk thought to himself as he burrowed further into his do-your-own-research rabbit hole. He googled "Chiang Rai" and found an article that said the northernmost major city in Thailand was the child sex trafficking capital of the world.

After less than an hour of googling, Musk reopened Twitter. Two days earlier, *Bloomberg Businessweek* had published an interview with the billionaire in which he had acknowledged his lack of impulse control on the platform. "I have made the mistaken assumption—and I will attempt to be better at this—of thinking that because somebody is on Twitter and is attacking me that it is open season," he told the magazine. "That is my mistake. I will correct it." But not yet. First, he had to respond to Unsworth.

Starting at 6:56 a.m. in Los Angeles, he began an online barrage of false accusations. "Never saw this British expat guy who lives in Thailand (sus) at any point when we were in the caves," he tweeted, before asserting in a second tweet that his team would make a video showing the submarine making it all the way to the cave system where the boys had been trapped. Never mind that Musk's submarine had not arrived in Thailand until the rescue was well underway, and eight of the twelve boys had already been freed.

"Sorry pedo guy, you really did ask for it," he tweeted.

The insult instantly ricocheted around the internet. The notion of pedophilia fueled bizarre online conspiracies. In 2016, internet rumors about a DC pizzeria that hosted a child sex trafficking ring, which later became known as Pizzagate, led to a gunman firing shots at the restaurant. The following year a crazed political movement known as QAnon began to foment around the idea that a cabal of child-molesting government officials were conspiring against President Donald Trump.

Musk was summoning a deluge of online conspiracists to harass an otherwise anonymous civilian whose expertise had just aided the successful rescue of children. But Musk's supporters had blind faith in him—after all, he had a sterling reputation as a businessman pushing humanity toward a future of a cleaner Earth and space travel. Musk's reputation was that of the smartest people on the planet. Surely he must know something about Unsworth that the average person did not?

"Bet ya a signed dollar it's true," he tweeted later that day, continuing his tirade against Unsworth.

Three days later, Musk tweeted an apology: "My words were spoken in anger." But he couldn't let it go, and continued to seed rumors about Unsworth months after his initial tweets. In September 2018, Musk doubled down on his accusation that Unsworth was a pedophile with more tweets and emails to a *BuzzFeed News* reporter* in which he wrote the Brit was a "child rapist" who took a "child bride who was about 12 years old."

A few weeks later, Unsworth sued Musk for defamation.

---

* That reporter was one of this book's authors, Ryan Mac, who was then a senior technology reporter at *BuzzFeed News*. Elon Musk emailed him several times, unilaterally declaring that his messages were "off-the-record," which in journalism parlance means that the information provided may not be publicly disclosed. Mac had not agreed to those terms with Musk, and *BuzzFeed News* chose to publish the email exchanges in full because of their newsworthiness in revealing Musk's state of mind and beliefs about Unsworth, all of which were untrue. In a deposition ahead of the trial, Musk would call his emails to Mac "one of the dumbest things I've ever done."

>>> ON DECEMBER 3, 2019, Musk was once again looking at his tweets about Unsworth. But this time, he sat on a raised stand above the gallery of the federal courtroom in downtown Los Angeles. Musk stared ahead vacantly and pursed his lips. He listened to Unsworth's attorney's questions, shifting his eyes as he gathered his thoughts, before issuing laconic replies that belied his swaggering online persona.

The reporters, fanboys, and critics who had packed the courtroom to catch a glimpse of the world-famous billionaire didn't know what to think. There he was, hunched over in a black suit, white shirt, and bluish-gray tie. Could this be the same person who built the world's most important electric car manufacturer, or proclaimed he would die on Mars? Musk was anything but commanding.

Defamation cases are historically hard to win in the U.S., but Unsworth's case against Musk seemed strong. Musk didn't deny tweeting the claims about Unsworth, and though he had initially apologized for his remarks, he continued to make allegations of pedophilia.

Faced with such an open-and-shut case, most members of elite society would have settled, offering a six- or low-seven-digit sum to make the lawsuit go away. After all, what was a few million dollars to a man worth around $20 billion? Why would Musk bog himself down with the embarrassing burden of legal discovery, depositions, and a trial when he had important companies to run?

Unsworth's lawyer, L. Lin Wood, a high-profile defamation attorney from Georgia, prodded him with questions about his online habits. He asked Musk what people posted on Twitter.

"They can assert facts, fiction, and anything that comes to mind," Musk responded.

"You told me, I believe, that it was a place where, in conversations, people could convey facts, state opinions, or even insult people. Right?" Wood asked.

"I mean, Twitter is a free-for-all where there's all sorts of things, you know, that sort of aren't true, untrue, half true, where people engage in sort of verbal combat effectively," Musk said.

"I mean, there's everything on Twitter."

This was Musk's defense. The plaintiff was simply a "creepy, old white guy expat living in Thailand," Musk said, and he did not mean that Unsworth was literally a pedophile. His lawyer, Alex Spiro, a smooth talker who completed a fellowship at the CIA before pivoting to become a high-profile attorney to the stars, built upon this strategy. His client was simply joking, he said. Never mind that Musk had tasked one of his most trusted staffers, Jared Birchall, with hiring a private eye to dig up dirt on Unsworth and reverse engineer the truth of his allegation.

If one tried to imagine a person antithetical to Musk, one might come up with Birchall. The former private wealth manager who operated in the shadows, standing behind the impressive people he worked for, safeguarding their money and interests. He didn't tweet, and old posts on his largely defunct Facebook were mostly thank-yous to people wishing him happy birthday, videos about his love for God, and photos of his wife and five children.

A tall, broad-shouldered man with a narrow nose and dimpled chin, Birchall was guided by his faith. He was a devout member of the Church of Latter-day Saints who didn't consume alcohol or caffeine and grew up in a touring family band called the Birchall Family Singers. Birchall graduated from Brigham Young University in 1999, and a decade later donated to oppose a 2008 California ballot measure to legalize same-sex marriage.

It was at Morgan Stanley, which he joined in 2010 after being fired from Merrill Lynch following a ten-year stint as a private wealth manager, that he first met Musk. The billionaire hired him from the bank in 2016, appointing him head of Excession LLC, his family office, named for an Iain M. Banks science fiction novel. Birchall catered to Musk's every financial need, pledging himself to his boss's every cause to gain an unprecedented level of trust.

Birchall set out on his assignment to prove Unsworth was a creep. He paid $52,000 to someone he thought was a private investigator. But that man, who faked his credentials, turned out to be a former convict and fed Birchall and Musk false information about Unsworth.

As the four-day trial wore on, Spiro ran circles around his legal counterpart.

Wood struggled to explain the basic mechanics of Twitter, and failed to goad Musk into admitting fault. His Georgia drawl and over-the-top southern charm also grated on the jury.

But Spiro spoke like a gregarious older brother back in town for Thanksgiving. Not yet forty years old, he was already a rising star at Quinn Emanuel, one of the country's top litigation firms. His friendly, fast-talking manner had won him famous clients like Robert Kraft, the owner of the New England Patriots, and Jay-Z, the music mogul, and it also won over juries. Tall and lean, with a boxer's nose, he had an athletic ease even as he clunked around the courtroom floor in a giant boot—he had broken his foot playing pickup basketball. The Boston-raised Harvard law grad couldn't deny what Musk wrote. But he could spin it. His client had posted thousands of tweets over the years on a platform that was known as a place where people could argue, insult, and brawl, he said. Twitter was a war zone.

"These tweets are not allegations of crimes," Spiro lectured the jury. "They are joking, taunting tweets in a fight between men."

The jury seemed moved by that. Unsworth testified he felt "humiliated, ashamed, dirty" by Musk's label and added, "I have good days and I have bad days." Those in the court felt for the Brit, but it was nearly impossible to assess the price of those "bad days." Then, Wood committed the fatal error. In closing arguments, he totaled what he believed his client was entitled in damages: $190 million.

When the number was announced, someone in the gallery snorted. Reporters looked up from their notebooks in amazement. Members of the jury shifted their gazes downward as if the eye-popping sum was something forbidden. It was a shocking number and, if approved by the jury, would have amounted to the largest award ever in the defamation case of an individual.

On the afternoon of December 6, the jury deliberated for less than an hour and came back with a decision. Musk was not liable for defamation. Spiro stood up as if in a trance. He—and Musk—had won. From then on, Spiro would enjoy a position of privilege in Musk's inner circle, working alongside Birchall to make the billionaire's every wish a reality.

"My faith in humanity is restored," Musk told the journalists on his way out of the courtroom. Anyone who tried to follow him was bodied out of the way by a phalanx of burly guards. The billionaire disappeared down an elevator, out a back door, and into a waiting Tesla Model S.

—

>>> A TRAIL OF CAMERAS befitting a rockstar followed Musk out of the courthouse, but he was not always a household name. The eldest child of three from a divided South African family, he came to the U.S. by way of Canada, eventually earning a bachelor's degree at the University of Pennsylvania. He was accepted into a PhD program for materials science at Stanford before dropping out to pursue a new consumer technology called "the Internet."

In 1995, Musk and his younger brother, Kimbal, cofounded Zip2, an internet city guides company with a $28,000 check from their father, Errol Musk, a figure Musk recalled as influential yet abusive. At twenty-three, the entrepreneur toiled away at the early web service, sleeping in the office, subsisting on Jack in the Box burgers and Cocoa Puffs, and showering at a nearby YMCA in Palo Alto, California. It was a yeoman's story Musk rarely failed to mention in early interviews with news outlets.

"I think the Internet is the superset of all media," a gangly, balding Musk told CBS in 1998. "It is the be all and end all of media. One will see print, broadcast arguably, radio—essentially all media folded into the Internet."

Even in those early years, Musk understood the value of a good story. He was a salesman at heart, developing narratives around his work ethic and visions of the future that would serve him in the decades to come. And he gravitated toward anyone with a tape recorder or camera that would give him airtime. He was also image-conscious, later correcting his receding hairline.

After four years of toiling on Zip2, he sold the company to Compaq for $305 million in 1999. He personally made $22 million in what was the largest all-cash transaction for an internet company at the time, and then went on to create X.com later that year. Positing that the internet would fundamentally change all industries, Musk spoke in recruiting interviews of a grandiose vision of how X.com could transform banking and displace entrenched players like Visa and Mastercard. To the thirty employees he hired to work out of an office on Palo Alto's famed University Avenue, Musk was a charismatic founder, with a convincing pitch for his online banking moonshot and a track record. At twenty-eight, he was one of the few people in the nascent era of the consumer internet to have already sold a company. Why not hitch yourself to his star?

But before X.com could take on the big credit card corporations, Musk found himself preoccupied with another competing start-up called Confinity. Its founders, Stanford University alums Peter Thiel, Max Levchin, and Luke Nosek, had developed a product called PayPal that allowed people online to email each other money. For a while, Confinity was located in the same Palo Alto building as X.com, leading some Confinity employees to believe that X had copied their product.

The dot-com bubble burst in 2000, and by March of that year, both X.com and Confinity decided to merge under the X.com name. Musk was named the company's chief technology officer, and later became chief executive, but immediately grated on the former Confinity employees. Obsessed with X.com after reportedly spending a small fortune on the web domain, he lobbied to change the name of the PayPal feature to X, despite the fact that PayPal already had brand recognition. It was seen as a mistake by the Confinity crew, which included Thiel, Levchin, and David Sacks, a law school pal of Thiel who came to lead the company's product efforts. They saw Musk as an obstinate leader who prioritized personal glory over company success.

Musk never got to implement his vision. That September, he and his wife, Justine, left for a two-week honeymoon in Australia. Musk had also arranged for meetings with potential investors who could potentially replenish the company's cash reserves. While he was away, the former Confinity team schemed for his removal. Sacks drafted a letter addressed to the board, outlining his lack of confidence in Musk's leadership. He pushed for Thiel to be named chief executive.

In the press, Musk played off the change, arguing that it allowed him to flex his entrepreneurial prowess. "One has to recognize where one's strengths lie," Musk told the tech news site CNET after his ouster in September 2000. "It's more interesting for me in the early stages of running a company, where the concentration is all on developing the product." But the move left Musk fuming. Thiel didn't only take Musk's job—he also rebranded the whole company as PayPal, completely erasing his beloved X.

Thiel and Sacks's gambit, whether by incredible foresight or sheer luck, paid off. Seven months later, eBay announced a deal to acquire PayPal in 2002 for $1.5 billion in stock. The acquisition would take Musk, who cleared more than $175 million, into another stratosphere of wealth. Despite his management failings, he had sold two internet companies and become one of the most successful entrepreneurs of a Web 1.0 era that would see plenty of companies rise to the heavens before exploding into flaming heaps of discarded ideas.

Before eBay had even completed its acquisition of PayPal, Musk founded Space Exploration Technologies Corp., or SpaceX, in May 2002. He committed $100 million of his PayPal windfall to space, with the goal of getting to Mars and making humans a multiplanetary species, a particular fixation of his from a childhood immersed in sci-fi novels.

Musk wasn't the only entrepreneur trying to reimagine travel. In 2003, two engineers named Martin Eberhard and Marc Tarpenning founded an electric car start-up called Tesla Motors. Their work caught Musk's eye and he invested $5.6 million the following year, turning him into Tesla's largest shareholder and earning him the chairman seat on its board. SpaceX occupied the majority of Musk's brain, but he sometimes dabbled in Tesla's affairs and weighed in on design decisions for the company's first car, the Roadster.

Ahead of the car's public unveiling in July 2006, Musk's ego took over. Hungry for credit, he emailed a public relations contractor the company had hired for the event, saying he would "like to talk with every major publication.

"The way that my role has been portrayed to date, where I am referred to merely as 'an early investor' is outrageous," he wrote. "We need to make a serious effort to correct this perception."

Following a *New York Times* story about the Roadster launch, in which he wasn't mentioned at all, he raged over email and said he was "incredibly insulted and embarrassed" that Eberhard had been mistakenly referred to as chairman of the company. He threatened to end Tesla's relationship with the communications firm.

By the following year, Musk got the attention he craved. As the company fell behind on its production of the Roadster, he fired Eberhard and took on the chief executive role himself. The company soldiered on toward building a luxury sedan, losing millions of dollars a year. SpaceX also struggled, teetering on the verge of bankruptcy after three failed launches. In 2008, its Falcon 1 rocket finally reached orbit, leading the company to win a $1.6 billion contract from NASA later that year. But the reprieve was short-lived—Musk's personal life crumbled as he divorced from his first wife, Justine.

In early 2010, Tesla filed to go public. With no advertising budget, the carmaker depended on media coverage to stir up excitement among investors and consumers. Ahead of the IPO, Musk tried to control the coverage, using Tesla's blog to rebut inaccurate stories and trashing a *New York Times* reporter who suggested his vehicles were only for the ultra-wealthy, calling him a "huge douchebag." When the Associated Press reported on the imminent registration of Tesla's initial public offering, he called to scream at a communications staffer, telling them that their inability to handle the story had ruined the company.

Although Musk appeared on the cover of *Wired* magazine with a prototype of his car's second model, the Model S, following Tesla's initial public offering, he fundamentally distrusted reporters and feared that a single bad story could destroy his company's prospects. News stories were existential threats. If Musk saw an article he thought was incorrect—whether in *The New York Times* or an obscure financial blog in the Netherlands—he would pressure his public relations staff, sometimes emailing them well past midnight, to correct the record. Reporters who constantly questioned him or wrote too critically of Tesla or SpaceX were personally blacklisted by Musk.

His craving for narrative control led him to Twitter. Celebrities like Oprah Winfrey and Ashton Kutcher had joined the 140-character microblogging service with 58 million users, and people were seizing available handles—the unique usernames that followed the @ sign—as they would distinct website domains.

An account called @ElonMusk started parodying the ambitious yet paranoid entrepreneur in early 2009. Featuring a profile picture of a man with a wide-brimmed hat covering his face, next to a donkey, the account claimed to be the "genuine ElonMusk." That March, the account tweeted that it was "plotting to take over the world." The account posted a handful of times, sometimes mocking the Tesla and SpaceX chief executive, but never garnered much attention. Eventually, whoever was behind it stopped, and the account had its posts wiped. A new owner took over that summer and began following other accounts, including those of professional skateboarder Tony Hawk, actor LeVar Burton, and an organization dedicated to breaking news. It remained silent for more than a year before Musk announced himself.

"Please ignore prior tweets, as that was someone pretending to be me :) This is actually me," @ElonMusk wrote on June 4, 2010.

It took him more than another year to post again, a photo of a sign for a Southern California ice rink that he had just visited with his kids in December 2011. Then the tweets started to flow. There were thoughts on philosophy, books he had read, and the occasional brag. ("Got called randomly by Kanye West today and received a download of his thoughts, ranging from shoes to Moses. He was polite, but opaque.") His early posts were earnest, if random, and while there were some updates about Tesla and SpaceX, most of his tweets gave off the vibe of a bored middle-aged dad.

"Not sure I can handle just doing 140 char missives," he tweeted, frustrated by the character limits on tweets.

Still, he stuck with it, tweeting at all hours and jumping between corny jokes, news articles, and photos from nights out. The posts gave a window into Musk's mind and he spoke without the manufactured corporate sheen typical of business leaders whose messages were workshopped by communications staffers.

"Feb is huge month. Model X world premier and public reveal of our LA design studio on the 9th," Musk wrote in January 2012 about his company's new sport utility vehicle. It was followed by commentary on Tesla's stock price, design improvements to a rocket, and a photo of a SpaceX engine test-fire. Twitter gave Musk the visionary a way of becoming Musk the salesman, keeping anyone interested updated on the progress of his companies. On the platform, he owned his own narrative, and he clearly explained his mission for both companies. For those who worked with him, it was one of his greatest superpowers: the ability to stay consistently on message, despite the criticism and uncertainty.

Twitter also accentuated an uglier side to Musk. Having fashioned himself the underdog, he had had no problem going after journalists and industry experts who claimed Tesla and SpaceX would fail. He used bursts of acid text to debate stories on a Tesla vehicle's short battery life or refute reporting on what beverages he drank at breakfast. He called stories "fake." It was a business communications style never seen before, with an extremely online chief executive willing to go to the mat over any perceived injustice.

The communications pros Musk employed had no power over his unfettered tweeting. But they also had to deal with the fallout of his unplanned pronouncements or deadlines. They regularly monitored his Twitter feed and had notifications on for his tweets, in an attempt to prepare for damage control in case reporters came calling with questions.

In January 2012, Musk hinted in one late-night tweeting session that he was separating from his second wife, the British actress Talulah Riley. "@rileytalulah It was an amazing four years," he wrote. "I will love you forever. You will make someone very happy one day." An enterprising reporter at *Forbes* saw the tweet and contacted Musk, who proceeded to explain why he was divorcing Riley. By the time Musk's communication staff had woken up on the West Coast, it was too late. A story was already in the works.

"I've made a mistake," Musk said to one of those staffers.

Musk was realizing just how effectively his tweets could be used to spin his version of events to the masses. Court filings later revealed that it was Riley, not Musk, who had filed to terminate the marriage. (Musk would later remarry Riley and divorce her a second time in 2016.)

As Musk grew richer and his companies more valuable, he attracted more media scrutiny, which, in turn, fed his paranoia. He suggested that the entrenched players, from traditional car manufacturers to Big Oil to Wall Street short sellers, worked in conjunction with traditional media outlets to sandbag his electric automaker. In 2013, when people expressed concerns about the Model S cars whose batteries spontaneously combusted, he blamed the media. "Why does a Tesla fire w no injury get more media headlines than 100,000 gas car fires that kill 100s of people per year?" he tweeted.

Musk's tweeting fed his legend. Tesla would state later in a financial filing that its chief executive's tweets were able "to generate significant media coverage of our company and our vehicles." Musk's Twitter account became a go-to platform for unplanned company announcements. He would state a goal for one of his companies, which would then force his employees to work as hard as possible to make it a reality.

—

>>> IN SEPTEMBER 2016, SpaceX prepared to take one of its most notable payloads into orbit. Facebook had contracted with Musk's company and spent $200 million to launch a satellite called Amos-6, which was intended to bring mobile internet—beamed from space—to parts of Sub-Saharan Africa. Mark Zuckerberg, Facebook's founder and chief executive, had spent years planning the move as part of his social network empire building.

As two masters of the universe, Musk and Zuckerberg were acquaintances and occasionally met to discuss larger topics like artificial intelligence. The Facebook chief sometimes hosted Musk for walking meetings around the social network's sprawling Menlo Park, California, headquarters. Musk was particularly concerned by the possibility that companies like Facebook would harness their reams of data and technological resources to create AI that could destroy humanity. During a 2014 dinner at his Palo Alto mansion, Zuckerberg urged Musk to stop railing about the potential dangers of AI at speaking engagements and on Twitter. It was all nonsense, Zuckerberg believed. Musk, enraged, refused to back down.

Despite Musk's disagreement with Zuckerberg, SpaceX did not turn down Facebook's business. It needed the money and attention. The company loaded the Facebook satellite onto a Falcon 9 rocket at Cape Canaveral, Florida, and prepared for a historic launch.

The satellite never made it to space. Two days before the scheduled launch, SpaceX's crew was running a test of the rocket's engines when it suddenly exploded in a ball of flames, incinerating Facebook's investment. SpaceX initially said little about the explosion, and its employees were still scrambling to find out what had happened, when Zuckerberg posted to his Facebook page.

"As I'm here in Africa, I'm deeply disappointed to hear that SpaceX's launch failure destroyed our satellite that would have provided connectivity to so many entrepreneurs and everyone else across the continent," the Facebook chief wrote, placing the blame solely on Musk's company.

After seeing Zuckerberg's post, Musk instinctively wanted to counterpunch. To him, Zuckerberg was a pretender, a man who focused on building dinky social apps, while he solved hard problems like building rockets and electric cars. But Musk held his fire, letting his grudge against Zuckerberg build inside him.

>>> AT TESLA AND SPACEX, Musk refused to take advice from anyone about his posts, and when one executive dared to point out that his tweets about deadlines made life harder for his employees, he demanded that they never speak to him about tweeting again. In other conversations, the boss explained that tweeting was his way to rebut and spin the press.

At his SpaceX cubicle in late 2016, he told one executive that he thought journalists were generally "idiots."

"They need to be spoon-fed!" Musk said, miming a baby being fed mushy peas. When asked what his plan was for several big SpaceX announcements on the road map, he shrugged and simply said he would tweet.

"With Twitter, we can talk directly to the people," Musk added. "Why do we need to go through journalists?"

By 2017, Musk's Twitter habit had become an addiction. That year, he tweeted 1,162 times, increasing his output by nearly 60 percent from the previous twelve months to average more than three tweets a day. Most people, even those who used Twitter daily, lurked and only read tweets, but Musk tweeted freely, a stream-of-consciousness that blended juvenile jokes with audacious bets on his own companies.

Elon Musk
@elonmusk

I love Twitter

9:50 AM Dec 21, 2017

💬 5.8K   🔁 34K   ♡ 168K

@redletterdave

You should buy it then

9:51 AM Dec 21, 2017

♡ 139K ⟲ 1.5K ♡ 8.4K

Elon Musk
@elonmusk

How much is it?

9:52 AM Dec 21, 2017

♡ 1.2K ⟲ 6.1K ♡ 31K

—

>>> BY MAY 2018, Musk's resentment of the media coverage boiled over. The billionaire was fed up with what he termed "the holier-than-thou hypocrisy of big media" and proposed a solution. In Musk's view, major media outlets were controlled by powerful people with agendas funded by special interests. Some, he suggested, used bots—or automated accounts—to manipulate public opinion on social media like Twitter. He thought that he and his companies were frequent targets.

"Going to create a site where the public can rate the core truth of any article & track the credibility score over time of each journalist, editor & publication. Thinking of calling it Pravda," he tweeted, a reference to the official newspaper of the Soviet Union's Communist Party.

It was an idea Musk had considered for a while, and Birchall had registered "Pravda Corp" with the California secretary of state the previous year. Pravda would be a "media" organization, Birchall wrote in paperwork that listed him as its leader. In truth, he was just the figurehead. When needed by Musk, Birchall would assume titles for the billionaire's charitable foundation; his brain-computer interface start-up Neuralink; and the Boring Company, his tunneling start-up. (Under Birchall, little would come of Pravda.)

With Tesla sputtering to produce enough cars to meet its leaders' projections that spring, Musk spent more time confined to the company's factory as his Twitter account became one of his few links to the outside world. That May, his tweet output quadrupled from April, as he cultivated a parasocial relationship with his followers. Twitter was home to many fandoms—from

English Premier League soccer teams to K-pop groups to Donald J. Trump—and the rabid following for Musk was no different. Fans created accounts to glorify his companies, attack his critics, and interact directly with the man himself, who, unlike many celebrities, would like or reply to tweets.

Bob Lutz, a longtime auto executive who had worked at Ford, Chrysler, and General Motors, compared Musk's supporters to "members of a religious cult."

"Just like Steve Jobs was worshiped at Apple, it's the same way with Elon Musk," he said in one 2016 interview, adding that the Tesla founder "is seen as a new visionary god who promises this phantasmagorical future, a utopia of profitability and volume."

>>> IN AUGUST 2018, while Tesla struggled through a difficult period that Musk called "Production Hell," its leader published another tweet that would haunt him legally for years to come.

> **Elon Musk**
> @elonmusk
>
> Am considering taking Tesla private at $420. Funding secured.
>
> 9:48 AM Aug 7, 2018
>
> 5.6K   20K   79K

he wrote, offering a 20 percent premium on Tesla's share price that also casually referenced stoner mythology.

It was an extraordinary post. Publicly traded companies took many precautions before making announcements that could influence their stock price and usually filed public disclosures with the Securities and Exchange Commission, particularly around mergers and acquisitions or matters of potential changes of control. Musk disregarded it all. Tesla's stock price shot up by 11 percent.

Musk would later claim that he had an agreement with Saudi Arabia's Public Investment Fund to privatize the company. Yet there was no formal agreement in place, and funding was nowhere near "secured" as Musk had claimed. Some of Tesla's own board members, a number of whom had grown tired of his tweeting and asked him to focus on building cars, had not even been informed of his efforts.

In an interview with *The New York Times* less than two weeks after his tweet, Musk wept as he acknowledged the stress he was under and discussed rumors of substance abuse. He denied using marijuana—though Tesla's board members were aware of other recreational drug use—but did admit to relying on Ambien to help him sleep. The next month, he went on a podcast with the comedian Joe Rogan and smoked a blunt, reigniting speculation about the causes of his odd behavior.

In late September, the SEC announced that it was suing Tesla and its chief executive for making false or misleading statements to the public. Tesla's leader "knew or was reckless in not knowing" that his statements were false or misleading, the nation's top securities regulator wrote. Two days later, Tesla and Musk settled. The company and its CEO would pay two $20 million fines, while Musk would step down as chairman. The agreement required that Musk enter a consent decree to appoint a "Twitter Sitter," a lawyer at the company who would approve his tweets with material information to Tesla before they could be posted. Musk did not have to admit to wrongdoing, though he could also not deny that he misled investors. From the SEC's perspective, the latter stipulation would prevent the billionaire from claiming he did nothing wrong.

Musk learned nothing. The $20 million fine he paid personally amounted to less than a tenth of a percent of his net worth at the time and he appointed a longtime board member and loyalist, Robyn Denholm, to chair Tesla.

Musk continued to tweet and tweet and tweet. It was unclear if there was anyone approving his tweets at Tesla headquarters, and the SEC would open investigations into two more of Musk's tweets, including one that misstated Tesla's production numbers and a Twitter poll in which the billionaire proposed selling 10 percent of his holdings. Little resulted from those inquiries. Musk could not and would not be contained—not by investors, not by his own board, and especially not by the U.S. government.

Beating his defamation case in 2019 further emboldened Musk to tweet whatever he liked. In July 2020, richer and more influential than ever, Musk laid out his sentiments in a giant middle finger to the regulator. He knew he was untouchable.

"SEC, three letter acronym, middle word is Elon's," Musk tweeted. It was neither his first nor last dick joke on the platform.

4 > OneTeam

Dorsey was late. His phone lit up with a barrage of texts from his personal assistant. *Where are you?*

On a balmy Houston evening in January 2020, Twitter's chief executive was supposed to be at a party at NASA's headquarters. But it had already started without him. Instead of his phone, Dorsey focused on the communal bowl of queso in front of him, scooping up melted yellow ooze with a few tortilla chips, as a handful of curious Twitter executives watched. Sitting around him at a posh bar in the downtown Four Seasons Hotel, they were savoring the moment. Their CEO was eating.

Consuming food might seem like a normal human activity, but for Dorsey it was anything but. Years in the spotlight had changed him. He was less relatable, and even less accessible, as he juggled the management of Twitter and Square. The snack represented his first meal of the day. He habitually skipped meals in an intermittent fasting exercise that had become popular among Silicon Valley engineers. The booze was also catching up to him. Sitting in the hotel bar—complete with low-slung leather couches, blazing fireplace, and mood lighting—Dorsey and his entourage of Twitter vice presidents and directors threw back top-shelf tequila.

Somewhat satiated and a little tipsy, he finally acknowledged his assistant's messages, saying he was on his way and peeling himself from the bar. His employees followed dutifully, piling into a black SUV that had been waiting for them outside.

"Is this an Uber?" one of them murmured. But the driver already knew Dorsey, acknowledging him with a slight tilt of his head. The executives realized the vehicle was Dorsey's own, manned and tracked by his personal security detail. They were whisked from the lights of downtown Houston, traveling twenty-five miles southeast to NASA's Johnson Space Center.

Dorsey, who was expected to hobnob with officials from NASA while his employees drank and milled around the cavernous museum among the rockets, tried to pull himself together in the car. But the inebriated gaggle of staff were having none of it. They roasted him for his dietary habits and office wardrobe choices, which sometimes included a pair of running sandals. They brought up the twenty-three-year-old *Sports Illustrated* swimsuit model he had dated, teasing him for buying her a house. She was two decades younger than the Twitter cofounder.

"They have to be at least thirty-five or older," one of the finance executives cracked from the back seat. Others laughed.

To any other CEO, this might have been out of line. But Dorsey welcomed the insubordination. These were his people, the folks who had been in the trenches with him running Twitter, the multibillion-dollar company and global town square he had built. Together, Dorsey and his team had survived years of doubt, criticism, and financial struggles. The few days in Houston were an opportunity to look back on troubled times and realize they had come out of it stronger.

Indeed that's how it felt to many of the employees at the celebration called OneTeam. It was the second-ever gathering of the sort, a laudatory summit with all the Silicon Valley tech giant trappings and excesses designed to give workers the feeling they belonged to one of the kindest companies in a cutthroat industry. Employees from around the world had flown in for three days of group meetings, corporate indoctrination, and plenty of drinking. Twitter had picked Houston, the site of the devastating 2017 Category 4 Hurricane Harvey, in part to give back. First responders, citizens, and government organizations had used the platform to coordinate disaster responses and communicate emergency information during the storm, showcasing Twitter's power.

Beyond the programming of speakers and group events and parties, everything—from hotels to food to bar outings—was covered. Some members of the C-suite even had access to a private jet that could whisk them back to their Bay Area homes if they didn't want to spend the night in Texas. In all, Twitter spent tens of millions of dollars on the event, which was viewed by leaders as a justifiable expense that would build relationships and reinvigorate workers at a somewhat stagnating company. And at the center of Twitter's indulgent culture-building exercise was Dorsey.

>>> DORSEY'S SECOND STINT as CEO had been a mixed bag. By 2019, the stock price had barely budged from where it had been three and half years earlier when he had replaced Dick Costolo. Wall Street had cooled on Twitter. The company was no longer the high-flying disrupter of the late aughts. Twitter had become squarely a midsize corporation that lacked revolutionary ideas, new products, or vigor. Its public market valuation hovered around $25 billion.

Twitter was growing, but modestly. That February, the company announced that for the first time ever, it had turned an annual profit, making $1.2 billion in what Dorsey said was "proof that our long-term strategy is working." It also disclosed that it had 126 million daily active users, up 9 percent from the same period the year before. In the land of giants, however, that wasn't impressive. Other players like Snapchat, and a fledgling video app called TikTok, were gaining ground. Investors and analysts dogged Dorsey with questions about whether or not he could keep Twitter relevant, particularly while he had split duties leading Square at the same time. The return of the company's original product mind was supposed to stimulate new ideas and Dorsey made some cosmetic changes—notably expanding the tweet character limit from 140 to 280—but nothing seismic was underway.

Dorsey's push into "healthy conversations" did move the platform toward publishing and promoting less hate speech and toxic content, via user suspensions and bans, or decreased visibility for certain tweets in the Twitter algorithm. Yet when crises took place, people still seethed on Twitter. That March, a white supremacist had gone on a shooting rampage, killing fifty-one people at two mosques in Christchurch, New Zealand. He had live-streamed it on Facebook. To archive the footage and keep it online, the attacker's supporters had used Twitter. It was yet another example of bad actors seeming to be one step ahead of the Silicon Valley tech behemoths.

President Trump also continued to flummox the company. Unwilling to penalize the leader of the free world, Twitter let its most engaging user share a constant daily stream of bombastic posts, which drove constant traffic to the platform but caused its leadership to feel ill at ease. In 2019, Trump had used his account to tell Democratic minority congresswomen to "go back and help fix the totally broken and crime infested places from which they came," riling up racist abuse. He had also tweeted 115 times in a single day ahead of his first impeachment.

"We have seen abuse, we have seen harassment, we have seen manipulation, automation, human coordination, misinformation," Dorsey said onstage that spring at the TED Conference, an annual gathering of business bigwigs. "These are dynamics that we were not expecting thirteen years ago."

In Dorsey's mind, Twitter was a public square, just like Washington Square Park, the 9.75-acre space in Manhattan's Greenwich Village where, even as the billionaire chief executive of two major companies, he would sometimes sit, take work calls, and watch the tourists, students, and buskers around him. The park was neutral to what happened on top of it and there was an expectation that people could express themselves freely there. Yet, if someone with a megaphone started harassing others, people could call him out or summon park police to maintain order.

In practice, however, finding the balance between free expression and safety became the crux of Twitter's problem. The users who were subjected to abuse slammed the company for not doing more to clean up the site. On the other side, suspended users accused the company of censorship. While Dorsey spoke openly about Twitter's struggles, he often came off as philosophical.

That criticism was not helped by Dorsey's regular interactions with high-profile right-wing users of his platform. He had long been a proponent of hearing from a diversity of voices, but in the polarization of the Trump presidency, his efforts to meet with these divisive figures only caused more consternation, even among his own staff. In March 2019, he went on Joe Rogan's podcast with Vijaya Gadde, who by then had been elevated to Twitter's chief legal officer. Another guest, conservative blogger Tim Pool, attacked the pair for their supposed bias against the right. The following month, Dorsey met in the Oval Office with President Trump, who accused Twitter of discriminating against Republicans and removing thousands of his followers. The chief executive kept in touch with figures like Ali Alexander, a right-wing conspiracy theorist, and conservative firebrand Candace Owens, who had made a name for herself by operating a doxing website.

To combat the impression that their CEO was out to lunch, Twitter comms arranged for Dorsey to tour each of the company's global offices, calling it the "Tweep Tour"—a reference to the company's nickname for its employees.

Dorsey was particularly moved by a visit to Africa that November, which included stops in Ethiopia, Ghana, Nigeria, and South Africa. Over three weeks, he met with dignitaries, engineers, and entrepreneurs, becoming particularly fascinated with the potential use of Bitcoin, a popular cryptocurrency, in the continent's developing markets. "Sad to be leaving the continent . . . for now," he wrote at the end of his trip. "Africa will define the future (especially the bitcoin one!). Not sure where yet, but I'll be living here for 3-6 months mid 2020. Grateful I was able to experience a small part."

Dorsey's tweet was news to Twitter's executives and board. There had been no planning to allow him to run the company from any post in Africa. Already facing concerns from investors and analysts that Dorsey was seemingly distracted by Square, Twitter's executives braced themselves for questions about how Dorsey could run the company from halfway around the world. The company's stock fell 1.3 percent on the news.

—

>>> IF THERE WERE any indications that Dorsey was distracted or checked out from his duties, they did not come at the 2020 OneTeam. Leslie Berland, the marketing chief who orchestrated many of Dorsey's public appearances, placed him front and center.

Berland, who joined Twitter in 2016, was among the most beloved personalities at the company. With a bright smile that flashed across high cheekbones, the New York–based executive was Dorsey's first major hire during a reshuffle of the company's leadership organization upon his return.

Her move at the time was a gamble. Berland was on track to be one of the youngest people to join American Express's C-suite but left her comfort zone to take a chance on Dorsey and a dysfunctional company that was seen to have never passed beyond its start-up phase. In her years there, she had worn many hats, eventually adding human resources to her marketing responsibilities and the less official title of "Jack whisperer." To employees, Berland was a maternal figure, moderating company-wide meetings with a warm personality that drew everyone in.

For OneTeam, Berland leaned into the weirdness that was quintessentially Dorsey. A church choir was flown in from Los Angeles to perform one song, Toto's "Africa." The joke poked fun at Dorsey's goal of living on the continent, but it fell flat for some employees who were fixated on how much it cost to cover flights and hotels for the performers.

Dorsey was receptive to his executive team's efforts to make him the driving force behind OneTeam. To kick off the event on January 14, he walked onstage at Downtown Houston's George R. Brown Convention Center in an astronaut helmet, white Moon Boots, metallic silver pants, and white trench coat with the conference logo emblazoned on the back. He then led a fifteen-minute meditation session. Dorsey's mom, Marcia, beamed from the audience as her son led them through the cultish mind-clearing exercises.

Beyond the woo-woo, Dorsey seemed bought in. Still in his Moon Boots, he ran through an extensive presentation of his vision for the company. With some of his talk written out partly in emojis on his iPhone, he showed slides featuring historical figures such as Martin Luther King, Jr., that read: "Everyone can be great, because everyone can serve." He spoke about his desire to work from Africa and his plan to decentralize the company's workforce so people would no longer have to work at the office if they didn't want to.

He also made a prediction that there would be a coming global incident that may force people to shelter in their homes and use an internet platform to communicate. He envisioned a world where Twitter—already important for the 152 million users who visited the site daily—would be an even larger influence. It felt like a Tony Robbins seminar where people down on their luck spent thousands of dollars to hear from a guru revealing to them life's secrets. Dorsey was selling them on Twitter and, more important, himself.

Other leaders took the stage to make their pitches, too. They included

Dorsey's top lieutenants: Kayvon Beykpour, a genial product whiz; Bruce Falck, a demanding executive who oversaw products for advertisers; and Parag Agrawal, a quiet machine-learning expert who led Twitter's technology teams. The three men were responsible for executing Dorsey's ideas for the platform, and they joined Dorsey's other deputies for a panel that discussed their plans for the company. While Beykpour and Falck bantered, Agrawal sat quietly with a slight slump in his shoulders that belied his large frame. His onstage presence didn't hint at the particularly close relationship he enjoyed with Dorsey, who viewed Agrawal as a kindred spirit with a bone-deep understanding of Twitter's mission and future.

Between talks from the likes of Olympian Simone Biles and NFL superstar J. J. Watt, employees spent copiously on their corporate expense accounts at bars and restaurants around Houston. One night, in an especially grand display of the company's largesse, Twitter rented out Minute Maid Park, the home of the Houston Astros, for a fireworks display that wouldn't have been out of place on the Fourth of July.

Dorsey relished the good vibes. At a block party on the second night of OneTeam, he walked around with Biz Stone, who had returned to the company as an executive with an undefined role in 2017. They took it all in, waiting in line with other employees for the Ferris wheel. They got into a pod together, rising and falling on a clear Texas night on one of the last rides of the evening.

—

>>> THE CONFERENCE'S THIRD and final day on January 16 was lighter on programming. Among the speakers was Chrissy Teigen, the model and television personality who ran one of Twitter's more popular accounts. She had come second in a poll of speakers that Twitter employees wanted most to see at OneTeam, following the first-place vote getter who remained a secret to most attendees until he dialed in via video call. He had been slated to come to the company's first OneTeam in 2018 but had been sidelined by work and a few controversies involving his Twitter account. Finally, at the coaxing of Dorsey, one of Twitter's most followed—and polarizing—celebrities was about to appear.

"I have no idea if this is going to work," Dorsey said, as he strode onto the convention center stage, fumbling with an iPad, his screen projected behind him on a giant display for the audience. A dial tone blared through the speakers for a few seconds. Then, with a smile and a wave, Elon Musk appeared.

There were cheers and applause. "We love you," one employee shouted. Others whipped out their phones to take pictures and video of the supersize projection of Musk, who was clad in an open black bomber jacket over a T-shirt that read OCCUPY MARS. Dorsey had previously called the Tesla and SpaceX chief the most exciting influential person on the platform for "sharing his thinking openly" as he solved "existential problems." Still, some in the audience wondered if Musk was an exemplary Twitter user, given his history of bombast and legal problems that stemmed from his online activity on the platform. Musk had emerged victorious in his "pedo guy" tweet lawsuit only a few weeks earlier.

Over the video call, Musk, seated behind his desk at SpaceX's headquarters, was unlike his online persona. He was reserved and sometimes barely audible, lacking the spikiness or juvenile humor that permeated his tweets. Dorsey tried to draw him out, asking him about his mission to establish colonies on Mars. "When do you think we'll see the first tweet from Mars?" Twitter's CEO asked.

Musk stumbled over words, taking pauses to think. At other points, he interrupted his own thoughts and seemed unable to string together sentences as quickly as his brain fired. He complained about bots and trolls.

"I'm sure you guys see it all the time, people trying to manipulate the system," Musk said. "They're trying to sway public opinion and sometimes it can be very difficult to figure out what's real public opinion and what's not, and what do people actually want, what are people actually upset about versus manipulation of the system by various interest groups."

He then leapt into discussion of the Mars Rover. Interplanetary tweets might flow within nine years, the billionaire supposed.

Then Dorsey asked him for feedback. "What are we doing poorly, what could we be doing better, and what's your hope for our potential as a service?" Dorsey asked.

Then added, "If you were running Twitter—by the way, do you want to run Twitter?"

The assembled employees laughed loudly at the idea of Dorsey offering his company to Musk. Both men paused for a moment. The laughter faded.

"What would you do?" Dorsey asked.

5 > An Invasion

The invader came in quietly. By mid-February 2020, Twitter's board heard rumors of big purchases of its shares, but the buyer's identity was concealed by equity swaps—financial instruments that did not trigger SEC requirements for public stock disclosures. The value of the purchases crept up and up, from thousands to millions, until, by late February 2020, it hit $1 billion.

The buyer was Jesse Cohn, a top partner for Elliott Management, the $71 billion fund founded by Paul Singer that was known for shaking up—or destroying, depending on who you asked—corporate boardrooms around the world.

With his stake secured, Cohn texted Omid Kordestani, Twitter's board chairman, and asked to talk. It was a phase shift in the secret campaign he had been waging for weeks to remove Dorsey once again from his perch atop Twitter. Kordestani, a suave Iranian American salesman who had led Google's business development during its heady dot-com days, knew the second that Cohn's name appeared on his phone that it couldn't be good news. On the call, Cohn confirmed his fears. Elliott Management owned a 4 percent stake in Twitter.

Cohn was brief and blunt. He told Kordestani to expect a letter detailing the stake he had accumulated in Twitter and the demands that he planned to impose. It was irresponsible to let Dorsey run the business while he was distracted by his dual role at Square, Cohn argued.

On February 21, a few hours after his call with Kordestani, Cohn sent his promised letter to Twitter's board, formally announcing his stake in the company. The letter stopped short of calling for Dorsey's firing but outlined the concerns about his part-time leadership. The board panicked.

Twitter was unlike Facebook and Snapchat, where the founder-chief executives had structured their ownership specifically to protect themselves from this kind of interference. Those companies had super-voting shares that gave the founders outsize control over their organizations, even after the companies had gone public. Dorsey owned only a thin slice of Twitter, a 2 percent stake worth about $531 million, and had no super-voting stock, leaving him vulnerable.

Cohn had waited until the last minute to announce his position, just two days before Twitter's February 23 deadline to nominate new board members. Twitter had a few open seats on its board that it had intended to fill with independent directors, expanding the breadth of knowledge and prestige in its boardroom. But the process of filling them had moved slowly. At one point, Dorsey had suggested adding Musk but had been rebuffed.

Cohn insisted that he be immediately placed on Twitter's board along with three of his loyalists. There were only three open seats on the board, but Cohn wanted to ensure he would control them and have a backup plan to grab a fourth seat if any of the current directors stepped aside. The board seats would allow Cohn to quickly get inside the company and start flexing his muscles. The board huddled over the final week of February without Dorsey present, trying to figure out what to do.

Twitter's stock price had fallen more than 20 percent in late 2019 after the company missed Wall Street expectations and revealed its ad service to be ridden with bugs and outages that prevented advertisers from properly targeting consumers. The numbers didn't make Twitter look like a company operating at peak performance or Dorsey like a chief executive with his eye on the ball.

But people inside the company believed in Dorsey, and he had a loyal and close-knit team of executives, some of whom might follow Dorsey out the door if they thought he was forced to go.

Dorsey was livid about Elliott Management's intrusion. He didn't want to be thrust into the spotlight for a public litigation of his successes and shortcomings—not again after being fired once before and dealing with the fallout from the 2016 election. He loathed the idea of out-of-touch finance bros in windowpane-check button-downs meddling with engineering and his vision for the product, and he did not want to be the focal point of a drawn-out battle.

He threatened to quit rather than wait to be pushed out in an agonizing repeat of his previous firing. Dorsey's message to the board was clear: "It's him or me." If Cohn was allowed to come in, he would storm out.

>>> ON A GRAY FRIDAY MORNING, Cohn's jet touched down at San Francisco International Airport. As the wheels hit the tarmac, the corporate raider readied himself for a showdown. At thirty-nine, the Long Island native had already led several activist shareholder campaigns against companies like eBay and AT&T, and earned enough money for a $30 million Wall Street penthouse apartment.

Cohn homed in on underperformers, quietly buying up stock. Once he had some financial leverage, he would force out corporate leaders or seek concessions to improve the bottom line. As the target's profits rose, so would the stock price. Then Cohn would sell, walking away richer, busted careers in his wake.

Cohn did not particularly care who replaced Dorsey as Twitter's chief executive—anyone full time would be better, he believed. The company was languishing under its part-time leadership and Dorsey would never make Twitter his top priority, since the vast majority of his wealth came from his shares in his digital payments company, Square. Worst of all, the board had given Dorsey an unreasonably long leash, ignoring the problems his lack of focus caused, Cohn thought.

Two of those perhaps overly lenient board members were on their way to confront Cohn. Kordestani and Patrick Pichette, a graying venture capitalist who had once served as Google's chief financial officer, shared a car that morning as they rode past the steely waters of the San Francisco Bay and plotted their counterattack. The pair were in agreement: Dorsey stays.

Cohn waited for them in a private conference room in the terminal that had been specially booked for the showdown. When the board members arrived, Cohn greeted them casually, despite the seriousness of the situation.

Pichette and Kordestani were put off, although they tried to stay poker-faced. Cohn's coup could upend one of the most important social media companies in the world, a hub for political speech that was sure to play a key role in the upcoming U.S. presidential election. Was he taking any of it seriously?

The battle for Dorsey's future could begin.

Cohn was quiet and polite, belying his reputation as a hard-driving CEO killer. He dressed casually, forgoing the Wall Street uniform of a suit and tie, and had brought only Marc Steinberg, an Elliott portfolio manager, not a bevy of lawyers.

Cohn invited Pichette and Kordestani to sit and talk. Despite his mellow approach, he was firm: Twitter's stock was underperforming, and it was Dorsey's inattention that was causing the pain. He needed either to leave Twitter or abandon Square. It was untenable for him to do both, Cohn believed.

Pichette leaned on his experience at Google as he made his case to Cohn. Sundar Pichai was technically the leader of only one company, Alphabet, he argued. But under the Alphabet umbrella was the search company Google, the video service YouTube, the corporate web suite Cloud, and moonshot projects like self-driving cars. Pichai ran it all without Wall Street worrying he was distracted.

Why not let Dorsey run just two companies, Twitter and Square? What mattered wasn't the number of businesses but having the right teams installed around a visionary leader to make sure everything stayed on track. Cohn should meet Dorsey, Pichette insisted. Only then could he understand what made Dorsey so special.

Cohn thought Pichette and Kordestani seemed defensive, and he and Steinberg tried to soften them with their friendly, informal approach. They had brought stacks of documents to argue his point of view. Investors deserved more from Twitter, given its role in public discourse. To Cohn, it didn't matter how many companies Pichai was theoretically capable of running—it just wasn't good governance to have a chief executive who was constantly pulled in different directions. Just because a tipsy driver might get home safely didn't mean that getting behind the wheel after a night at the bar

AN INVASION

was a good idea. Twitter needed a leader who was focused on fixing its performance problems and could dedicate all of his time to the company, Cohn said.

By the following weekend, February 29, word of Elliott's intervention had leaked. Twitter employees immediately erupted in outrage at the idea that their quirky yet beloved chief executive might be ousted. Many of them credited Dorsey with creating Twitter's unique culture that put people and speech over profit and speed. They tweeted support for Dorsey under the hashtag #WeBackJack, sharing stories about his leadership and praying he would stay.

Even some tech leaders stood up for Dorsey. "Just want to say that I support @Jack as Twitter CEO," Elon Musk tweeted.

>>> SEVERAL TWITTER EMPLOYEES came home from Houston after OneTeam feeling incredibly ill. Maybe the aches and chills were caused by the constant drinking and partying, but dozens of workers developed fevers and stayed away from their offices in the weeks following the event. They joked that they had the "OneVirus."

By late January, news reports were circulating on Twitter of a mysterious virus spreading rapidly across China. First detected in Wuhan, the capital city of the country's central Hubei province, the COVID-19 coronavirus recorded its first U.S. case near Seattle on January 20, 2020. *Had they perhaps contracted something other than the flu at OneTeam?* Twitter employees wondered.

Just as his employees rallied around him, Dorsey sent them away. On March 2, the company encouraged everyone to work from home. Twitter was one of the first major American corporations to do so—unsurprisingly, given that its chief executive already encouraged remote work. As Twitter workers scattered to their homes, they were told nothing about Elliott's maneuver.

Dorsey still seethed. He was required to recuse himself from board discussions about potential changes, and resented being in the dark. From the outside, it seemed like no one on the board was standing up for him, and he became furious not just with Cohn but with his own board members.

Twitter's bankers at Goldman Sachs, meanwhile, mounted their defense. The best Dorsey could hope for, they said, was to fight money with money. Dorsey needed a white knight investor to save him. Dorsey immediately suggested Laurene Powell Jobs, the widow of Apple's founder Steve Jobs. The blond mogul managed her husband's multibillion-dollar trust and was one of the few people with the financial firepower to come to Dorsey's aid. Dorsey deeply admired her late husband, and knew that Laurene supported founders and was fond of him. She also tended to invest in media companies, climate projects, and other vehicles with a clear public benefit; Twitter fit comfortably into that niche.

Dorsey pushed the board to pitch Powell Jobs, and she met with its members to discuss a potential investment. But despite her wealth and friendship with Dorsey, she declined to back him in the fight against Elliott. The investment seemed wrong for her portfolio for several reasons, among them that a messy boardroom battle didn't suit her polished image.

Agonizing over what to do, Dorsey received an unexpected phone call during the first week of March. On the other end of the line was Egon Durban, the cohead of Silver Lake, an investment firm that specialized in technology and entertainment. Unlike Cohn, Durban, a Texas native with a deep tan from his golfing and surfing excursions around the world, had a long history in Silicon Valley and understood what Dorsey wanted to achieve at Twitter. He had made a name for himself by brokering the sale of Skype to Microsoft for $8.5 billion in 2011. The Silver Lake leader may have agreed that Twitter wasn't living up to its potential—he had considered taking Twitter private several years earlier—but he was a friendly face who had gotten to know Twitter when he probed the company for a potential acquisition.

Durban had read the news about Elliott Management and was calling with an offer. Cohn had come to Twitter with a $1 billion battering ram. To head him off, Durban would match him, funneling $1 billion into Twitter himself. The money would be enough to force a truce with Cohn, keeping Dorsey in place and buying him time to improve Twitter's stock price. It would turn into a high-stakes game of tug-of-war, but Dorsey had few options. He accepted Durban's offer.

Pichette continued to urge Cohn to meet with Dorsey himself. He flew to New York to make his case to Cohn in person, walking through an eerily empty airport as more and more travelers abandoned their plans amid fears of the virus. Twitter's bankers at Goldman hosted the second meeting, and Pichette insisted that Cohn should see firsthand what Dorsey had to offer.

Cohn agreed, and Dorsey sullenly accepted the meeting, understanding he would need to charm the investor in order to keep his job. Cohn flew back to the West Coast and met Dorsey at Goldman Sachs's office in Menlo Park on Sand Hill Road, the famed Silicon Valley corridor where the biggest venture capitalists set up shop to hear pitches from the most promising start-up founders.

While Pichette and Kordestani had insisted that Cohn would recognize Dorsey's genius once he saw him in person, Cohn felt as though he already knew reclusive founders like Dorsey—young men who believed they were uniquely capable of changing the world and clogged Sand Hill Road as they traipsed to investor meetings. While the venture capitalists might be convinced that all it took to succeed was a laptop-toting visionary, Cohn was more old-school. Businesses needed structure and discipline, he thought.

Dorsey eyed Cohn suspiciously, trying to put on a friendly face. He spoke in his usual thoughtful cadence, trying to explain Twitter's importance to Cohn. But the activist had come ready to poke the bear.

"How can you run both companies?" Cohn asked Dorsey.

Dorsey responded that he had a management process, explaining the way he relied on Agrawal, Beykpour, and Falck, the deputies who had recently joined him onstage at OneTeam, to make decisions in their areas of expertise. The same went for Square, Dorsey said—he had people he trusted running the show. When he needed to step in and make the final call, they knew where to find him.

Cohn didn't like what he was hearing from Dorsey in the meeting. He worried that Dorsey surrounded himself with yes-men. Many of his trusted executives, including Agrawal, had spent most of their careers at Twitter and didn't know how a normal business would be run. There was no replacement for face time with a chief executive, and Dorsey's tweeps just weren't getting enough.

At the end of the meeting, Durban headed over from Silver Lake's office, just yards away from the Goldman suite. His sudden presence—and his promised investment in Twitter—took Cohn by surprise. The activist left with his worries about Dorsey reinforced. But his role was to improve Elliott's investment, and a drawn-out fight with Silver Lake wouldn't help.

Cohn set about making a deal that he could live with. Pichette and Kordestani shuffled between the parties, ironing out a truce. A frustrated Dorsey bemoaned Twitter's lack of super-voting shares. Cohn insisted Twitter expand revenue and attract new users, and pressured the board to set up a new committee focused on governance, an attempt to tame the freewheeling company. With $1 billion apiece on the line, Cohn and Durban each had seats on Twitter's board. Cohn dropped his demand for additional board seats and only took one for himself.

By the end of the week, Dorsey had been rescued. "We are deeply proud of our accomplishments and confident we are on the right path with Jack's leadership and the executive team," Pichette said in a March 9 statement announcing the truce. "While our CEO structure is unique, so is Jack and so is this company."

There was also some reshuffling. It was clear to Cohn that the arrangement with Kordestani let Dorsey run amok. Pichette, an owly Canadian who favored Patagonia pullovers, had more management experience and an ease about him that made him a better candidate to rein in Dorsey. He was swapped in as chairman.

Pichette also snagged the job of running the governance committee Cohn had demanded. His first task, according to the deal Cohn struck with Silver Lake and Twitter, was to evaluate the effectiveness of Dorsey's leadership, "given that the Company's chief executive officer has another chief executive officer role."

The agreement brokered with Elliott was clear—the investment firms could meddle in corporate governance and try to tether Dorsey to his desk, but they would keep their noses out of product and policy decisions. The arrangement would give Dorsey some leeway to pursue his lofty goals.

>>> ON MARCH 19, little more than a week after the truce with Elliott, California issued its first stay-at-home order, banning residents from leaving their homes except for essential tasks. As the pandemic began to rage, Twit-

ter's headquarters, an Art Deco high-rise on San Francisco's Market Street, lay empty.

Dorsey initially retreated to his mansions in San Francisco's Sea Cliff neighborhood and in Big Sur. But he soon hit the road, traveling to Hawaii, Costa Rica, and French Polynesia. His direct reports complained that they could sometimes hear roosters crowing in the background of his conference calls, a reminder that he was on an island, while they were trapped in their apartments. In May, Twitter became the first tech company to announce that its employees would work remotely, forever—a decision driven by Dorsey's desire to stay away from the office.

By the summer, he stopped chiming in during meetings entirely. He routinely kept his camera and microphone off, leaving employees wondering if he was still sitting near his laptop at all or if he had wandered off to surf.

As Dorsey faded away, Cohn and the other directors discussed a plan for Twitter to double its revenue by 2023 and increase its daily audience by at least 64 percent, to 315 million people. Cohn pushed for even higher targets, but Twitter executives resisted his demands, fearing that they were unrealistic. If they committed to goals that were too lofty, the stock would certainly spike and give Cohn an opportunity to sell his shares for a win. But the Twitter team would be left holding the bag.

In late October and early November, a gaunt-looking Dorsey videoconferenced into Congress to testify several times about content moderation. Hidden behind a scraggly gray-and-white beard that descended past his chest, Dorsey responded patiently to the senators' questions about decisions the company had made to delete certain tweets and allow others to remain on the platform. He walked them through his nascent plans for decentralizing social media—inspired by the technology behind Bitcoin, he believed social media could run on public code governed by regular people rather than corporations. Dorsey told the senators he wanted to give users more control, empowering them to choose their own algorithms that would govern the kinds of content they saw and excusing Twitter from its moderation responsibilities. This was the path forward that all social media companies would eventually follow, he insisted.

Although Dorsey took pains to set up his laptop with just a simple white wall as his backdrop, some of his employees who had been on conference calls with him recently recognized the location—Dorsey was dialing in from a rental property in Hawaii.

Despite Elliott's insistence that Twitter could not survive with a part-time CEO, Dorsey was freer than ever. Not even a public hearing in front of Congress could confine him to a particular home or desk. A few days after his November testimony, paparazzi captured him strolling on a secluded Hawaiian beach with Sean Penn and Jay-Z.

6 > A Polynesian Vacation

On March 25, 2021, Dorsey, along with Facebook chief executive Mark Zuckerberg and Google CEO Sundar Pichai, dialed in to testify virtually in front of the U.S. House of Representatives' Energy and Commerce Committee for a hearing on combating disinformation and online extremism. It was the fifth such hearing that Twitter's leader had been summoned to in front of Congress, and he was sick of it. The hearings felt theatrical, with the senators or representatives grandstanding on the culture war topic du jour.

Dorsey attended the hearing—still virtual because of COVID concerns—from his Sea Cliff mansion kitchen, calling in from an iPad stacked on top of a pile of books. His dishes, glassware, and a blockchain-based clock could be seen in the shot framed behind his buzzed head and graying beard.

He did his best to explain how Twitter had arrived at its content moderation decisions, trying to walk the lawmakers through the process. But they kept interrupting him, demanding that he give yes or no answers that left little room for nuance. Annoyed, Dorsey tweeted out a poll. "?," he wrote, asking his followers to vote yes or no.

Word of the tweet quickly reached the representatives who were questioning him.

"Mr. Dorsey, what is winning, yes or no, on your Twitter account poll?" Representative Kathleen Rice, a Democrat from New York, asked, raising her eyebrows behind thick tortoiseshell glasses.

"Yes," Dorsey replied, the corners of his lips poking upward from behind his beard as he stifled a smile.

"Hmm, your multitasking skills are quite impressive," she replied.

But Dorsey wasn't looking to multitask for much longer.

—

>>> NO LEGISLATION EVER came out of the hours-long grilling sessions. The press pilloried Dorsey and Twitter, regardless of what he said. And most lawmakers seemed like they wanted to ask only about specific content moderation decisions: *Why was this taken down? Why was this left up? Why was my account being shadowbanned?* They didn't appear to care about the future of technology or solving the problems they complained about.

Dorsey firmly insisted that world leaders should not be subject to the same rules as other Twitter users and refused to allow their accounts to be suspended when they posted threats that would get a regular person booted off the platform. He also staunchly defended Alex Jones, the far-right podcaster and Sandy Hook school shooting truther, after he was banned from Facebook, YouTube, and Apple's podcast network, saying that Jones hadn't violated any of Twitter's rules. Eventually, as pressure mounted and Jones used Twitter to tell supporters to prepare their "battle rifles" against the mainstream media and harass a reporter, Dorsey relented, banning the conspiracy theorist and his company, Infowars.

Vijaya Gadde took on Twitter's critics with a firmer hand. She had joined Twitter in 2011, while Dorsey was still trying to find his way back into the company's day-to-day operations. When he secured his second CEO term in 2015, harassment was becoming rampant on the platform. Under his watch, she crafted Twitter's rule book, drafting rules with her team and securing Dorsey's sign-off. To other Twitter executives, Gadde, like Leslie Berland, was one of the few trusted voices who could speak Dorsey's language. She could convince him to adopt her suggestions, and translate his desires to lawyers and investors who might otherwise find him inscrutable.

Born in India, Gadde moved to the United States with her family when she was three. She grew up in Beaumont, a small city in southeastern Texas where her father had to ask permission from local Ku Klux Klan leaders before he went door-to-door to sell insurance. After high school, she studied law at New York University. She had worked for a decade at the Silicon

Valley legal powerhouse Wilson Sonsini Goodrich & Rosati, where she focused on complex acquisitions and corporate governance matters. At Twitter she avoided the limelight, and when she was forced to speak publicly, her voice would sometimes crack. She was fiercely protective of her connection with Dorsey, and some colleagues complained that she didn't give them enough insight into his movements and opinions.

In March 2020, Gadde realized that Twitter would need new rules banning misinformation about COVID so it could take down tweets that promoted ineffective treatments or spread conspiracy theories like the idea that 5G technology caused the virus. But there was a problem—Twitter had no framework for tackling misinformation and no staff responsible for policing it.

Unlike at other social media companies—which would simply swoop in to remove a post without explanation, and then retroactively justify the decision—at Twitter, there had to be a rule first, and the rule had to be public. Sometimes, Twitter published drafts of new rules and then gathered feedback over months—even years—before finalizing them and enforcing them. It was this bureaucratic approach that allowed Jones to stay on the platform much longer than he was allowed to remain on others. At Twitter, they had waited for Jones to break enough rules on the site to merit a ban.

But COVID changed things. Gadde turned to Yoel Roth, one of her deputies who had helped identify and root out Russian disinformation accounts after the 2016 election.

Roth, an earnest, baby-faced man, had spent most of his career at Twitter, fighting misinformation. He joined the company in 2015, soon after completing a PhD focused on online communication and communities, and had written the closest thing Twitter had to a misinformation policy not long before COVID struck. The new rules prohibited users from sharing photos and videos that had been altered by artificial intelligence. The imagery, known as "deepfakes," included fake videos of politicians making fictional statements or artificial pornography that placed someone's face on a different person's body, and were becoming commonplace with the more widely available and easy-to-use AI and video editing software. After consulting with his team, Roth decided people who shared deepfakes on Twitter would be banned in the most egregious cases, but if users created them to parody someone, Twitter would simply label the tweet, warning viewers it contained manipulated media.

Twitter executives were taken with the idea of labeling, no one more so than Dorsey, who was fed up with the pressure from Congress and the public over misinformation. His company had no business becoming an arbiter of truth, he thought, and it was in an untenable position in a polarized political climate where the idea of truth itself was contested. Labels allowed Twitter to avoid taking down anyone's speech wholesale. Every day felt like another fire drill, and Dorsey was becoming inured to it.

The approach of labeling some risky tweets let Twitter protect its reputation—it was doing *something* about misinformation—without needing to fact-check every single questionable post or overstepping into censorship of unverified content. Dorsey worried about censorship and wanted to err on the side of leaving tweets online.

Then the pandemic struck. Usage of Twitter spiked as people sheltered at home around the globe, totally reliant on their phones for contact with the world, and Twitter became a hub for doctors to share tips on avoiding the virus, for scientists to write long thread breakdowns of the latest research, and for an army of armchair experts to swap theories about the virus's origin, debate the efficacy of masks, or make their own predictions about the length and severity of the pandemic.

Twitter flooded with misinformation about COVID, and its team of content moderators were overwhelmed. The only tools they had were deleting tweets and suspending users, and they tried to reserve those punishments for the most dangerous posts—like ones encouraging fake "cures" that were, in fact, harmful.

Among the do-your-own-research experts on Twitter was Musk. On March 6, he tweeted, "the coronavirus panic is dumb." Two weeks later, on a day when the U.S. reported 2,000 known COVID cases, he posted that the country would be "close to zero new cases" by the end of April and that "kids are essentially immune" from the disease. (Children were not immune from the virus, and the U.S. averaged more than 20,000 daily COVID cases by the end of April.) Armed with seemingly little information aside from what he read in his Twitter timeline and the random links he clicked on, Musk argued with virologists and doctors on the site, cast doubt on the virus positivity rates in the country, and advocated for a drug—anti-malarial treatment hydroxychloroquine—that had no verifiable impact on the disease.

In some ways, Musk was simply mimicking the means by which he learned about rocketry or automobile development, fields in which he had little prior experience. He consumed publicly available information, made quick judgments, and posited in his typical contrarian fashion that he could do better than the experts. With rockets and cars, he had been wildly successful. So why couldn't he be right again, about COVID?

Musk also had a financial incentive to doubt the severity of the virus. A pandemic would destabilize the economy and, more important, the operations of Tesla and SpaceX. The people who worked on his manufacturing lines would be forced to stay home. Musk had always required his employees to show up for in-person collaboration and hands-on work, and COVID lockdowns would prevent that. On March 13, Musk sent out a company-wide email encouraging SpaceX's workers to continue coming in to the office because he had seen data that he claimed showed the illness "is *not* within the top 100 health risks in the United States.

"As a basis for comparison, the risk of death from C19 is *vastly* less than the risk of death from driving your car home," he wrote. "There are about 36 thousand automotive deaths [per year], as compared to 36 so far this year for C19."

While SpaceX stayed open—declared an essential business because of its government contracts—Tesla was obligated to follow California's stay-at-home orders, sending Musk into a rage. The shelter-in-place orders meant his workers weren't in the factory. Musk became fed up, tweeting at 11:14 p.m. on April 28, "FREE AMERICA NOW." He then replied to a far-right activist, who had declared that the scariest thing about the pandemic was how easily Americans "bow down & give up their blood bought freedom to corrupt politicians," and called her assessment "true."

Musk continued his tirade on a Tesla earnings call the next day, when he admitted the company was worried about not being able to quickly resume production of cars in Fremont. "The extension of the shelter-in-place, or frankly I would call it 'forcibly imprisoning people in their homes against all their constitutional rights'—that's my opinion—and breaking people's freedoms in ways that are horrible and wrong and not why people came to America or built this country," he said. "What the fuck? Excuse me. Outrage. It's an outrage.

"If somebody wants to stay in their house, that's great," Musk continued. "They should be allowed to stay in their house and they should not be compelled to leave. But to say they cannot leave their house and they will be arrested if they do, this is fascist. This is not democratic. This is not freedom. Give people back their goddamn freedom."

—

>>> AT TWITTER, the tsunami of tweets about the pandemic was unlike anything Gadde and her team had dealt with before. New details about the virus emerged daily, so a tweet posted today could be misinformation tomorrow.

Roth appealed to Dorsey. Twitter could use the labels it had built for deepfakes to add warnings to misleading tweets about the virus, he said. There was only one problem: the message cautioning viewers about manipulated media was hard-coded into the label Twitter had built, so it could only be used for one thing. Dorsey declared a "code red" and demanded the company's product and engineering teams build a label that could include any kind of warning Twitter's content moderators desired. Roth seized the new tool and quickly began adding labels to COVID misinformation.

The company did remove thousands of tweets under a policy that assessed "demonstrably false or potentially misleading content" that had "the highest risk of causing harm" but left up hundreds of thousands of others, including those from Musk, which questioned the official response to the pandemic but didn't cross Twitter's line into harmful misinformation.

With no end to the pandemic in sight, Dorsey doubled down on the label strategy. Heading into the 2020 U.S. presidential election, the company was under fire to do something about President Trump's endless cascade of reckless tweets. It had a long-standing policy of ignoring rules violations if they came from world leaders, arguing that the public interest outweighed potential harm. But maybe Twitter didn't need to delete messages from Trump, Roth thought. Maybe his tweets could be labeled, too. Dorsey gave the go-ahead.

Trump had started to rail against the electoral process, warning of fraudulent results and complaining about mail-in ballots, which were expected to fall in favor of his opponent, Joe Biden.

> **Donald J. Trump**
> @realDonaldTrump
>
> There is NO WAY (ZERO!) that Mail-In Ballots will be anything less than substantially fraudulent. Mail boxes will be robbed, ballots will be forged & even illegally printed out & fraudulently signed. The Governor of California is sending Ballots to millions of people, anyone. . . . .
>
> 8:17 AM May 26, 2020
>
> ○ 34K   ↻ 30K   ♡ 82K

> **Donald J. Trump**
> @realDonaldTrump
>
> . . . . living in the state, no matter who they are or how they got there, will get one. That will be followed up with professionals telling all of these people, many of whom have never even thought of voting before, how, and for whom, to vote. This will be a Rigged Election. No way!
>
> ○ 12K   ↻ 11K   ♡ 47K

The intervention Twitter undertook was small—just a link underneath Trump's tweet that suggested users "get the facts about mail-in ballots" by reading a CNN article. The reaction, however, was massive.

Republicans decried "censorship" and Trump blasted Twitter in more tweets later that day. The company was "interfering" in the election and "completely stifling FREE SPEECH," he ranted. While Dorsey took most of the heat, Trump's supporters also started to target other employees. Kellyanne Conway, one of the president's most senior advisers, trained her ire on Roth, after people started to dig into his old tweets looking for any proof of left-wing bias among Twitter's employees.

They quickly found what they were looking for. "I'm just saying we fly over those states that voted for a racist tangerine for a reason," Roth had tweeted in November 2016, when he was still a junior employee. In a 2017 post, he wrote there were "ACTUAL NAZIS IN THE WHITE HOUSE." The tweets were screenshotted and spread across the site, making Roth, one of the main people at Twitter who helped handle abuse campaigns, the direct target of one. In interviews, Conway decried Roth as a censor, and Trump piled on by posing for photos with a copy of the *New York Post*, which featured a story about Roth and his supposed censorship on the cover.

Twitter, of course, had seen abuse campaigns before, but few employees besides Dorsey had been targeted at this scale: Roth was suddenly an ant under the beam of light through a magnifying glass. As Twitter did damage control, Roth went dark. Twitter assigned a security guard to sit outside his home as death threats streamed in online. Dorsey also attempted to step in. "There is someone ultimately accountable for our actions as a company, and that's me. Please leave our employees out of this," he tweeted on May 27.

Twitter's labels emboldened Trump to tweet more, leading Gadde and her team to build a process for labeling Trump's tweets. Only a few people could make the final call to slap a label on the president's posts, including Gadde herself; Del Harvey, the head of safety; and Roth. When Twitter's content moderators flagged a tweet from Trump that seemed to violate Twitter's rules, Harvey and Roth would receive an urgent page. If they didn't respond, Gadde would receive one next.

The executives lived in the Bay Area, near Twitter's headquarters, and were routinely jolted awake in the wee hours of the morning by another alarming message from Trump, who habitually spent time in the morning tweeting from his bedroom in the White House before beginning his day. The process quickly shifted from a monumental one of historical consequence—a Twitter employee deciding to curtail statements by the leader of the free world—to a minor irritation squeezed in between morning alarms, coffee, and getting kids ready for school. Sean Edgett, the company's general counsel, was added to pager duty for Trump's wild tweets, but Dorsey was purposely left off the list. His global travel and routine absences sometimes made him tough to reach.

To some of his employees, Dorsey seemed increasingly disillusioned with Twitter and its daily free-speech fights. Over time, Twitter had created a strike policy that booted users off the platform after they received several warning labels. Dorsey began to wonder if Gadde was going too far by banning people, making Twitter into a censor again—the thing he had hoped labels would prevent. Dorsey also hated the requests from government officials, including the FBI and White House, who flagged tweets to be taken down that they believed were in violation of the company's policies. It felt like the company was doing the bidding of too many outsiders and deviating from its mission to be an open public square.

Dorsey's disagreements with his top policy deputy burst into public in October, just weeks before the U.S. presidential election. Ahead of the November vote, the FBI had repeatedly warned Twitter and other Silicon Valley companies to be prepared for hack-and-leak campaigns similar to those seen in 2016, when Hillary Clinton's emails had been published in part on Twitter by a hacker.

As Gadde's team received more warnings from the FBI, they steeled themselves for something similar. Two years earlier, as the company took stock of Russian meddling on its platform, it had instituted a new rule forbidding users from sharing information gleaned by hacking. If such an occurrence were to happen again, Twitter would block ill-gotten documents or information from appearing on the platform and freeze any accounts that tried to share them.

It didn't take long for that policy to be put to the test. On October 14, 2020, the *New York Post* published an explosive article featuring emails extracted from a laptop belonging to Hunter Biden, presidential candidate Joe Biden's son. The messages showed the younger Biden had brokered a meeting between his father and a Ukrainian executive he worked with, in contradiction to Joe Biden's claim that he never involved himself in his son's business dealings. There were also nude images of Hunter, a violation of Twitter's policy forbidding explicit pictures shared without the subject's consent, and images of him using drugs.

The provenance of the emails was murky. The *Post* reported that Hunter's laptop had been turned over by a repair shop.

To Twitter's executives, the story seemed to have all the telltale signs of a hack-and-leak: embarrassing emails leaked from an unknown source, just in time to tip the scales of a presidential election. Gadde quickly made the decision to block the link to the article from being shared on Twitter. She also greenlit the suspension of the *New York Post*'s official Twitter account, a move that blocked the publication from sharing any other stories until it deleted its tweet that featured the Hunter Biden article.

The backlash was immediate and furious. Republican lawmakers and Trump campaign officials accused Twitter of censorship, and even some Democrats questioned whether Twitter, by cracking down on a mainstream media outlet, had overstepped. Even Dorsey objected. "Our communication around our actions on the @nypost article was not great," he wrote. "And blocking URL sharing via tweet or DM with zero context as to why we're blocking: unacceptable."

His statement seemed confusing. Who was Dorsey criticizing? Internally, however, employees knew where the message was being directed. While Dorsey had been a proponent of empowering the executives he put into positions of power to make their own decisions, he was now effectively admonishing Gadde. She took his tweet personally, particularly because he was more direct and harsh in what he said publicly than the criticism he shared in private, she told people close to her.

On October 16, Twitter decided to limit the hacked materials rule only to prevent hackers themselves from sharing information they stole, and to add contextual labels to other tweets about the Biden emails, warning viewers that they came from an unknown source. The *Post* would once again be able to share its links, but the controversy continued for weeks.

—

>>> ON ELECTION DAY, November 3, Twitter had a team tracking misinformation and voting results around the clock. As it became clear that Biden was going to beat Trump, it monitored attempts to undermine trust in the electoral process. The company labeled some 300,000 tweets over a two-week period covering the election and its aftermath. Nearly 40 percent of Trump's election tweets in the four days after the election received labels, warning that their content "might be misleading about an election or other civic process."

Dorsey had gone along with the labeling or the outright removal of misleading information about COVID in the pandemic's early days. But with the advent of newly created vaccines from major pharmaceutical companies like Johnson & Johnson and Pfizer, he balked at the idea of taking action against

A POLYNESIAN VACATION

tweets that questioned the efficacy of the shots. He had given Gadde free range to make decisions previously, but by the spring, he began micromanaging the moderation policy process around vaccine and pandemic content.

He demanded to be added to an internal email thread used by moderators that sent out an automated message when a tweet was removed for violating the COVID misinformation policy, and often forwarded them to Gadde, second-guessing the moderators' decisions. Twitter's CEO later objected to labels applied to the account of Alex Berenson, a former *New York Times* reporter who became a vocal opponent of the vaccines. Dorsey also questioned the removal of tweets that discussed vaccination, bickering over whether they truly violated Twitter's policy against spreading misinformation. Dorsey's emails, often with few words, landed in Gadde's inbox on a weekly basis, a steady source of paper cuts to her relationship with her boss.

She and others couldn't understand why he had suddenly involved himself. Still, Gadde tried not to make decisions about COVID policy without speaking to Dorsey. "I need to talk to Jack about this, and we'll hear from him when we hear from him," she would tell her team. Plans about how to handle incendiary tweets about the virus often stagnated.

Some employees speculated that Dorsey was skeptical of the vaccines. Over time, Dorsey's closest staffers would assume he was unvaccinated, although because Twitter's offices remained closed and Dorsey did not come in, they never knew for sure.

>>> IT WAS AROUND NOON on January 6, 2021, in the middle of the South Pacific, when Dorsey got the call from Gadde, recommending he put the president of the United States in time-out. He was on Tetiaroa, a pristine atoll in French Polynesia, in a luxury resort called the Brando. It was from here, on volcanic-formed land, shielded from the wild open ocean by a natural fortification of living coral, that Dorsey had decided to run his two businesses as most of the world remained closed during the pandemic.

Most employees had no idea where Dorsey was and some executives tried to keep his whereabouts secret. It would be horrible for morale if workers—many of whom were cooped up working from their beds or couches—knew their leader was on an island with thirty-three private villas, kayak tours, and a vast spa with personalized massage treatments. A few months earlier, Kim Kardashian had spent $1 million to rent the whole resort for her fortieth birthday, drawing widespread condemnation for partying in the pandemic.

Working from the Brando, three hours behind San Francisco headquarters, Twitter's chief executive dialed into a Twitter leadership meeting to go over the company's goals for the year. It was at that moment that executives started to get alerts about a large group of Trump supporters amassing in Washington, DC. They had been drawn there by the president with the promise of an epic rally. Trump had tweeted repeatedly that his vice president, Mike Pence, could overturn the outcome of the election.

Dorsey had spent more than four years resisting the idea of booting Trump off Twitter, digging into his position the more he was criticized for it. Twitter's leaders decided to finish the goals meeting, though some kept one eye on their platform and developing news as Trump's supporters made their way to the U.S. Capitol.

To Twitter's trust and safety employees, the potential for violence was glaring. Harvey and Roth had warned for months that Trump would use his account to stir up trouble and argued for his removal from the platform. A few weeks after the election in November, Roth had drafted a document he titled "Post-Election Protests and Calls for Violence" that outlined the kind of activity Twitter might see in the wake of Trump's loss, and how the company should respond.

If there were tweets calling for protests based on misinformation—that the election had been rigged or ballots had been altered—those tweets would be labeled. Calls for violence from extremist groups would be removed. Roth also wrote code that scanned Twitter for the coded phrases Trump had used to rile up supporters, like "locked and loaded." Tweets that contained the phrase would be triaged by the company's content moderation team within thirty minutes, a rapid response in case they contained specific election-related threats.

As the riot unfolded, however, Roth found there was little he could do. He kept the television in his home office on mute, aghast as the mob flooded through the Capitol, and tried to focus on Twitter. But few in the crowd were tweeting about their exploits in real time. Most of the posts about the riot were coming instead from journalists, news outlets, or terrified online observers.

Roth also watched @RealDonaldTrump. After giving a speech near the White House to his supporters, the president retreated to his residence to egg them on from there as they proceeded to the Capitol. "Mike Pence didn't have the courage to do what should have been done," he tweeted, raging that his second-in-command had not stopped the certification of the vote and that the election was being stolen from him. His "sacred landslide election victory" had been "unceremoniously & viciously stripped away," he wrote.

Roth and his manager, Harvey, agreed that it was time to suspend Trump. He was clearly undeterred by the hundreds of warning labels that had been added to his tweets. His post about Pence posed a threat to the vice president's safety and warranted a permanent ban, Roth argued. He and Harvey drafted an extensive letter outlining their rationale and presented it to Gadde in a video call.

But the idea of banning a sitting president was too much for Gadde, and she worried Dorsey wouldn't approve even if she went along with the suggestion. For regular users, there was always an escalatory process before they got banned. They would get warnings, and then be blocked from tweeting for a "cooling off" period. If they continued to break the rules after that, they were finally suspended. "We haven't done that yet," Gadde argued during the call. "We haven't timed him out yet."

After Gadde signed off, Roth and Harvey sat in silence for a moment, just staring at each other. There was nothing to say.

Roth and Harvey rewrote the recommendation for Gadde, asking for Trump to get a temporary time-out instead. "Any further policy violations will result in suspension," Roth wrote. He stared at the words on his screen, then went back and bolded, italicized, and underlined them to hammer home his point:

"*Any further policy violations will result in suspension.*"

As Trump's supporters ransacked the Capitol, he released a video on Twitter, repeating his claims that he had been cheated out of the presidency.

Each of his tweets seemed to manifest more violence in the real world. As long as he had his favorite megaphone, there wasn't an end in sight. The safety executives expanded their document to include Trump's video.

Gadde took the document to Twitter's most senior executives, who threw their support behind it. Beykpour took notice of Roth's emphatic text. "Does this mean we would be suspending him for anything else?" Beykpour asked. "Good."

"This is the right thing to do," Matt Derella, the company's chief customer officer, commented at the top of the document. Roth watched as the executives' comments appeared in the margins—but he didn't see anything from Dorsey.

From his villa, Dorsey watched the events unfold on Twitter and joined various emergency calls that sought to deal with the madness. Gadde called him to tell him the news. The steady refrain of false claims—and the violence that stemmed from them—were finally enough. On the phone, Gadde told Dorsey that she had decided to lock Trump's account for twelve hours. From his retreat halfway around the world, Dorsey assented.

Dorsey sent a company-wide email to his employees, saying it was important for Twitter to follow its own rules and allow any user, even Trump, to return after a temporary suspension.

Roth felt shell-shocked and confided in Gadde. "I feel like I have blood on my hands," he told her.

"I think you're taking too much of this on yourself," Gadde responded. "It's not just your decisions, it's not blood on your hands. We are making these decisions together."

Employees were livid. Their leaders had not done enough. Many had long believed that Trump had no place on the platform. They watched as their peer companies, like Facebook, unilaterally banned the former president. Meanwhile, they were giving the president back his favorite megaphone after a night's sleep.

More than three hundred employees signed a letter directed to Dorsey and other company leaders. "Despite our efforts to serve the public conversation, as Trump's megaphone, we helped fuel the deadly events of January 6th," read the letter, which was delivered to executives on the morning of January 8. "We must learn from our mistakes in order to avoid causing future harm. We play an unprecedented role in civil society and the world's eyes are upon us. Our decisions this week will cement our place in history, for better or worse."

The letter called for Trump to be immediately suspended. But some senior engineers at the company went even further. They began discussing what actions they would take if Dorsey refused, and decided they would stop working by the end of January if Trump remained on the platform—either holding a strike or resigning en masse.

Trump returned from his twelve-hour time-out unrepentant. He declared he would not attend Biden's inauguration. In another tweet on January 8, he referred to his supporters as "great American Patriots" whose voices would be heard long into the future. "They will not be disrespected or treated unfairly in any way, shape or form!!!" he wrote.

Harvey believed the tweets were an invitation for violence at the inauguration. Gadde and Edgett were resistant—the messages were somewhat coded, and could be interpreted as benign by some readers. Several lawyers at the company leaned on Edgett to change his mind, while Harvey showed Gadde the responses to Trump's tweets. His supporters were interpreting it as she feared—an incitement to attack.

That afternoon, Harvey and Roth sat down again to write a new recommendation: Trump had to go. As they drafted it, they stayed on the phone and argued their position with Gadde, who eventually caved. As they continued to work, Gadde signed off to call Dorsey and Twitter's board.

The board agreed with Gadde's recommendation. But Dorsey had one demand—if Trump was going to be removed, Twitter had to publish its reasoning publicly for the world to see.

Suddenly, the document that Harvey and Roth were throwing together had to serve as the official explanation for the historic decision to silence America's leader, cutting him off from his 88 million followers.

They quickly cleaned up the document for publication. World leaders "are not above our rules entirely and cannot use Twitter to incite violence," the two policy executives wrote. Trump's new tweets "are likely to inspire others to replicate the violent acts that took place on January 6, 2021, and there are multiple indicators that they are being received and understood as encouragement to do so."

Harvey had no time to contemplate the decision. She was already worried about the consequences for Twitter itself. Removing an account like Trump's, which occupied a large chunk of Twitter's social graph—the web of following and follower connections between users—could easily crash the entire service. Freezing the account was only part of the process. It also had to be removed from the social graph—other users' follower and following lists—and from the thousands of block lists of people who had gotten fed up with seeing Trump in their feeds. She called several engineers who worked on the social graph and warned them that Trump was about to go, and that he might take the entire site with him.

Having faced the backlash of Trump supporters once before, Roth was worried about safety. He moved the document he and Harvey had written to a burner account and took out the names of all the other employees who had worked on the draft. If it were leaked, the employees who had made the decision to ban Trump would be anonymous.

Then, just after 3:00 p.m., it was time.

On the internal dashboard that governed all Twitter accounts, there was a big red button: "PERM-SUSPEND." If clicked, it would permanently terminate a user's account, erasing their social graph. Harvey decided she would be the one to click it.

Roth stood up from his desk and walked upstairs to his living room, where his husband sat watching the news. "Something is about to happen," Roth said. Moments later, the news broke: @RealDonaldTrump was gone.

The days of collective deliberation didn't stop Dorsey from once again second-guessing Gadde and her team in public. Dorsey painted the Trump decision as one he wasn't responsible for and didn't wholeheartedly defend. "I believe this was the right decision for Twitter. We faced an extraordinary and untenable circumstance, forcing us to focus all of our actions on public safety," he tweeted on January 13. "That said, having to ban an account has real and significant ramifications. While there are clear and obvious exceptions, I feel a ban is a failure of ours ultimately to promote healthy conversation."

In the aftermath of the Trump ban, Harvey grappled with anxiety. She had seen the underbelly of the internet, but the Capitol riot weighed on her more heavily as some commentators rushed to trivialize the event. She believed a political game was being played with people's lives, one that glossed over a societal failure. She scheduled a leave of absence from Twitter for that fall. In June, when the Nigerian president Muhammadu Buhari tweeted a threat to regional separatists, invoking the Nigerian Civil War and promising to treat his opponents "in the language they understand," Harvey acted quickly, temporarily suspending Buhari's account and requiring him to delete the call for violence.

When Gadde realized what Harvey had done, she was frustrated that she hadn't been consulted. The Nigerian government responded by banning Twitter in the country, an embargo that would take months to resolve and cause tension between Gadde and Harvey. Trump cheered the decision in a statement, saying, "Congratulations to the country of Nigeria, who just banned Twitter because they banned their President." In October, just one day shy of her thirteenth anniversary working at Twitter, Harvey resigned.

To some close to him, the Trump ban also broke Dorsey. It was the red line he had spent nearly half a decade promising not to cross. Twitter wasn't an idyllic, free, and open town square. "It was like he was the kid who built the robot, which then destroyed the world," said one former Twitter executive.

7 > Resource Plan

On February 25, 2021, Twitter staged a flashy event called Analyst Day to introduce investors to the new corporate goals it had outlined with Cohn—doubling its annual revenue to $7.5 billion by the end of 2023 and attracting 315 million daily active users by the same deadline.

Kayvon Beykpour, Twitter's consumer product lead, and Bruce Falck, who led the product team dedicated to advertising, were largely responsible for the projections and were confident in their ability to hit them in about three years. Beykpour would attract millions of new users with innovative features, and Falck would drive up revenue with increasingly precise ad targeting.

But Falck's figures seemed especially aspirational to other executives. He had claimed he would be able to get Twitter to $10 billion in revenue by the deadline. Ned Segal, a dark-haired and dimpled former Goldman Sachs banker who was the company's chief financial officer, crunched the numbers and scaled back Falck's figure before allowing Dorsey to present it to investors, warning the board that it was unreasonable.

Even the scaled-back figures seemed outlandish to some executives—the equivalent of a longtime NBA benchwarmer saying he would average thirty points a game and become an All-Star in three years. They meant the company would need to increase its 2020 revenue of $3.7 billion and dramatically improve its user figures, which stood at 192 million daily active accounts.

The projections made leaders across the company incredibly uncomfortable. "Some of us looked at each other knowing there was no way we'd hit those targets," one said.

Still, Dorsey trusted Falck and Beykpour to get the job done. To help them, Dorsey had asked the board to support what he referred to as the "Resource Plan," which amounted to a spending spree. Like most tech companies, Twitter had come through the height of the pandemic less bruised than its leaders had anticipated. There had been huge bumps in engagement as people glued themselves to their phones during lockdowns. Dorsey wanted to use the extra cash on Twitter's balance sheet to hire his way out from under Elliott's scrutiny.

Dorsey's plan was to hire and acquire as many people as possible. The new hires—and several key start-ups that Twitter would snap up—would build new products and bring new innovations to Twitter, which in turn would bring new users. And as the new users flocked in, so would the advertisers. That would all lead to a massive increase in revenue, Dorsey believed. The board rubber-stamped Dorsey's plan, giving him freedom to spend as he desired.

At Analyst Day, Dorsey wore a long, graying beard and a slicked-back ponytail of hair that had grown long during the pandemic. He was surprisingly candid about Twitter's failings.

"We're slow, we're not innovative, and we're not trusted," Dorsey said, assuring Wall Street that all that was about to change. It was a victory lap for Cohn, who had successfully pressured Dorsey to focus on revenue. Cohn's other obsession hadn't faded—he still believed that Dorsey was being pulled in too many directions to diligently lead Twitter and that he needed to step down. But they were in a stalemate. Dorsey refused to leave while Cohn was on the board, and Cohn didn't want to relinquish control until he was sure that Dorsey would go.

While the governance committee agreed Dorsey could remain Twitter's CEO, Cohn pressed the board to come up with a succession plan, and they put together a list of executives who would be able to fill Dorsey's shoes when the time came. The governance committee saw potential in Agrawal, a brilliant engineer and Twitter's chief technology officer, who had an intensity that Dorsey lacked.

Although he tended not to draw attention to himself, Agrawal had wisely cultivated his relationships with key board members. He rarely missed a board meeting and spent time carefully walking the directors through the logistics of Twitter's infrastructure. Crucially for Cohn, Agrawal also had the trust of Dorsey and many key technologists at the company. He was a successor everyone would be able to stomach.

Dorsey grew distant as Cohn's meddling continued. During one company-wide meeting, he tried calling in, but his internet failed. In another, he dialed in from a Costa Rican getaway and gave a lecture on the power of Bitcoin, his iPad camera pointing upward at him at an unflattering angle. Employees shifted uncomfortably in their seats. *We don't really do anything with Bitcoin here,* some thought. After the meeting, some placed bets on when Dorsey would be on his way out of the company.

—

>>> BY THE END OF MARCH 2021, Dorsey was fed up with Cohn. A month earlier, he had stepped in front of a camera and committed to the goals that Cohn demanded. It was time for Cohn to go, Dorsey told his board. The board began negotiating for Cohn to exit its ranks. It was a delicate conversation—Cohn and Dorsey each wanted to emerge looking like they had bested the other.

Cohn could depart on a high note, the board suggested. Twitter's stock was performing beyond anyone's wildest expectations—its price was over $70, a whopping 95 percent higher than when Cohn had joined the board a year earlier. And the succession plans were in place for when Dorsey was ready to exit. On March 31, Cohn agreed to leave the Twitter board once the company found a director to replace him, and by June, a substitute had been identified.

Twitter replaced Cohn on June 9 with Mimi Alemayehou, a Mastercard executive who focused on development and finance in Africa. In a terse statement, Dorsey briefly acknowledged Cohn's role in redirecting the company's goals. "We continue to build upon our strengths and are proud of our progress. We are appreciative of Jesse's input and support during an important year for us," Dorsey said.

Having rid the company of Elliott, Dorsey kept up his global travels as Twitter's offices remained shuttered. In August, he and his good friend, the music producer Rick Rubin, traveled to Boca Chica Village, a small Texas town on the Gulf, to marvel at Starbase, SpaceX's rocket launch facility. Dorsey snapped a photo of a massive stainless steel test rocket to further cement his bromance with his host. "Grateful for @elonmusk & @SpaceX ♥," he tweeted. Behind the scenes, Musk and Dorsey maintained contact. The SpaceX CEO sometimes messaged Twitter's leader about his frustrations with the platform, notably about an account called @ElonJet, which tracked the whereabouts of his private plane using public flight data. Musk pressured Dorsey to ban the account, though Gadde and her team determined that the account did not violate any of Twitter's rules.

Travel distractions and frustrations with Elliott aside, Dorsey still cared about Twitter. That September, Esther Crawford, a director of product management who had come to the company after it acquired her start-up, met Twitter's leader for the first time. She had prepared to explain her vision for allowing content creators to make money on the platform, a potential new stream of revenue for Twitter, which had been looking to expand its business from its dependence on advertising. If creators chose to accept cryptocurrency payments, her strategy might also be a way for Dorsey to bring Bitcoin into Twitter.

Staring piercingly from the other side of the video call, Dorsey was more philosopher king than chief executive.

"We don't want to turn Twitter into a casino," he said. To Crawford, Dorsey felt like a protective parent guarding his creation. Then he veered into a critique of capitalism.

"The corporation is bigger than the individual and the corporation just wants to make money at all costs," Dorsey continued. "Even if you have good intentions, the power of the corporation will corrupt. All of us could be corrupted. I could be corrupted."

While Dorsey had made a fortune worth billions of dollars off Twitter, he was anxious that his product, a communications platform meant to connect people, could survive and grow only because it was a business. Perhaps this was Twitter's original sin, attempting to make money in the first place. He continued to discuss with Crawford the ethics of implementing creator monetization.

"There are mothers in Africa who will literally spend money on Twitter rather than give it to their children," he said.

The feature, a subscription service for Twitter creators to allow subscribers to access premium content, launched a few weeks later. But @Jack would not be around long to witness its progress.

8 > Parag

On the morning of Thanksgiving in 2021, one of Twitter's board members posted an odd tweet.

Martha Lane Fox, a British businesswoman who had sat on Twitter's board since 2016, wished a happy holiday "to all the people im lucky enough to work with stateside."

A dot-com-era entrepreneur, philanthropist, and baroness member of the House of Lords, Lane Fox added a bit of international flavor to Twitter's collection of directors. While she didn't celebrate the holiday, she rattled off a list of Americans she was thankful to work with at Twitter: Leslie Berland; Sean Edgett; Vijaya Gadde; Twitter's leaders for human resources, Jen Christie and Dalana Brand; and Ned Segal.

One name was conspicuously absent: Jack Dorsey.

For employees who kept tabs on the board, it was a potent signal that something might be amiss with their chief executive. Some of them wondered: What has Jack done now?

Four days later, on November 29, Dorsey was gone from Twitter's top job. The news leaked to the press before Dorsey had the chance to announce it internally, and Twitter employees woke up to find news articles on Dorsey's exit plastered across their timelines. They frantically swarmed into their workplace chat app, Slack, and into their group chats, searching for any scrap of confirmation.

But Lane Fox, a few other board members who had been privy to the succession talks, and a handful of Twitter executives in whom Dorsey confided had been bracing themselves. Earlier that month, Dorsey had dialed in to the board's quarterly meeting and somberly announced that he planned to step down as chief executive. Several members of the board, who had not been aware of the governance committee's plans, were shocked at the sudden proclamation.

As the news sank in, it began to make more sense. Dorsey had been increasingly disillusioned. When he bothered to come to meetings, he seemed sullen and taciturn. He had one final demand for the board: His successor would be Parag Agrawal.

At first, a few board members balked. They knew Agrawal as a thirty-seven-year-old soft-spoken engineer who toiled in the background to rehab the company's decaying infrastructure while its product, finance, and policy leaders made headlines. Sure, he had been inked in Twitter's succession plan as the person to temporarily take the reins should Dorsey unexpectedly go out of commission—but some of them thought that Agrawal needed more development before he could become the boss. Most major companies did contingency planning and picked replacements for key leaders. But it was one thing to imagine an inexperienced leader taking over for an interim period, and another to hand him the job permanently. Some thought a formal search should be conducted to seek out Dorsey's replacement.

But Dorsey was adamant and left the board little time to consider any alternatives. By departing immediately, he was forcing the board's hand. They could go with Agrawal or be left with no permanent CEO, making it seem as if the directors were unprepared. The direction of the company and its stock price would be uncertain.

The board, which had backed Dorsey over and over again, decided to comply, partially, with his demands. There would be no recruiting firm hired, no formal search conducted, no candidates interviewed. Instead, members of the board could toss out names of executives they thought were capable of the job, and the others would consider them. David Rosenblatt, a long-tenured board member who had run the ad platform DoubleClick before it sold to Google in 2008 for $3.1 billion, had overseen Dorsey's return to Twitter in 2015. He suggested interviewing leaders from other social media companies. Others suggested offering the top job to Agrawal on an interim basis while they searched for a permanent leader.

Their backs against the wall, the board members agreed to elevate Agrawal. An offer letter was hastily drawn up and presented to the introverted technologist on November 29, the Monday after Thanksgiving. "The Board believes that you will add tremendous value as the Company's CEO," Bret Taylor, Twitter's chairman, wrote in the letter. It was mainly boilerplate language, but it outlined his compensation package, which, typical for a top Silicon Valley job, was a pretty penny.

He would earn a million-dollar salary and be granted $12.5 million in stock that he would earn out over time. He would also be protected in the event he was suddenly fired; in that case, his stock would be paid out on an accelerated schedule and he would receive severance payments. Taylor also used the letter to push back on some of Agrawal's doubters.

"The Board wishes to reiterate our excitement at your accepting this new leadership role," Taylor wrote.

After rumors of his departure had swirled online since the early hours of November 29, and on the home pages of every major American news site, Dorsey finally addressed them. "Not sure if anyone has heard, but I resigned from Twitter," he tweeted sarcastically just before 8:00 a.m. He also shared the victory lap email he had sent to employees moments earlier, announcing his departure and Agrawal's appointment. The subject line was: "Fly."

"Parag started here as an engineer who cared deeply about our work and now he's our CEO (I also had a similar path . . . he did it better!)," Dorsey wrote. Like Taylor, Dorsey made sure to seed the narrative that Agrawal had the board's unwavering backing.

"The board ran a rigorous process considering all options and unanimously appointed Parag," Dorsey told employees. "He's been my choice for some time given how deeply he understands the company and its needs. Parag has been behind every critical decision that helped turn this company around. He's curious, probing, rational, creative, demanding, self-aware and humble."

His praise for Agrawal seemed sincere to those who knew both men well, though Dorsey's description of the board's "rigorous process" wasn't accurate. It had been rushed, precisely because he wanted it to be.

Employees reeled. Dorsey's departure, announced just after the holiday, seemed abrupt and ill-timed. Was another activist investor like Elliott creeping into the stock, forcing Dorsey out before he was ready to go? Much of Twitter's rank and file didn't know how absent Dorsey had become over the past year.

Twitter's workforce was even more surprised to see Agrawal rise to the top job. He wasn't a visible leader within the company, and many workers, had they known Dorsey was planning his exit, would have placed their bets on Beykpour, Segal, or Gadde. Beykpour had a handle on the product, Gadde had been Dorsey's right-hand woman for years and taken on the company's toughest political issues, and Segal seemed well equipped to run the business after keeping it mostly in the black during the tumult of the pandemic.

Most employees hardly knew Agrawal—if they knew him at all. And while Dorsey credited him with turning Twitter around, his fingerprints weren't visible to many within the company.

Agrawal was eager to change that. In mid-November, soon after he had been offered the top job at Twitter but before the announcement, he called a trusted adviser and told them Dorsey would soon be stepping down. Agrawal would become Twitter's chief executive. And he wanted to discuss who to fire first.

Agrawal wanted to restructure and streamline the company, getting rid of several leaders in the process. Some of those leaders were his closest friends and Twitter's longest-serving employees, but Agrawal wasn't sentimental. He believed he needed to do what was best for Twitter, and that meant radical change to invigorate the company.

"I need you to help me," Agrawal told his friend. He wanted advice, and he liked to reason through difficult decisions by having a trusted deputy argue each of the pros and cons. Agrawal liked to puzzle through the counterarguments to his own opinions, and wanted to debate getting rid of allies as well as enemies. In his mind, no one person—not even himself—was more important than Twitter.

He also needed to figure out what to do with his own small team. Despite being the company's CTO, Agrawal had less than forty people reporting to him. It was an unusual setup that left him with little management responsibility and generated accusations from detractors that he sat in an ivory tower, disconnected from the company's actual employees. In a few days, Agrawal, who had always avoided managing people, would be responsible for more than 7,000 workers. The management of his small team of employees would need to be taken over by someone.

As he shifted people around, he began thinking of another bold idea: Why not restructure the entire company? Rebuilding Twitter's corporate leadership could dissolve the blockages that had plagued Dorsey's second tenure as its leader and allow the company to move faster. Twitter suffered from having three men responsible for various bits of its product—there was Beykpour, who oversaw the side of the platform that its users saw; Falck, who led products for advertisers; and Agrawal himself, who oversaw its infrastructure. All of them funneled into Dorsey, vying for his attention and trying to get him to back their decisions even as he drifted further away from Twitter. The engineering team, led by longtime tweep Mike Montano, and Twitter's design team, run by a former Facebook executive named Dantley Davis, drifted in between, each also reporting to Dorsey but dispatched to work on various bits of the product. The convoluted structure tangled the chain of command. Agrawal considered it a five-headed monster.

Agrawal didn't need to stroke egos by handing out C-suite titles, and he didn't want to waste time sorting out who was making bad choices. He wanted a few effective lieutenants, and clear accountability for screw-ups. He suggested thinning out Twitter's top ranks to just three executives, including himself.

The person on the other end of the phone call was overwhelmed. The timelines Agrawal proposed for firing executives and revamping the company were aggressive. Agrawal had a lot he wanted to do, and he was eager to act.

>>> AGRAWAL HAD BEEN working toward the role for more than a decade. He was raised in Mumbai by an academic mother, who had taught economics, and a science-minded father, who worked for the country's Department of Atomic Energy. At first, he followed in his father's footsteps, studying at the Atomic Energy Junior College in Mumbai, but soon gravitated to computer

science. He graduated at the top of his class at the Indian Institute of Technology Bombay, and immigrated to the United States in 2005 to enroll in a PhD program at Stanford University. India's technology institutes were revered in Silicon Valley for churning out exceptional engineers, and Agrawal, a strikingly confident and relaxed student, enjoyed a warm welcome at Stanford. He considered several professors before joining a research team focused on data management, led by Jennifer Widom.

In the third year of Agrawal's studies, Widom took a sabbatical to travel. She was gone for the whole year, isolated from her students by distance and unreliable internet in some of the remote locales she visited. Without her supervision, some of the students struggled. But not Agrawal, who "blossomed in his independence and capabilities," Widom recalled.

Before he could finish his dissertation, Agrawal got an intriguing offer to head north to San Francisco for a job at Twitter, a nascent social media company that had been founded a year after he arrived in America. It wasn't the kind of tech giant that typically snapped up Stanford graduates, many of whom were lured by the more lucrative salaries offered by Google or Facebook, but it did give Agrawal the kind of database work that occupied his attention at school.

"Oh don't worry, I'm going to finish my thesis," he promised Widom. Then, in October 2011, he vanished into his work at Twitter. It took him another year to complete the dissertation, and he ended up returning to Stanford for a few months to focus on the work. When Widom signed off in the summer of 2012, Agrawal celebrated with a rare tweet: "Finally," he wrote, sharing a photo of the finished, signed dissertation.

Agrawal rarely participated in the combative, goofy, and high-velocity discourse on Twitter. If he tweeted at all, it was usually to reshare travel photos from his Instagram or snippets about his relationship with Vineeta, a fellow Stanford student whom he would eventually marry, or to celebrate Twitter milestones, like a 2013 blackout that wiped out power during the Super Bowl and sent hordes of users rushing to the site to chatter about the outage. He also occasionally fanboyed over Elon Musk, retweeting his posts about rocket launches or his revelations of new Tesla models. In September 2015, he shared a photo from the launch event for the Model X, with Tesla's chief executive standing front and center as he presented the new white electric SUV. To a certain subset of Silicon Valley engineers who were Agrawal's peers, Musk represented the ideal leader: a visionary technocrat guided by science and engineering who listened to his own instincts above all else.

In 2018, Twitter gathered its entire workforce together in San Francisco's Moscone Center for its first-ever OneTeam. Dorsey took the stage to wild applause from his employees. He was deep in his guru phase and spoke about drinking his salt juice for maximum hydration. Then he told his employees to check under their seats.

They reached under their chairs to find gift bags filled with bottled water, a packet of lemon juice, and a little baggie of salt. Agrawal dumped all the lemon juice and salt into his water bottle, gulping the concoction before Dorsey could finish walking through the recipe. "Just a pinch of salt," Dorsey advised.

>>> DORSEY VIEWED AGRAWAL as a confidant, someone with a holistic vision for Twitter who was also fluent in its code. The two sometimes met on weekends, with Agrawal finding ways to draw the taciturn Dorsey out of his shell and into debates about the future of the platform. Dorsey seemed to appreciate Agrawal's bluntness—he wasn't afraid to disagree with Twitter's founding father and didn't seem cowed by his celebrity status like other employees, who would stop the chief executive in the office to ask for selfies. While other executives tiptoed around him, Dorsey could count on Agrawal for an honest take.

But Agrawal was blunt with his teammates, too, and his hard-charging nature sometimes clashed with other engineers who worked alongside him. He could be intractable, and when his team disagreed with him, he would sometimes keep pushing what he wanted anyway. Before he became CTO, he began arguing that the company had to stop running its own servers and move everything into a cloud service like Google Cloud or Amazon Web Services.

Google and Amazon offered flexibility, allowing companies to expand quickly without having to buy and build more server space, giving startups and Fortune 500 corporations alike the ability to be more agile and not worry about the upkeep of infrastructure. That was the world Agrawal wanted to operate in. At Twitter, no one could get servers fast enough to launch their projects, and he was annoyed by the slow pace.

There was just one problem with his plan: cost. Twitter's in-house way of piecing together its infrastructure in its own data centers came at a fraction of the cost the company would be charged by Google or Amazon. Still, Agrawal wouldn't drop it. Years later, when he became Twitter's chief technology officer, Agrawal finally executed his plan, inking multimillion-dollar deals with Google and Amazon, even while Twitter still ran its own data centers. Agrawal had a knack for turning Dorsey's idealism into concrete plans. In 2019, Dorsey tweeted he would ban political advertising, catching his policy specialists by surprise. He asked Harvey to craft a new rule based on his tweets but said Agrawal would have the final say—Dorsey was leaving for a silent retreat.

Dorsey was fascinated by Bitcoin and wanted to find ways to incorporate the blockchain, the public ledger technology upon which the cryptocurrency was built, into the social media service. Both agreed Twitter had to stop moderating content by hand, picking and choosing which tweets violated its bespoke rules. Eventually, most tweets would need to remain live on the platform, with algorithms choosing which should be circulated widely and which should be choked off from the company's powerful distribution systems.

They also wanted Twitter to be more transparent, so its users would understand its inner workings. In meetings with other engineers who believed in the power of decentralization, or the idea that technology shouldn't be controlled by one person or company, they concocted plans to build a new version of social networking with publicly available code. Instead of hiding away its proprietary secrets, Twitter would tell everyone how it functioned. The new Twitter would rely on a protocol, which was a way of saying that it would be a web service that anyone could build on, similar to the one that served as the backbone of all email services, allowing a Google mail account to connect with a Yahoo one.

It was an audacious plan. If Dorsey and Agrawal succeeded, they could end the era of empires in social media. There would be no more walled gardens lorded over by Mark Zuckerberg or others who owned pieces of people's online identities. Instead, users could easily flit from Twitter to Instagram to TikTok, posting to each platform as they saw fit with a single identity. The plan could someday rid Twitter of its persistent annoyances— shrill questions from members of Congress about why this tweet or that one was allowed, advertisers and Wall Street bankers pressuring the company about revenue. Twitter's guts would match its mission to be a public utility for conversation.

They called the project Bluesky. The name was intended as a symbol of freedom for Twitter's bird logo, referencing the open firmament where the bird would someday fly.

"We're facing entirely new challenges centralized solutions are struggling to meet. For instance, centralized enforcement of global policy to address abuse and misleading information is unlikely to scale over the long-term without placing far too much burden on people," Dorsey tweeted in December 2019, announcing the project. "The value of social media is shifting away from content hosting and removal, and towards recommendation algorithms directing one's attention."

9 > Bluesky

On January 21, 2020, Agrawal ventured into the fog of San Francisco's Richmond district. He'd recently returned to the city after Twitter's OneTeam festivities in Houston and was wearing only a thin gray sweatshirt, not enough for the wind and chill. He braced himself against the cold as he dashed up the marble steps of the Internet Archive, a digital library founded in 1996, which occupied a former Christian Science church bolstered by commanding white columns, to attend a series of talks on the future of decentralized social media.

Agrawal found a seat in the wooden pews. In the church of online knowledge, Agrawal was much more at home than at the glitzy parties of OneTeam.

He watched as Jay Graber, a slight developer dressed cozily in a pullover and black tights, took the stage and rattled quickly through a slide presentation on decentralized social media. She had been researching the ecosystem for several years and saw potential in decentralized technology, but also flaws—none of the current offerings could manage to meet the basic expectations of average social media users, she noted.

At the end of the evening, the host called Agrawal to the mic. Behind his wood-framed glasses, Agrawal appeared flustered. Standing at six foot two, he seemed to shrink into himself, tensing his shoulders like he was a turtle peeking out of its shell. Twitter wasn't properly addressing misinformation and harassment, he said, and the company's job was moving from simply hosting content to directing attention. "What we see is, controversy and outrage rather than healthy conversation gets more attention," Agrawal admitted.

He hoped that starting over with Bluesky could solve Twitter's problems. "Decentralization is not an end, it's a means to an end," he said. His primary job was to find people to staff the team and to appoint a much-needed leader, Agrawal added.

In the crowd, Graber listened closely. Bluesky was everything she wanted to build—an elegant decentralized platform that could introduce radical new technology to everyday users without asking them to learn clunky blockchain systems or compromise on the ease of mainstream services like Instagram and Twitter. After the talks ended, she approached Agrawal.

Agrawal appeared more relaxed in the small discussion than he had been onstage—he seemed entirely different, less like an engineer with his head lost in the code and instead warmer and more personal.

As they chatted, Agrawal and Graber felt each other out. She wanted to understand what his motivations were for building Bluesky, while he was curious about her approach to decentralized tech. Was she, like him, a nerd? Was she a crypto acolyte, or a protocol fanatic?

"No one has an inherent right to reach millions," Agrawal told Graber. He and Dorsey had started using the catchphrase "freedom of speech, not freedom of reach" to describe their plans for Bluesky, he explained. They envisioned it as a place where anyone could say anything—but not everyone would be able to tap into the viral potential of Twitter's recommendation algorithm.

Graber liked what she heard. Agrawal invited her to join a chat room dedicated to the project. It was filled with a dozen or so other developers who had heard Dorsey's call to action and wanted to contribute to Bluesky. The group began to hold conference calls every few weeks to hash out what their future might look like. Agrawal sometimes joined the calls, and he hung out in the chat room to size them all up, looking for the right person to lead the group. Dorsey joined the chat, too, but was quieter.

It became clear to Agrawal that Graber was the best choice. He invited her to come visit Twitter's headquarters. It seemed as if momentum was building for her to lead Bluesky. But then twin hammers dropped. Elliott Management charged into Twitter's stock, and COVID hit. The chaos threw Bluesky's very existence into question. Agrawal had championed the project, but without Dorsey to guarantee funding and make it a priority, Graber feared the entire vision could collapse. And as the pandemic set in, Agrawal and other Twitter employees were increasingly distracted, navigating remote work and battling wildfires of misinformation that flared incessantly across the platform.

Graber became certain: Bluesky would need to be fully independent of Twitter if it was going to succeed. She couldn't rely on a controversial and easily toppled CEO for money and protection. She also didn't want to lean on Twitter's engineering teams to pick up her slack. Bluesky would need its own money and its own engineers. Also, while it could take feedback from Twitter, it would need the authority to break away from its parent company at any time. Rumors swirled among Twitter employees and Bluesky contributors that the project would be killed before it even began, slashed in another ruthless maneuver by Elliott. Even after Dorsey was rescued by Silver Lake, Graber wasn't confident he could protect Bluesky forever.

Dorsey approved of Bluesky being formed as a fully independent company in late 2021. All he wanted was a form of social media that could evade capture or control altogether, be it by advertisers, Wall Street raiders, politicians, or shareholders. A month after Graber's lawyers signed off on the Bluesky paperwork, Dorsey handed Twitter over to Agrawal.

>>> IN HIS FIRST message as CEO to employees on the morning of November 29, 2021, Agrawal thanked Dorsey, calling him a mentor, and stressed the need for performance above all else. Then he turned his attention to the more than 7,000 workers he now led.

"I joined this company 10 years ago when there were fewer than 1,000 employees," he wrote. "While it was a decade ago, those days feel like yesterday to me. I've walked in your shoes, I've seen the ups and downs, the challenges and obstacles, the wins and the mistakes."

It was time for everyone to throw their full weight onto the gas pedal.

"The world is watching us right now, even more than they have before," he said. "Let's show the world Twitter's full potential!"

Agrawal believed that fixing the way Twitter moderated content was one of the most important things he could do to demonstrate that potential. In his conversations with Dorsey, he had become certain that Twitter had gone too far in policing online speech; he wanted to open the platform up more. The company's only tool for controversial tweets was to ban the user who posted them. It was a guillotine where Agrawal wanted a pair of tweezers.

With deep experience in artificial intelligence, he thought he could use the technology to solve many of Twitter's problems—the ever-accelerating harassment, the fears about speaking up, the sluggish growth—with one master stroke. Soon after he took over as CEO, Agrawal tapped Jay Sullivan, one of his product leaders, to help flesh out the plan.

He called the sandy-haired Sullivan into the office on a weekend in early 2022. The pair took over a conference room and Agrawal explained his vision. He wanted to slash the dependence on human reviewers and allow machine learning to take faster action against content that violated its rules. Agrawal planned to stop removing content from Twitter altogether, except in the most egregious situations.

In his typical fashion of requiring his deputies to reason through all sides of an argument, Agrawal asked Sullivan to define "free speech." He thought the phrase had lost all meaning. He wanted to hear Sullivan argue for what it was, and what it wasn't.

The two sat staring at the blank whiteboard. "How can we enable freedom of expression?" Agrawal asked.

Before joining Twitter, Sullivan had worked at Facebook and dealt with some of the platform's most challenging content moderation problems, like child exploitation. He'd also advocated for Facebook to encrypt direct messages, a move that lawmakers disliked because it would allow criminals to hide their messages from subpoenas. He was deeply aware of the tradeoffs a massive social media company had to make as it juggled privacy, safety, and speech.

The product manager felt Twitter had already gotten a lot right. He told Agrawal that the platform should keep its policies that prevented societal harm, taking down tweets that promoted terrorism, threatened violence, or sexualized children.

Agrawal agreed. It was the gray area that concerned him—tweets that made Twitter feel like a toxic cesspool but didn't break its rules. "How do you make it more fun and lively, like you want to be there?" the chief executive mused.

The two men discussed what parts of Twitter should be inaccessible to the site's infamous troll armies. Accounts that shared ugly but permissable content shouldn't be promoted by Twitter's algorithms or allowed to monetize their accounts—this would give the company some distance from its worst offenders. Advertisers would have to be protected, of course. The major brands that fueled Twitter's business often balked when they saw harassment or hate speech on the platform, pausing multimillion-dollar campaigns until the scandal of the week flamed out.

Agrawal and Sullivan began to sketch their ideas on the whiteboard. Instead of playing whac-a-mole with bad tweets, each account on the platform would earn a ranking of 1 through 5, based on the kind of content it produced. If the account consistently posted tweets that violated Twitter's rules, it would get shoved down the rankings to a 5, where it would lose access to promotion by the company's algorithms and couldn't be shared. If an account posted valuable content, was verified, and followed Twitter's rules, it would be ranked at 1 and get the most reach.

The system ought to be transparent, Agrawal told Sullivan, letting users and the world know where their account fell on the spectrum and what they would need to do to move up a level. That change would lessen the conspiracy theories that always swirled around the company, as people across the political spectrum claimed Twitter was censoring or shadowbanning them.

The whiteboard sketch looked like a series of concentric rings, with the lowest-ranked accounts at the gravitational heart of Twitter. The outer reaches represented the company's miscreants, and their content would be visible only to people who chose to follow them.

Sullivan was intrigued. He started copying down Agrawal's vision in his notebook. "What is this thing?" he asked Agrawal.

Agrawal wasn't sure what to call it yet. All he knew for certain was that Twitter had to change.

Sullivan stared at the drawing on the whiteboard. With Twitter at the center and its users orbiting around it, the sketch looked like a planet. "What about Saturn?" Sullivan suggested.

10 > Twitter in Trouble

Agrawal was worried about the state of Twitter's business. In early February of 2022, he and Segal had reassured their investors that the company could still reach the ambitious growth goals laid out in response to Elliott Management's intrusion—$7.5 billion in total revenue and 315 million daily active users by the end of 2023.

Despite those promises, Twitter netted a $220 million loss on $5 billion in sales in 2021, setting the company up for a tense 2022. The company would need to run flawlessly that year to stay on track with its revenue and user targets, and Agrawal, a first-time chief executive, knew he had little room for error. He was already operating with a reputation deficit. The stock market reacted to his appointment with little more than a yawn, and the prevailing question from most was summed up by one *New York Times* headline: "Who is Parag Agrawal?"

Four days after he was named chief executive, on December 3, he went through the dramatic restructuring he had promised. He fired Montano, the engineering head, and Davis, the design leader, in an attempt to create "clear decision-making, increased accountability, and faster execution," he wrote in an internal email.

While Davis had caused internal strife among some employees during his tenure, Montano was seen as a loyal soldier who, like Agrawal, had spent more than a decade at the company rising through the engineering ranks. His exit was seen as an early example of the chief's ruthlessness. Agrawal would not be afraid to show some of his closest friends the door.

Privately, he fretted that Twitter hadn't transformed enough since Elliott's attempted coup against Dorsey. Agrawal knew that another hedge fund or private equity firm could crunch the numbers and realize Twitter was a target. Silver Lake had been willing to bail out Dorsey, but Durban, its deep-pocketed leader, was emphatic that Twitter needed to improve its numbers and that he wouldn't double down with another rescue.

Agrawal scuttled hiring plans for 2022 and looked for other ways to make Twitter more lean. Product and engineering teams, who had just been asked to present plans for growth that year, were taken aback. The hires they had been told they could make would no longer happen, and the plans they'd presented before the holidays were trashed.

To prove he was still confident in Twitter's progress, Agrawal and Segal announced a $4 billion share buyback program in early February 2022. "It represents confidence in our strategy and execution," Segal said. "We are putting our money where our mouth is."

While share buybacks typically lift the price of a company's shares—as an organization spends cash from its balance sheet to decrease the supply of its stock—Twitter's shares barely budged in the weeks after. But that may have been due to another announcement a few days later.

On February 16, Twitter announced that its newly appointed chief executive was taking a "few weeks" of parental leave so that he and his wife, Vineeta, could welcome their second child, a baby boy. It was a progressive move in keeping with Twitter's people-first equity ethos, where employees, regardless of gender, received up to twenty weeks of parental leave for new children. But it was still radical for a CEO, much less one who had been in the job for less than three months, to make such a move, particularly in the grinding, male-dominated culture of Silicon Valley.

Some tech leaders shared the views of Musk, who told *The New York Times* in 2020 that he didn't have to spend much time with his newborns, including one he had recently welcomed with his then-partner, the musician Grimes.

"Well, babies are just eating and pooping machines, you know?" he said. "Right now there's not much I can do. Grimes has a much bigger role than me right now. When the kid gets older, there will be more of a role for me."

Twitter's executives attempted to head off any criticism. Agrawal planned to continue working a few days a week, and official statements leaned into the idea that all parents should have a role in their children's early development. "Thank you @paraga for leading by example and taking paternity leave," Segal tweeted. "I wish leaders did this when I was early in my career and becoming a father."

The market remained unforgiving. By the end of February, Twitter's stock was down more than 54 percent from the year prior. Agrawal could do no right.

—

>>> MUSK, MEANWHILE, started 2022 flying high. With a net worth of $219 billion, he ranked at the top of the *Forbes* billionaires list for the first time, propelled by the public market performance of Tesla and the ever-increasing private valuation of SpaceX. Musk had feared that the pandemic and its economic standstill would stifle Tesla's momentum after surviving the uncertain days of 2018, but the automaker thrived, shipping cars at an increased pace, constrained only by its ability to meet voracious global consumer demand.

The company had enjoyed a sixfold increase in profit to $5.5 billion as its market valuation hit more than $1.2 trillion in November 2021. The company had been worth a little more than $60 billion when Musk proposed taking it private three years earlier. Apple, founded in 1976, took forty-two years to become the first publicly traded American company to hit the $1 trillion market cap milestone. Tesla did it in eighteen years.

But many people worried the company was overvalued, including Musk himself. Tesla's shares seemed to have decoupled from any financial reality or traditional metrics. The advent of mobile trading apps like Robinhood made it easy for a new class of millennial and Gen Z first-timers to buy and sell stocks, and Tesla, with its bombastic, always-online chief executive, became an attractive pick.

Musk had at times chafed at Tesla's unrealistic value. In early May 2020, after Tesla's stock price had more than tripled over the previous twelve months, he tweeted, "Tesla stock price is too high imo." The stock price fell more than 12 percent in thirty minutes, though it would recover in the days after and soar to new heights, bolstered by unparalleled fandom.

And in November 2021, with Tesla's shares near an all-time high, Musk had asked in a Twitter poll if his followers supported his "selling 10% of my

Tesla stock" because of criticism around "tax avoidance." While it seemed magnanimous, he already knew he'd have a massive tax bill—nearly $11 billion—due to the exercise of a stock bonus plan that gave him $24 billion worth of Tesla shares. By the end of that December, he would sell off 15.7 million Tesla shares to cover his tax obligation, and give another $5.7 billion worth of stock to his own foundation to incur further savings.

>>> LESS THAN TWO WEEKS after Agrawal told employees that he'd be taking some time off for parental leave, Russia invaded Ukraine on February 24, igniting a conflict that gripped the world as imagery of bombings, bodies, and blood spread across social networks like Twitter. The war led advertisers to halt their spending on social media. Brands like Ford and Visa didn't want to see their ads plastered next to a gory image of battle on Twitter, as the war complicated an already darkening global economic situation marked by inflation and rising interest rates.

Within a month, the bleak truth was clear. Twitter's advertising had plummeted and Agrawal realized the company was not going to be able to deliver on the targets. His focus flipped quickly from success to survival.

He began to search for other areas to cut fat. During the pandemic, many Twitter executives worked remotely from Hawaii and proposed to hold the next #OneTeam on the islands. When they realized the price tag for bringing the entire company to Hawaii for a few days, they scaled back and settled for Disneyland in Southern California instead. The budget for the lavish affair, which would bring Twitter employees back together for the first time since the pre-pandemic times of January 2020, was $37 million.

To Agrawal, it was an egregious expense. While the company had already put down deposits for hotels and conference space, he decided that the cultural celebration would have to be canceled. It was all too easy to imagine photos of Twitter employees taking in the fireworks landing on people's timelines alongside pictures of Ukrainian corpses.

He also began planning a layoff. The move, a consequence of overhiring during the pandemic as all online businesses received boosts, was set to be Twitter's biggest staff reduction in years. Agrawal hoped to cut between 20 and 25 percent of Twitter's workforce, dismiss many of its contractors and consultants, and shut down office spaces that had gathered cobwebs during the pandemic. He code-named the cost-cutting measures "Project Prism" so execs could chatter about it in the office without alerting their employees.

The austerity proposed in Prism would be extreme by Twitter standards. The layoffs alone would save the company $750 million, and by slashing parties, travel, marketing budgets, and the cloud contracts he had pushed for, Agrawal hoped to save more than $1 billion. The days of freewheeling spending and lax oversight were over.

11 > Musk's Shopping Spree

While Musk's businesses seemed to be booming, his personal life was another matter. He was increasingly erratic, mixing Ambien and late-night Twitter binges. He had a reputation for using drugs including LSD and ecstasy at parties, and would later say publicly that he had a prescription for ketamine, a dissociative anesthetic that can ease depression, and used it to get out of negative states of mind. Close family members became so concerned with Musk's mood swings and shifting behavior in early 2022 that they began discussing a possible intervention that could make him aware of his issues. The billionaire, however, was unreceptive and evaded his family's attempts to interfere in his personal life.

Musk's Twitter feed had become faster, louder, and increasingly unhinged. By the end of 2021, Musk was averaging more than 250 tweets a month, and they had taken a distinct rightward turn. He always fashioned himself a libertarian with liberal tendencies, a business scion who backed Barack Obama and deigned to join the Trump administration's business councils, albeit holding his nose. The pandemic changed him.

He had rarely entangled himself in politics, but his tirades against COVID lockdowns and his crusade to keep Tesla factories open drew him into political sparring matches on Twitter. He believed his critics were infected by a "woke mind virus" that sought to destroy the country and limit progress by focusing on divisive issues of racial and social justice. He was disdainful of the social justice movements that had gripped the U.S. in 2020, following the murder of George Floyd, and hated the resulting pushes for diversity, equity, and inclusion that swept corporate boardrooms in the months after.

Like many white male tech CEOs, he saw his companies as pure meritocracies, where hard work and smarts rose to the top. To Musk, diversity initiatives were antithetical to success, watering down his workforces with underqualified female or minority candidates and discriminating against white or male applicants.

Railing against the ails of liberalism, he fought against efforts to unionize workers at Tesla's factories. In 2021, a federal court found that a Black former lift worker at Tesla's Fremont factory had been racially abused at work, awarding him $137 million. While a judge later reduced that payout on appeal, the decision opened the floodgates for a host of other discrimination lawsuits against the electric automaker.

"traceroute woke_mind_virus," Musk tweeted in December 2021, referencing a diagnostic computer command that allowed engineers to track the path of data across the internet and discover where it originated and how it was spreading.

Musk had gone fully anti-woke. On Twitter, he regularly engaged with right-wing influencers, slammed Democratic politicians over initiatives to tax the wealthy, and decried transgender people's insistence on being referred to by their chosen pronouns, a demand he viewed as contrary to free speech. He was uninhibited, and continued to use the platform to lash out.

He also used Twitter to flirt. He had met Claire Boucher, the musician known as Grimes, over Twitter, and the couple appeared together at the 2018 Met Gala. She was seventeen years his junior, but the elfish singer, known for her futuristic style and shock value antics, was a kindred spirit. She, too, dreamed of dying on Mars and was fascinated by the possibility of a techno-utopian future governed by AI.

Musk rarely sought peace in his romantic life. He had previously dated the actress Amber Heard, in a relationship that seemed to swing wildly between extremes of passion and anger. Their bust-ups, which mostly stayed out of the major gossip rags, were infamous among his family and staff.

His romance with Boucher could be erratic, too. In 2020, shortly after the singer announced she was pregnant with Musk's child, he unfollowed her on Twitter, the second time in two years he had done so during a rough patch. The mess didn't seem to bother Musk—after all, he had used Twitter to announce his divorce from his second wife.

Musk could have falling-outs with Kimbal, too. He subjected his brother to the same unfollowing treatment on Twitter, which made it awkward given Kimbal served as a director at Tesla and SpaceX. Kimbal, who sometimes had concerns about Musk's mood swings, eventually resigned from SpaceX's board in January 2022. And while the brothers made up after their arguments, Musk's willingness to cut Kimbal and other kin off when he was angry reaffirmed to his family that he could be prone to bouts of emotional instability.

>>> IT WAS IN THIS ENVIRONMENT that he decided, seemingly on a whim, to make his move. As Agrawal was slashing at Twitter, Musk was spending. His spree was enabled by Tesla's share price, which had come down a bit from the heights of the end of 2021 but still translated to a public market valuation of more than $1 trillion by early January. That gave Musk a stunning amount of purchasing power.

Unlike the average person, Musk did not have to sell his shares to get cash. Indeed, he had sold some of his Tesla stock at the end of the year to cover his tax obligations, but he preferred to hold on to as much of his equity as possible to give him more control over Tesla's decision-making. Instead, to get liquid cash, Musk used his Tesla shares to obtain loans. Tesla disclosed in financial filings that by the end of 2021, Musk had pledged more than 92 million shares, or nearly 40 percent of his stake in the company, "as collateral to secure certain personal indebtedness." At the time, the shares pledged were worth more than $32 billion.

This unfathomable well of wealth made Musk extremely confident. Tesla was clearly doing well, as was SpaceX, which completed thirty-one launches in 2021 as the market leader in the private launch business. As private share sales at the end of the year valued SpaceX at $100 billion, Musk's mind began to wander on to the next problem he wanted to solve for humanity. "So I asked myself what product I liked, and that was an easy question," Musk later told his biographer, Walter Isaacson. "It was Twitter."

Musk spent hours on Twitter every day and his shifting politics and online filter bubble made him particularly sympathetic to the idea that the company was silencing speech. He saw himself as the only man who could stem Twitter's continued leftward slide. By the end of January, he deputized Jared Birchall to begin quietly buying up chunks of Twitter's shares.

Birchall had his former firm, Morgan Stanley, execute the trades, which began on January 31 when the stock opened trading on the New York Stock Exchange at $35.43. The instruction was simply to buy as much as possible, as quickly as possible.

In their haste to execute Musk's command, the traders blew past a deadline imposed by the Securities and Exchange Commission, which requires any entity that accumulates more than 5 percent of a publicly traded company to disclose their stake. Twitter's board never caught wind of the transactions.

By March 14, Musk had accumulated 9.2 percent of Twitter's outstanding shares. The stake, which was worth nearly $3 billion, made him the single largest shareholder in the company, bigger than mutual fund company Vanguard, and Dorsey, who held only 2.3 percent. Few besides his innermost circle knew about the move. He was required by law to disclose what he owned by March 24, but Musk did not publicly announce his purchases until April 4, in a belated SEC filing.

Up until that point, Musk behaved fairly normal online—at least by his standards. He tweeted about environmental, social, and governance (ESG) corporate initiatives ("the Devil incarnate") as well as the mainstream media (full of "groupthink"), and inserted himself in the crisis of the period, the Russian invasion into Ukraine.

At first, he showed unwavering support for Ukraine and played the hero, saying that SpaceX's satellite internet service, Starlink, would be available in the country. "Hold Strong Ukraine," he tweeted in the early weeks of the invasion, with six Ukrainian flag emojis. Later, he posted in a mix of Russian and English to challenge Vladimir Putin to "single combat."

Musk, however, grew increasingly uncomfortable with his initial position. In early March, he tweeted that he refused requests by some foreign governments to block Russian news sources that had been used to spread propaganda about the war from Starlink's internet.

"Sorry to be a free speech absolutist," he wrote, angering Ukrainian supporters who saw his move as tacit endorsement and enabling of the Kremlin's information warfare. The criticism stung Musk, who began to see the full-throated support of Ukraine as an orthodox position held by some of the "woke" liberal elements that he had grown to hate.

Ten days after his post about Ukraine holding strong, he had developed a new opinion and shared two memes to underscore his reversal. One featured a photo of a cartoon drawing of a non-player character, or NPC, a derogatory term for people who are incapable of forming their own independent thoughts. "I support the current thing," read the text on the image, with the cartoon framed by LGBTQ symbols as it held a Ukraine flag. Another meme, taken from the Netflix show *Narcos*, showed a man staring off into the distance. "Netflix waiting for the war to end to make a movie about a black ukraine guy falls in love with a transgender russian soldier," read the inscribed text.

Musk hardened his position as he spoke with his closest associates. "I saw your tweet re free speech," Antonio Gracias, a technology investor, former Tesla board member, and close friend, texted Musk on March 5. "Wtf is going on Elon . . ."

The EU passed a law banning Russia Today, and internet providers like Starlink had been told to deny access to their sites, the SpaceX chief replied. "Actually, I find their news quite entertaining," he continued. "Lot of bullshit, but some good points too.

"Free speech matters most when it's someone you hate spouting what you think is bullshit," Musk added, continuing his defense of Russia Today.

"I am 100% with you Elon," Gracias replied. "To the fucking mattresses no matter what. . . . . this is a principle we need to fucking defend with our lives or we are lost to the darkness."

Musk's conversations with Gracias and other friends gave him purpose. By then, his Twitter share acquisition spree was well underway, but he had a renewed justification for the object of his desire. If liberals and their media and tech industry cronies couldn't be trusted to uphold free speech, maybe he could do it himself. Twitter, after all, was a company based in one of the most liberal cities in the U.S. It must be compromised, he believed. "I'm worried about de facto bias in 'the Twitter algorithm' having a major effect on public discourse," Musk tweeted on March 24.

That day, his phone started to buzz with messages from his ex-wife Talulah Riley. Unlike his first partner, Riley stayed in longing contact with

Musk, who sometimes tweeted about his cherished memories of their relationship. On this occasion, the actress was incensed about Twitter's recent suspension of the account of the Babylon Bee, a right-wing satire site that Musk and Riley thought was funny. The publication had received warnings from Twitter in the past for breaking its rules, but had its account suspended after it violated a policy against misgendering people, when it named Rachel Levine, a Biden administration official and transgender woman, its "man of the year."

"Can you buy Twitter and then delete it, please?! xx," Riley texted Musk.

"Or can you buy Twitter and make it radically free speech?" she wrote moments later, peppering her messages with more kisses. "So much stupidity comes from Twitter xx." Musk liked her message, appending a thumbs-up to it, but didn't respond further.

Instead, he lashed out in a late-night Twitter binge. "Free speech is essential to a functioning democracy. Do you believe Twitter rigorously adheres to this principle?" he asked his Twitter followers in a poll. "Please vote carefully," he cautioned.

By March 26, the poll had accumulated more than two million votes. More than 70 percent of Musk's followers had an overwhelmingly negative view of the platform and its handling of speech. "Given that Twitter serves as the de facto public town square, failing to adhere to free speech principles fundamentally undermines democracy," Musk tweeted. "What should be done? Is a new platform needed?"

More responses poured into his Twitter notifications, congratulating him on his tweets and egging him on. Twitter was a censor, a cesspool, a plague on open discourse, they said. "Buy twitter," replied Mike Cernovich, a far-right influencer who had once encouraged Pizzagate. He wrote that Twitter only respected those who were "leftist making death threats against conservatives."

"Doesn't sound very balanced," Musk replied.

The billionaire also received encouraging text messages from friends and acquaintances, among them Oracle founder Larry Ellison, who agreed for the need of a new Twitter; Mathias Döpfner, a German media executive who called for Musk to buy Twitter and then hand it over to him to run; and

MUSK'S SHOPPING SPREE

Sam Bankman-Fried, the founder and leader of crypto exchange FTX, who wanted to invest.

But one message in particular caught Musk's attention. It came from a typically quiet correspondent who had been entered into the Tesla chief's phone contacts as "jack jack."

"Yes, a new platform is needed," Dorsey texted Musk. "It can't be a company. This is why I left."

Although he had resigned as chief executive months earlier, Twitter's cofounder had yet to disentangle himself fully from the social media platform and remained on its board. Still, despite his position and the obligations that came with it to protect Twitter, he was encouraging Musk to build a competitor.

"Ok," Musk wrote back. "What should it look like?"

Dorsey didn't hesitate to describe the vision he'd laid out to Agrawal and other executives before his resignation. "I believe it must be an open source protocol, funded by a foundation of sorts that doesn't own the protocol, only advances it," Dorsey responded. "It can't have an advertising model. Otherwise you have a surface area that governments and advertisers will try to influence and control."

"Super interesting idea," Musk replied.

Dorsey was pleased with the response and went even further, trying to woo Musk. Twitter was too compromised to salvage, he suggested.

"I'm off the twitter board in mid May and then completely out of the company. I intend to do this work and fix our mistakes. Twitter started as a protocol. It should never have been a company. That was the original sin," Dorsey texted. "I wanted to talk to you after it was all clear, because you care so much, get it's importance, and could def help in immeasurable ways. Back when we had the activist come in, I tried my hardest to get you on the board, and our board said no. That's about the time I decided I needed to work to leave, as hard as it was for me."

The two billionaires were courting each other. But both of them were keeping secrets, too. Musk didn't divulge that he had recently acquired more than three times as much Twitter stock as Dorsey. At one point during the exchange, Dorsey griped that his ownership stake was only 3 percent—

not even remotely close enough to give him the power to make unilateral, paradigm-shifting decisions needed to resuscitate the social network.

Dorsey didn't mention that his protocol work on Bluesky had begun years ago, or that Twitter's board had gone to the mat to save him from Elliott. He didn't disclose his fiduciary duties and conflicts as a board member at Twitter and Bluesky. And he didn't note that he had hand-selected Agrawal as his successor—even as he undercut him.

But what Dorsey said got Musk's juices flowing. He started to fixate on how he could improve Twitter. Over the past few weeks, he had used his account to slam Twitter for its focus on cryptocurrency-related features, its lack of ability to handle bots, and its nonpublic algorithm. He also made jokes about how terrible the platform made him feel, despite the hours he spent on it every day making and consuming content. He was a shareholder now, and a large one at that. Maybe he could be the one to agitate for change?

An hour after chatting with Dorsey, Musk texted Durban. He had worked with Durban before on his botched deal to take Tesla private in 2018. Silver Lake had agreed to back Musk in the deal, only to have the transaction fall apart when the Saudi money Musk had claimed was guaranteed turned out to be anything but. The relationship between Musk and Durban, once close, had become chilly in the aftermath of the Tesla disaster.

"This is Elon. Please call when you have a moment," Musk texted. "It is regarding the Twitter board."

On the call with Durban, Musk was more direct than he had been with Dorsey. He was Twitter's largest shareholder, Musk said, having snapped up more than 9 percent of its stock. And he wanted to use his position to force changes at the company, perhaps joining the board as Durban had done. It came spilling out of the mouth of the world's richest man.

No one in Twitter's leadership had heard whispers of a possible intrusion. Birchall's contacts at Morgan Stanley had been able to avoid detection by Twitter as they amassed Musk's stake, which was almost three times larger than Elliott's had been.

The Silver Lake head quickly phoned Martha Lane Fox, who oversaw the board's governance decisions; Bret Taylor, Twitter's chairman; and Agrawal. Twitter had a problem, Durban told them. The group agreed to keep its interactions with Musk positive—he was a much better ally than enemy.

>>> DURBAN CONNECTED MUSK with Lane Fox, Taylor, and Agrawal to organize a face-to-face meeting. (Musk, who often tweeted boob jokes and appointed few women to leadership roles in his companies, saved the two men by their full names in his phone but referred to Lane Fox as "Martha Twitter NomGov" in his contacts.)

The intrusion could have been Agrawal's nightmare come to life—another outsider meddling in Twitter's stock. But he didn't see it that way. He had imagined that he and Twitter might be vulnerable to another activist investor who would come in, squeeze him for a few months, and then leave. But Agrawal was a Tesla fan and thought Musk's big investment in the company showed he was committed to Twitter's success.

With Musk came chaos. But if it meant Agrawal, a green CEO, could lean on the perspective of one of the industry's most successful people, it might benefit Twitter. He and Taylor had a quick call with Musk on the night of March 27, after the Tesla chief finished a meeting with his autopilot engineering team.

Agrawal found that he had things in common with Musk. Both executives felt Twitter had gone too far in restricting speech on the platform and wanted to find ways to loosen the company's grip on content moderation. They tended to be hard-driving, and impatient for change. And each saw himself as a top engineer, able to solve the complex problems Twitter faced that had stymied former leaders.

But Musk came to the call with a new bombshell that unsettled the two Twitter leaders. He would consider taking a seat on Twitter's board, he told Agrawal and Taylor. But he was also mulling the other options he could pursue with his new stake in Twitter, including taking it private or spinning up an entirely new competing social media service. Durban had already warned them that Musk could be a wild card, and he was proving it on the call. He might be willing to work with them, but he could just as easily turn their lives upside down.

They met in person for the first time on the last day in March, with assistants booking an Airbnb near the San Jose, California, airport to stay out of the public eye and accommodate Musk's travel schedule.

Taylor, who looked like Prince William if the royal had been raised on coding boot camps instead of Eton etiquette, had a veteran's résumé in Silicon Valley. He helped develop Google Maps and launched a start-up that he later sold to Facebook, where he rose to become chief technology officer. But he had never been in a position like this.

Calm and analytical, he was being groomed to lead Salesforce by Marc Benioff, the billionaire chief executive of the sales software company. Benioff made him co-CEO in the same week that he was elevated to be Twitter's board chairman, as Dorsey handed the reins to Agrawal. Taylor was a rising star and a steady, inoffensive choice. He rarely tweeted, aside from occasional posts about his beloved San Francisco 49ers and Golden State Warriors.

He preferred to do business in person and spent much of his time globetrotting to meet face-to-face with Salesforce customers. While he was respectful of Twitter's remote work policies and went along with Twitter's custom of holding most board meetings over video calls, Taylor didn't want to pass up the opportunity to meet Musk in person. He found that his willingness to show up often charmed clients and thought it would get him on Musk's good side.

Taylor flew into San Jose that night. He was the first to arrive at the Airbnb. "This wins for the weirdest place I've had a meeting recently. I think they were looking for an airbnb near the airport and there are tractors and donkeys," he messaged the group.

The men shared a meal at the house and discussed what Musk could bring to Twitter as a board member, interrupted occasionally by noisy police helicopters that flew overhead. Agrawal and Taylor had agreed to view the meeting as a job interview, and questioned Musk about what he would bring to the boardroom, probing his views on Twitter and asking about his management of his other companies.

Musk was interested in joining the board, he told them. But he was still weighing his options and thought he might buy the company outright, or start a competing social media platform. Musk's disclosure was bizarre. If he really intended to start a competitor, why would he need to accumulate a massive stake in Twitter to do it? And why would he give them advance warning of his plans? Surely the billions he had already invested in Twitter would be better spent on his own start-up, if he truly intended to found one. It seemed to Taylor and Agrawal that Musk's two realistic options were to stage a hostile takeover or join the board. It would be up to the two men to convince Musk to take the latter route.

"Great dinner :)," Musk wrote in the group chat. "Definitely one for the memory books haha."

—

>>> MUSK WOULDN'T STAY at the strange Airbnb or in San Jose for very long. He had a planned weekend of debauchery in Berlin and headed to Germany's capital soon after his dinner with Agrawal and Taylor. By Friday evening, he was spotted at KitKatClub, a legendary club founded by an Austrian pornographer in the early '90s and once known for its latex parties and encouragement of public sex. On Saturday, there was a stop at another club, Sisyphos, for the birthday party of his good friend and college roommate, Adeo Ressi. Musk donned a Zorro mask before he and a group headed to Berghain, the techno mecca that is regarded as Berlin's most famous—and hard to access—club.

Located in a former heating plant built in the postwar years, the club straddles Berlin's East-West divide and can be imposing for outsiders, who are very much not welcome. The line, guarded by gruff bouncers with face tattoos, can stretch for hours around graffitied walls and chain-link fences; those who are permitted to enter know the etiquette. Dress down and preferably in black. Don't speak, but if you must, do so in German.

Musk arrived in the witching hours of Sunday, and followed none of the customs. He strode up to the bouncers, announced who he was, and asked to come in. They looked him up and down and shook their heads. "Not today." Unused to rejection, Musk took out his phone and turned to Twitter in a less than sober state. "They wrote PEACE on the wall at Berghain! I refused enter," he tweeted at 4:53 a.m.

As Musk crashed after his weekend of partying, Twitter's board gathered on a video call to talk about him later that same Sunday, April 3. Most of the board members were wary. Adding a famous, loudmouthed director would change the chemistry of the group and could be distracting, especially as they tried to steer a novice CEO, who also attended the meeting. And of course, some of Twitter's employees wouldn't be pleased about it, since

Musk used Twitter to spread misinformation about COVID and push anti-transgender rhetoric.

Lane Fox argued that getting Musk to agree to join the board would be a huge relief. His musings about staging a take-private or founding a competitor would destabilize Twitter if he brought them to fruition. Adding him to the board was the easiest way to neutralize him, since it would harness him with a duty to act in the best financial interests of Twitter.

Dorsey chimed in with a rare comment. "I consider him a friend," he said of Musk, adding that his technical accomplishments would be valuable for Twitter.

Eventually the board agreed to welcome Musk to their ranks. The directors decided that the best way to keep Musk from running off the rails would be to limit his ownership of Twitter to below 15 percent and require the erratic billionaire to sign a cooperation agreement that would dictate how he could behave as he sought to influence the company.

Dorsey quickly shared the board's decision with Musk. "I heard good things are happening," he texted.

As Musk prepared to fly out of Berlin, Lane Fox scrambled to get the necessary paperwork in front of him. The sooner he signed, the better. Musk connected Lane Fox to Birchall and asked the head of his family office to deal with the "important paperwork" needed for him to join Twitter's board.

But when the documents came through, Musk wasn't particularly happy with the arrangement. The agreement Lane Fox sent him capped the amount of shares he could buy and forbade him from making critical public statements about the company and its leaders. The document was corporate boilerplate, virtually identical to the agreement that the board had required Jesse Cohn to sign when he agreed to halt his hostile campaign two years earlier.

Most directors wouldn't flinch at the idea of being required not to publicly dump on the company they managed. But Musk wasn't like most directors, and he hated the idea of being muzzled. Birchall rejected the document on his behalf.

By Monday morning, April 4, Twitter's board was in a bind. Musk had belatedly filed a notice with the SEC disclosing his ownership stake in the company, and it was now public, leading to a flood of media coverage. Twitter's lack of public communication on the matter made it unclear if he was friendly or hostile. The uncertainty was causing turbulence for Taylor and the other directors, who were still trying to wrangle the billionaire into signing the board agreement. With Musk groggy from his flight, Lane Fox offered a revised agreement that was "even more pared back."

It was still not to the liking of Musk, who held all the power. With his ideals about making more irreverent speech possible on Twitter, he couldn't agree to silence himself. Twitter's directors needed him to sign this agreement. Without it, there was no way of knowing what the erratic billionaire could do. But Musk remained difficult. It was his way or no way.

"Thank you for considering me for the Twitter board, but, after thinking it over, my current time commitments would prevent me from being an effective board member," he wrote on a text thread with Birchall, Taylor, and Lane Fox. "This may change in the future."

The text led Taylor, Lane Fox, and Agrawal to hit the phones. They had to give Musk what he wanted or risk another Elliott situation. Twitter's CEO, barely in the hot seat for four months, pushed for appeasement, and eventually got it. Twitter's board quickly returned to Musk with a revised agreement. The stock ownership limit remained, but there were no restraints on what he could or couldn't say about Twitter.

> The Company and the Board will take all action necessary so that as Mr. Musk will be appointed to the Board.

On Monday night, Musk signed on the dotted line. With his problem resolved, Agrawal sent Musk the language of a tweet he was going to post making the announcement. It was something Musk would never do—ask someone for approval of his posts—but he played along.

"I'm excited to share that we're appointing @elonmusk to our board!" Agrawal tweeted the next morning. "Through conversations with Elon in recent weeks, it became clear to me that he would bring great value to our Board. Why? Above all else, he's both a passionate believer and intense critic of the service which is exactly what we need on Twitter, and in the Boardroom, to make us stronger in the long-term. Welcome Elon!"

The congratulations rolled in, but few well-wishers were more effusive than Dorsey, who became very emotional when Musk's appointment was confirmed. "Thank you for joining!" he wrote. "Parag is an incredible engineer. The board is terrible. Always here to talk through anything you want."

Dorsey, who had watched as frantic emails rolled in from his fellow board members about how to handle the Musk situation, had remained silent. Although he had told them about his friendship with Musk, the directors had no idea that Dorsey had been encouraging the entire maneuver in private messages with him.

After years of trying, Dorsey had his favorite tweeter on the board of the company he helped create. It was a board he hated, despite the protection it had offered him over the years. With Musk around, perhaps there was a way to implement the radical changes he had dreamed of. Musk was his opportunity.

"When is a good time to talk confidentially?" Musk texted back.

Dorsey responded in fourteen seconds: "anytime."

Musk rang him soon after.

12 > An Offer

Since taking over as Twitter's chief executive at the end of November 2021, Parag Agrawal had been afforded few breaks. There had been the internal reorganization, his attempts to resuscitate Twitter's stock price, and the board dance with Musk. Even his part-time paternity leave, which was supposed to give him time with his newborn son at the start of the new year, had been cut short by the Ukraine war. By the first full week of April 2022, however, the tide seemed to have turned. He had announced Musk as his newest board member and was ready to relax.

Twitter's CEO wasn't the only one breathing a sigh of relief in early April. As COVID restrictions loosened around the world, employees went globetrotting. Hawaii was a popular destination, and some leaders on the finance team headed to Maui. They were still far away from their home base in San Francisco when Musk broke the news: He would not be joining Twitter's board.

Instead, he intended to buy the whole company.

—

>>> THE INITIAL REACTION among employees to Musk's board appointment was mixed. The billionaire had developed a brash, trolling reputation on Twitter, and few people were better acquainted with his online tendencies and persona than those who worked on the platform. Indeed, that reputation had garnered him hordes of followers—Dorsey among them—who saw Musk's post-first-ask-later approach to Twitter as refreshing and aspirational. To them, he was a builder who couldn't be tamed and whose rough edges could and should be ignored because he had achieved so much.

Over the years, however, Musk's posting had alienated a lot of people. Twitter employees immediately began to surface concerns on the company's Slack channels. After all, one of the company's core values was to "communicate fearlessly to build trust." Dorsey had always encouraged his workers to speak truth to power, which in practice meant that employees tweeted about company concerns, spoke up at internal meetings, and challenged decision-making in internal chat rooms. When Musk was appointed, Twitter's employees did what they had always done.

Their concerns included his treatment of workers at Tesla and how factory workers had sued the company over allegations of racial discrimination. They brought up the misinformation he spread during the early months of the pandemic, as well as his false accusation of pedophilia on their very platform. Many knew their protestations would likely be fruitless—the decision had been made. But they felt obligated to say something.

"We know that he has caused harm to workers, the trans community, women and others with less power in the world," one tweep wrote on Slack. "How are we going to reconcile this decision with our values? Does innovation trump humanity?"

One employee pointed out that if they or any of their colleagues posted some of the things Musk tweeted, they would be the subject of a human resources investigation. "Are board members held to the same standard?" they asked.

Others at the company, however, didn't agree. To them, Musk was a once-in-a-lifetime entrepreneur and the possibility of learning from a man who had built a multibillion-dollar fortune across disparate industries was alluring. And now that he was on the board, he wouldn't bad-mouth the company anymore, they argued. Agrawal, who read the dissenting messages on Slack, was firmly in this camp. No reasonable board member would harm the company, and bringing Musk onto the board was one of his most significant changes as CEO. He was ready to defend it, even as it crumbled before him.

—

>>> MUSK'S RESERVATIONS about joining Twitter's board of directors began almost immediately, even before Agrawal's celebratory tweet. As a director, his voice would be one among ten others. He would have no

AN OFFER

operational control. For his whole career, Musk had succeeded by being the lead decision-maker. He followed his gut and couldn't be shackled by any corporate structure or opposed by dissenters. His leadership style was completely undemocratic. The board at SpaceX, for example, consisted of friends who were early backers, company leaders, and an executive at Google, which had invested in SpaceX.

Within hours of his delayed SEC filing disclosing his stake in Twitter on April 4, he was inadvertently reminded of the limitations of his power by friends and acquaintances who buzzed his iPhone with congratulations, including hedge fund billionaire Ken Griffin ("Love it!!"), Döpfner ("Fast execution 😎"), and libertarian venture capitalist Joe Lonsdale ("Hope you're able to influence it"). Some railed against what they saw as left-wing influence run amok, a progressive ooze that had seeped into the company's executive functions and affected decisions around content moderation and product development.

"Are you going to liberate Twitter from the censorship happy mob?" wrote Joe Rogan.

"I will provide advice, which they may or may not choose to follow," Musk replied.

Musk's inability to call the shots ate at him. He told a Tesla board member that he had not really wanted to join the board but gave in only after Twitter pushed hard. And when others asked him to pull favors—including the Babylon Bee editor Kyle Mann, whose publication's account had been suspended for violating the platform's harassment rules—he conceded he wasn't in a position to make such calls.

"So can you bust us out of Twitter Jail now lol," Mann asked.

"I do not have that ability," Musk said tersely.

In another text with CBS host Gayle King, who asked him to bring an edit button to Twitter, he wrote, "Owning ~9% is not quite control."

On April 7, Musk was preparing for another major Tesla event in Austin with Twitter still on his mind. Named the "Cyber Rodeo," by Omead Afshar, one of Musk's most trusted associates, the gala was the grand opening for the car manufacturer's new Texas factory, whose development had been overseen by Afshar. Since Tesla and SpaceX had moved more operations to the Lone Star State, Musk had started to think of Texas as "home," or whatever the approximation of home was for a jet-setting billionaire. But as he flew to Austin from Colorado Springs, where he had given a talk to cadets at the Air Force Academy, he remained more concerned about Twitter than his upcoming speech at the Tesla party. "I have a ton of ideas, but lmk if I'm pushing too hard. I just want Twitter to be maximum amazing," Musk texted Agrawal. He reminded Twitter's chief that he "wrote heavy duty software for 20 years" and that he still wanted to see Twitter's codebase.

Agrawal was thrilled that one of Silicon Valley's idols wanted to go so in-depth. "In our next convo—treat me like an engineer instead of CEO and lets see where we get to," he responded. He highlighted his engineering experience and said he has spent long hours in the codebase as chief technology officer. He also told Musk not to worry about leaks to the press that suggested employees were unhappy about the prospect of Musk joining the board—the "silent majority" supported him, Agrawal promised.

Musk was welcomed back to the Gigafactory that night by a drone light show, in which flying robots with LEDs formed the company logo and various Tesla vehicles in the evening sky near the 10-million-square-foot building. Inside, thousands of adoring fans waited for the arrival of their "techno-king," who drove onstage in a black Tesla Roadster through the *Blade Runner*-styled party. There were cheers, raised cell phones, and shouts of "You're my hero!" and "I love you!" Musk exited the car, wearing a black cowboy hat, Aviator sunglasses, and a too-tight Cyber Rodeo T-shirt. The rock star adoration was accentuated by Musk's late arrival. "We needed a place where we could be really big," Tesla's chief said of the new factory. "And there's no place like Texas!"

More than twenty-five minutes later, he paused his halting, partly ad-libbed speech for the orgiastic screams, which only got louder when a prototype of the Cybertruck—the Musk-described "magnum opus" of his company—drove onstage. "Here at Tesla, we believe in throwing great parties, so let's get this party started!" he exclaimed before clambering into the driver's seat as fireworks shot into the night sky.

Despite the adulation and sycophancy, however, Musk's mind remained restless. As he moved around at his own Giga Rodeo and after-party, all anyone wanted to talk about, it seemed, was the other company that he

AN OFFER

didn't own. Sure, a new Tesla factory was noteworthy, but there was more excitement about Twitter and his involvement. *What was he going to do?* Everyone wanted to know, and even those closest to him had no clear idea.

Neither did Musk.

>>> THE NEXT DAY, Friday, April 8, the Tesla chief flew to Lanai, the Hawaiian island owned by Larry Ellison. The trip was meant to be for rest and relaxation with his latest love interest, Natasha Bassett, a twenty-nine-year-old Aussie actress, but Musk continued to stew about Twitter. He had a call with Agrawal to discuss engineering processes at the company and he voiced his gripes about fake users and spam that flooded his replies. Agrawal listened, and tried to assuage his newest board member by acting as a sounding board. When Musk sent him a screenshot of a bot he had come across on Twitter after the call, Agrawal texted back within the hour. "We should be catching this," he responded.

Musk's initial hours on the small Hawaiian island were anything but restful. He stayed up late into the night browsing Twitter and at 2:37 a.m. local time reshared a chart showing graphs of the differences in how people of different political parties trusted different news outlets. Then he trained his attention on another set of stats. An account called @stats_feed tweeted the top 10 most followed accounts, placing his own in eighth place, with 81 million followers. Ahead of him were @BarackObama (131.4 million followers), soccer player @Cristiano Ronaldo (98.8 million followers), and singer @LadyGaga (84.5 million followers), but none of the accounts posted at the volume that he did—some hadn't tweeted in days—and he wanted to know why.

"Most of these 'top' accounts tweet rarely and post very little content," he wrote in the witching hours on Lanai. "Is Twitter dying?"

It was an observation that might have felt innocuous from someone who was new to the platform. Of course celebrities posted less. They had teams of social media experts and communications people dictating, editing, and vetting what they could or could not say, and for most of them, posting was about self-promotion or the pushing of products (#ad). Musk was one of the few celebrities who controlled his own account entirely and tweeted with reckless abandon. He found it incomprehensible that he was atypical, a celebrity with a massive platform shitposting, replying to fans, and duking it out in the marketplace of ideas. He observed that Taylor Swift had not posted for three months and Justin Bieber had tweeted only once in 2022— this was a travesty to a man who couldn't go a few hours without jabbering away online.

From a board member, however, Musk's post was an attack on the company. Board members had a fiduciary duty to shareholders and asking if Twitter was falling apart wasn't exactly the greatest endorsement of the platform.

It fell to Agrawal to do some damage control. In their earlier conversations, he and Musk had agreed to be candid with each other, and Agrawal felt no hesitation in bluntly telling Musk off. Without thinking much of it, he composed a long, typo-strewn text message to Musk and sent it off, not bothering to ask the board's advice. He was the chief executive after all, and wanted to be firm.

"You are free to tweet 'is Twitter dying?' or anything else about Twitter -but it's my responsibility to tell you that it's not helping me make Twitter better in the current context," he wrote to Musk, an hour and a half after the billionaire's tweet. "Next time we speak, I'd like to you provide you perspective on the level of internal distraction right now and how it hurting our ability to do work."

Musk responded forty-nine seconds later.

Bzzzt. "What did you get done this week?"

A seven-word slash at Agrawal's heart. Never mind that he had spent much of his week corralling Musk, explaining Twitter's operations, fielding his questions, and announcing his appointment to the board. Dealing with Musk felt like a full-time job, but the billionaire had taken aim at his work ethic. He had little time to process it before the next bomb dropped forty seconds later.

Bzzzt. "I'm not joining the board. This is a waste of time."

Another fifteen seconds.

Bzzzt. "Will make an offer to take Twitter private."

The board was bureaucracy, and no one was going to tell him what to do or how to do it. Certainly not a guy who had been chief executive for four months.

He hadn't given up completely on the idea of creating his own separate social media platform either. Earlier that morning, he had been texting with Kimbal about a Twitter-like site built on blockchain. Users would spend tiny amounts of money to post messages, cutting out bots. As Musk slung arrows at Agrawal, he was simultaneously texting his brother about his idea. "I think a new social media company is needed that is based on blockchain and includes payments," he wrote to his younger sibling.

Still, to Musk, Twitter was unique. After all, it was where he had built a loyal following and become one of the platform's most prominent voices. Other people, including Donald Trump, had tried and failed to build their own popular social networks. Owning it was more than attractive, and with Tesla's shares still flying high, he had the means.

Agrawal was shell-shocked. He read Musk's one-two punch over again, then fixated on the knockout blow.

"Can we talk?" Agrawal typed back. Musk didn't text back.

As he waited for a response that would never come, Twitter's chief executive called Taylor, Twitter's chairman. He quickly laid out what had happened and said he hadn't been expecting Musk to go from 0 to 100 over such an innocuous text. Taylor read through the text screenshots that Agrawal sent, equally surprised. Until then, the chairman had felt that his and Twitter's relationship with Musk had been going swimmingly. Since the Airbnb meeting, they had worked in lockstep to meet Musk's demands and announce his appointment to the board. A few days earlier, Taylor had even sent the Tesla chief a link to a tweet of "Twitter's next board meeting" featuring a photo of Musk, from his appearance on *Saturday Night Live*, in which he was dressed as a character from the Super Mario video game franchise, sitting at a conference table with normally dressed young professionals. Musk reacted with "Haha."

That Saturday morning, Taylor was wondering how the man whom he had welcomed with open arms was now holding a gun to his head. "Parag just called me and mentioned your text conversation," Taylor texted Musk at 5:10 a.m. in Hawaii. "Can you talk?"

Musk scanned the message and drafted a reply. He didn't change his tone. Between texts from his brother he sent his response. "Please expect a take private offer," he wrote.

"I saw the text thread," Taylor responded, his fingers racing across his iPhone screen. "Do you have five minutes so I can understand the context? I don't currently."

His phone lit up with four fast texts from Musk in a two-minute span.

"Fixing twitter by chatting with Parag won't work."

"Drastic action is needed."

"This is hard to do as a public company, as purging fake users will make the numbers look terrible, so restructuring should be done as a private company."

"This is Jack's opinion too."

Taylor understood Musk's concerns about bots—he had brought it up in the Airbnb meeting and with Agrawal—but why did Musk think Twitter needed to be private to fix the issue? Even less clear was the comment about Dorsey. Taylor knew Twitter's cofounder had been deeply affected by the Elliott process, but Dorsey hadn't suggested taking Twitter private. Still, Taylor knew that arguing with Musk would only further enrage the billionaire. He turned down the thermostat.

"Can you take 10 minutes to talk this through with me?" he texted a few minutes later. "It has been about 24 hours since you joined the board. I get your point, but just want to understand about the sudden pivot and make sure I deeply understand your point of view and the path forward."

Musk demurred, saying he was busy and would talk to Taylor another time instead.

The billionaire spent the rest of the day playing the hits from his Twitter account, tweeting about his array of enemies, including the media, U.S. senator Bernie Sanders of Vermont, and short-sellers who bet against Tesla. By Saturday evening, Musk turned to his new perceived enemy, Twitter's management, and sharpened his knives. He started a Twitter poll and shared it with his followers, asking if he should "convert Twitter SF HQ to homeless shelter since no one shows up anyway."

To his followers, it looked like Musk, a newly christened board member, was just getting to work. They had become accustomed to this trolling behavior, and the tweet, which took shots at both Twitter's supposedly lazy workforce and San Francisco's social issues, was red meat to them. To Taylor and Agrawal, however, it revealed Musk was unserious about being a director. The cooling-off period had done nothing for his hostility. They had to act.

"Hey—can you speak this evening?" Taylor texted Musk. "I have seen your tweets and feel more urgency about understanding your path forward."

Musk didn't text back. But a few hours later, he shot off another poll. "Delete the w in twitter?" he wrote, a change that would flip the company's name from a bird reference to a joke about female anatomy. More red meat.

By then, Musk was already making moves. He had informed Birchall of his plan to leave the board and engage in a takeover. Birchall contacted some former colleagues at Morgan Stanley, who got down to business to engineer a plan for the boss.

13 > Poison Pill

As they had during the Elliott challenge, Twitter's board enlisted their bankers at Goldman Sachs to mount a defense. If Musk truly intended to snatch the social media company out of their control, he had two options for doing so. He could bring an offer to the board for them to review, and they could choose to accept if the price was right. Or he could stage a tender offer, naming his price per share and inviting Twitter's current shareholders to sell off their stock to him. The latter option was riskier for both Musk and the board. A public tender offer could throw Twitter's stock into seizure, with wildly fluctuating prices as the market tried to bet on the deal and its probability of closing. Musk would have to make more disclosures to the SEC and they would govern the process—inviting his most despised regulator to take a magnifying glass to his plans. And for the board, it would rip control out of their hands. Musk could name any price, bash the company in any way, and create chaos at every turn.

Popularized during the bull market of the 1960s, tender offers became the weapon of choice for hostile buyers as they laid siege to vulnerable companies. A bidder would announce their intentions by placing an ad known as a tombstone in a national newspaper. In the ad, the bidder would name their price, which was usually higher than the stock's market value, to entice shareholders into selling. A purchase offer would also be mailed to a company's stockholders, informing them of the opportunity to sell. Such bids often came against a company's board's wishes, but shareholders sometimes saw the chance to make some quick money, and they took the bait.

As opposed to a tender offer, a direct offer from Musk was theoretically more appealing for both the billionaire and the board. Musk would face fewer regulatory hoops, and the board could negotiate. Its members could set terms for the deal and attempt to rein in Musk until the transaction could close.

Prolonged negotiations could also buy Twitter's board more time. As they had during Elliott's attack, board members could shop for friendlier investors and explore a sale to a more stable buyer. But in order to put the company up for sale, they needed to decide what Twitter was worth. Its bankers at Goldman Sachs set about developing financial models to determine a reasonable price.

By the next day, Taylor wanted to move on from the board seat offer. Musk was a pain in the ass, and there was no sense in keeping someone on the board who didn't want to be there. Moreover, he and Agrawal were now sitting on information that was inherently material to Twitter's stock and its shareholders. If Musk wanted out, they would have to formally tell the SEC. And they would have to disclose his apparent overture to buy the company.

"Acknowledging your text with Parag yesterday that you are declining to join the board. This will be reflected in our 8-K tomorrow. I've asked our team to share a draft with your family office today," Taylor wrote to Musk, using a term for an SEC financial filing that Twitter was obligated to share with the public whenever major changes occurred at the company. "I'm looking forward to speaking today."

It took nearly two hours for the billionaire to respond.

"Sounds good," he wrote. "It is better, in my opinion, to take Twitter private, restructure and return to the public markets once that is done. That was also Jack's view when I talked to him."

Again, Taylor was blindsided by conversations that a sitting Twitter board member had been having with a loud public critic and potential bidder. Musk kept mentioning his private conversations with Dorsey as justification for his takeover. Why was Dorsey encouraging him?

Taylor thought the board might have time to round up other potential buyers and set off an auction. But Musk moved with characteristic swiftness—and uncharacteristic silence. He kept as quiet as possible for three days, leaving Agrawal, Taylor, and the rest of the board guessing at his next moves. His Twitter account, an expression of Musk's id and a typically reliable barometer for what he was feeling, remained relatively inactive. His only tweet during the period came on the evening of Sunday, April 10. It was a single emoji of a blushing face with a hand over its mouth. (The post was later deleted.)

On Wednesday, Musk flew from Hawaii to Vancouver for the TED Conference, the annual confab where willing people gave inspirational talks in front of rich people under the motto "ideas worth spreading." Agrawal crossed paths with him in the air as he headed to Hawaii for some time off with his family. Musk left one girlfriend, Bassett, for another, as Boucher joined him in Canada and used the opportunity to take their son, X, to visit her grandparents in a nearby town. But Musk didn't join them. Instead, he worked with Birchall from room 1001 at the five-star Shangri-La, hammering out an offer for Twitter.

It was an unusual transaction for Musk. The billionaire and his companies were not known as big spenders. SpaceX rarely made acquisitions, and Tesla had completed only about a dozen deals in its lifetime, the most notable being its $2.6 billion acquisition of solar energy company SolarCity in 2016. Even that transaction was highly suspect. Musk had been SolarCity's chairman, and it was speculated that he had approved the deal to bail out a floundering company founded by his two cousins, Peter and Lyndon Rive. Acquisitions typically went against Musk's ethos. He sought to run his companies as efficiently and cheaply as possible, and usually pushed Tesla and SpaceX to find in-house solutions rather than purchase other firms.

Even taking into account the handful of deals he'd orchestrated at Tesla and SpaceX, Musk's push for Twitter simply had no comparison. He wasn't buying the social network to glom it on to an existing company. He wanted it for himself. If there was any precedent, it was Jeff Bezos's $250 million purchase of *The Washington Post* in 2013, widely perceived as a goodwill project to save one of the great institutions of American journalism. Musk buying Twitter was more than two orders of magnitude larger than that deal, and would give him control over one of the most trafficked and influential sites of the modern internet.

That Wednesday afternoon, Musk broke his monastic silence and texted Taylor. Musk typically made decisions impulsively and enjoyed stepping on the occasional rake, but this time he carefully chose his words. His message seemed to have been vetted by advisers, with a tone that was far too formal to have been tapped out by his own thumbs. Immediately, Taylor understood that Musk was completely serious about going hostile.

"After several days of deliberation—this is obviously a matter of serious gravity—I have decided to move forward with taking Twitter private," Musk wrote. "I will send you an offer letter tonight, which will be public in the morning. Happy to connect you with my team if you have any questions. Thanks, Elon."

The two executives had a tense phone call after Musk sent the message. Musk was cold and blunt. He would not tolerate any negotiations, he told Taylor.

"I am not playing the back-and-forth game," Musk said. "I have moved straight to the end." His offer would be firm and final, he said.

If the deal didn't work, Musk said he would consider unloading his Twitter stake, potentially sending the stock's price spiraling downward. "This isn't a threat," Musk said. He simply had no confidence in Agrawal, and the investment was junk if he couldn't call the shots and make changes as he saw fit.

Taylor played it cool. "Send the offer," he told Musk. Twitter would give it a fair consideration and get back to him. That was all Taylor could commit to, and he refused to be sucked into a debate about the qualifications of current management. They ended the call with Musk directing any further questions to Birchall.

Musk had tapped Birchall's network of Morgan Stanley bankers, and they had rallied to meet his demand for a quick acquisition of the social media company. For such a complex transaction, Musk's letter, which was sent that evening, was shockingly brief, a mere 145 words. It was also a radical departure from the belated disclosure Musk filed with the SEC announcing his investment, in which he declared that he would remain a passive shareholder, not using his stake to sway the company. Everything had changed.

> Bret Taylor
> Chairman of the Board,
>
> I invested in Twitter as I believe in its potential to be the platform for free speech around the globe, and I believe free speech is a societal imperative for a functioning democracy.
>
> However, since making my investment I now realize the company will neither thrive nor serve this societal imperative in its current form. Twitter needs to be transformed as a private company.

> As a result, I am offering to buy 100% of Twitter for $54.20 per share in cash, a 54% premium over the day before I began investing in Twitter and a 38% premium over the day before my investment was publicly announced. My offer is my best and final offer and if it is not accepted, I would need to reconsider my position as a shareholder.
>
> Twitter has extraordinary potential. I will unlock it.

The letter was serious, but had a callback to Muskian lore. The price, $54.20, was a pothead reference and a callback to his 2018 attempt to take Tesla private. It would value Twitter at about $44 billion. Musk shared a laugh with Birchall over the weed reference, before settling in for the night with Boucher.

At 4:23 a.m. while hanging with the pop singer in his Vancouver hotel room, he tweeted a link to an SEC filing.

"I made an offer," Musk wrote.

—

>>> TAYLOR CALLED AN emergency board meeting on Thursday morning, April 14. The discussion was sobering. Musk had accelerated rapidly from bizarre threats to what seemed to be—despite the trolling joke in his proposed price—a real offer to buy Twitter.

They needed to give Musk serious consideration, Taylor instructed. It was their duty to shareholders to weigh the offer carefully, especially because it represented a premium on Twitter's current stock price. But it wasn't clear to the directors that Musk's offer was one of conviction, or if he might leave them twisting in the wind.

Twitter would need to fend him off, at least for a little while, the board decided. They had to survey other buyers and suss out whether another competitive bid was possible. And even though Musk had firmly said he wouldn't negotiate, he was the richest man in the world. Maybe they could squeeze him.

They also decided that Taylor, Lane Fox, and Pichette should be responsible for managing the deal-making with Musk on a special transaction committee. Each of them were independent directors, without the same biases and ties to the company as Agrawal. And none of them had had extensive previous dealings with Musk like Durban, who had consulted with Musk on his botched take-private of Tesla in 2018.

The board said publicly that it would "carefully review" Musk's proposal. But privately, its members agreed to implement a "shareholder rights plan," or a company defense better known as a "poison pill," a tactic developed by company boards in the 1980s to shield themselves from the corporate raiders. A poison pill would mean flooding the market with new Twitter shares that existing shareholders would be able to buy at a steep discount. If implemented, the move would likely damage Twitter's share price by increasing the supply of outstanding shares. But it would also make it incredibly expensive for Musk to keep buying up shares to accumulate a controlling position.

Musk's offer was certainly a premium on Twitter's share price, which, after recovering in recent weeks, hovered around $45. But many Wall Street analysts believed Twitter was worth far more—after all, its stock had been above $70 just a year earlier. The social media company could reach those heights again, some firms insisted, urging the Twitter board to seek a higher price from Musk.

While Twitter's board huddled on a video call, Musk was on the move. He strode onto a glossy stage in Vancouver that morning, wearing a crisp white shirt beneath a black overcoat, plopped down in an armchair and pulled his ankle up across his knee, revealing shiny motorcycle boots. These would be his first public remarks about his Twitter offer, and the audience at TED sat with rapt attention.

"I do think this will be somewhat painful, and I'm actually not sure I'll be able to acquire it," he said to his interviewer.

In his discussions to join the board, Musk had talked with Agrawal about his concerns of free speech, content moderation, and the supposed liberal influence on internal decision-making. At the conference, he ratcheted up that rationale even further, positioning himself as Twitter's savior. He didn't care about the money. He was doing this to wrest control of the internet's town square from its censorious overlords.

"This is just my strong, intuitive sense that having a public platform that is maximally trusted and broadly inclusive is extremely important to the future of civilization," he said. "I don't care about the economics at all."

Once he stepped off the TED stage, Musk upped the ante, goading the board with a tweet. "Taking Twitter private at $54.20 should be up to shareholders, not the board," he wrote, hinting at his nuclear tender option.

>>> INSIDE TWITTER, many employees were aghast. Could Musk really buy the company? What changes would he make?

But some cheered the idea. Musk had fans throughout Silicon Valley, and Twitter was certainly no exception. They celebrated the idea of a decisive leader. Others believed that Twitter had caved too easily to liberal ideals and lost sight of its moral mission—allowing all conversations to thrive.

Sensing their voices were not the majority, Musk's internal cheerleaders broke away from the frenzied chatter in the company's Slack channels and formed their own group, titled #I-Dissent. The name was a nod to their opposing views on Musk, and the channel quickly became a battleground. Employees bickered and went back and forth about the pros and cons of Musk buying the company.

The fear and anger brewed into a toxic stew and Agrawal scrambled to address his troops. He took a red-eye from Hawaii on the night of April 13— the same day he'd arrived in the islands—and landed back in the Bay Area early the next morning. Once on the ground, he called into the board meeting, then prepared for an emergency all-hands meeting with his workers, in front of whom the chaos was playing out publicly via frantic tweets.

To many tweeps, Musk was antithetical to the people-first, empathetic culture they had signed up for. Some tweeted jokes to mask their concerns, with one employee quipping that they had moved a therapy appointment to attend the Agrawal-led all-hands meeting. Others began updating their résumés. Later, one engineer would create a widely shared internal document titled "Reasons Not to Work for Elon Musk," with two dozen bullet points of the billionaire's character failings, among them his early COVID denial, extreme wealth, and online bullying.

The pushback about Musk's purchase on internal chats was loud and furious. Employees clung to the news of other potential acquirers. Some workers made impassioned pleas for their bosses to reject Musk's bid, while others glued themselves to their Twitter timelines and began a crash course in corporate finance. What was a poison pill and how would it work? Would it change the price of their own stock grants? What grounds could the board use to reject Musk's offer, and what would it take to make another offer viable? Employees from around the world dialed in to the video meeting, which started at 2:05 p.m. in San Francisco.

Agrawal tried to rally them, telling them he believed "everything would work out as it should." His workers pressed him to say more about the deal and asked why he hadn't been more transparent with them when Musk first started buying up stock in the company.

"This provides all of us with this moment where we feel distracted, where we feel a loss of control," Agrawal told employees. "I am personally going to spend my time focusing on things I can control, and I believe it will matter."

One employee said Musk had placed them in a hostage situation, essentially cornering the company's management into selling Twitter.

"I don't believe we are being held hostage," Agrawal said to them all.

The comment didn't instill confidence. Twitter's staff wanted leadership. Some wondered why Agrawal didn't use his Twitter account to fight back while Musk bashed them. It was the kind of thing they could have counted on Dorsey to do. Trying to encourage calm, Agrawal seemed rigid and his answers rehearsed, as if he was reading off a script.

"I believe Twitter stands for way more than one human, any human, me or anyone else, and their values. Twitter stands for open, public conversation," Agrawal said. "It is best defined by everyone who uses the service."

Given his answers, some employees thought Agrawal was already resigned to the outcome. He told them to ignore the noise Musk was making and focus on their jobs, since that was the only thing within their control, and that he would do the same.

What did he just say? Employees frantically messaged each other. Was their chief executive really just going to bury his head in the sand while the richest man in the world lambasted them on Twitter?

It wasn't Agrawal's first testy all-hands with employees, but it left many wanting more. Dorsey, for all his faults, projected stability and knew how to wield his influence. Agrawal seemed unprepared for the battle ahead.

ACT II >>

14 > "Bring the cattle"

As soon as Musk tweeted on April 14 that he had made an offer for Twitter, everyone on the platform began taking sides. A contingent of Trumpers, right-wingers, and libertarians, who believed the company had inappropriately censored accounts that shared their political beliefs, cheered the move. Musk was their savior, the man who nobly sacrificed his own money to rescue society.

"Elon Musk is willing to do what only he can do," wrote Dinesh D'Souza, a right-wing political commentator known for denying the results of the 2020 U.S. presidential election. "He is putting a substantial chunk of his vast fortune on the line to save free speech in America."

Others were less comfortable with the idea of one human controlling one of the most important pieces of digital real estate. Fred Wilson, an early Twitter investor and former board member, called the platform "too important to be owned and controlled by a single person."

"The opposite should be happening," he wrote. "Twitter should be decentralized as a protocol that powers an ecosystem of communication products and services."

But unlike email, Twitter wasn't a protocol. It was a business and a public one, making it possible for a motivated individual with the highest net worth on earth to put in an offer.

—

>>> AFTER NEWS OF his offer broke, it didn't take long for Musk's phone to light up with messages. Gayle King, the prominent television host, called

the proposed takeover a "gangsta move" and told the billionaire he was "not like the other kids in the class." Marc Merrill, the cofounder of video game developer Riot Games, called Musk "the hero Gotham needs" while Adeo Ressi sent a prayer hands emoji with hopes that his former college buddy could fix the platform. The level of hero worship that Musk—who often cast himself as the savior for the world's problems—received gave him confidence. This was the right move.

Congratulations also came from members of his inner circle. These were some of his closest friends and company lieutenants—men he knew he could rely on. And like courtiers, they lined up to kiss the ring.

"Thank you for what you're doing," wrote Omead Afshar, the de facto leader of Tesla's Austin operations whose official title was Project Director for the Office of the CEO. A former ski instructor and medical device company engineer, he worked his way through the ranks of Tesla and had ingratiated himself with Musk by never saying no. At the automaker, he became known among employees for his ability to manage Musk's expectations or translate bad news to the boss, and was given the responsibility of overseeing big projects like the build-out of the Gigafactory.

"We all love you and are always behind you!" Afshar texted. "Not having a global platform that is truly free speech is dangerous for all."

Steve Davis, head of the Boring Company, also chimed in. Like Afshar, he idolized the entrepreneur, though people who knew him said he took it even a tad further. Davis saw Musk as his North Star, and his life's mission was to help Musk achieve his idol's dreams. As Birchall would tell others: "If Elon asked Steve to jump out of a window, he would do it."

Musk had texted his tweet announcing his offer to the Boring Company's head early on Thursday morning along with ideas for a "Plan B" to build a "blockchain-based version of Twitter."

"Amazing!" Davis wrote back after waking up and seeing the messages from his boss. "Not sure which plan to root for. If Plan B wins, let me know if blockchain engineers would be helpful."

Neither Afshar nor Davis, however, were quite as jazzed as Jason Calacanis, a tech entrepreneur who pivoted to investing and hosting podcasts. One of his early companies had funding from Musk, and over the years, Calacanis, a beady-eyed man with a penchant for name-dropping his famous friends, became one of the staunchest defenders of Tesla and its leader. He often made it known that he owned the first Model S ever created.

Over a thirty-one-hour period starting on the day of Musk's announcement, the investor sent eight texts in a row before his friend deigned to respond. Calacanis said that removing bots and spam from Twitter would be far easier than the engineering problems faced by Tesla's self-driving team. He argued that Twitter's verification system of blue check marks should be extended beyond members of the media and celebrities. And after getting his first reply from Musk, he suggested that Musk immediately fire employees to cut costs.

"Sharpen your blades boys 🔪," he wrote. He then sent Musk some of his own tweets outlining product ideas.

In return, Musk gossiped. He believed Agrawal, who had already returned to San Francisco by Thursday morning, was still on vacation. He impugned the Twitter chief's work ethic to Calacanis, while failing to mention that he, too, had been in Hawaii earlier that week. Twitter's chief executive and its hopeful buyer had, by coincidence, left the islands on the same day.

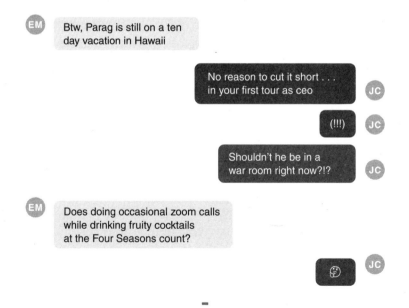

>>> ONE MEMBER OF Musk's inner circle who knew better than to leave a paper trail of texts was Birchall. Perhaps having learned from the "pedo

guy" trial, which exposed his embarrassing communications, he limited important communications with Musk to phone calls or messages on Signal, an encrypted app that could be configured to delete text history.

Birchall didn't necessarily worship Musk unconditionally. He did, however, see him as a generational leader who could usher humanity into a new era of progress with his cars and rockets. He had relocated his family from California to Texas in 2020, as his boss warred against the Golden State's COVID policies, and settled in a mansion outside of Austin to be closer to Musk.

"Elon lives on the rails and he always comes close to the edge," Birchall would sometimes tell people who came to work closely with Musk. "It is our job to keep him on the rails."

It was Birchall who authorized Musk's Twitter share purchases—flaunting SEC rules in the process—and handled the paper pushing during the board-seat rodeo. And it was Birchall who snapped into action to backfill a plan when Musk rashly decided he didn't want the seat but rather the whole company. As his boss laid out his vision on the TED stage in Vancouver, the financier was already sussing out investors who might put money into a Musk-led Twitter. The world's richest man would have control and would finance the deal mostly with his own cash and debt commitments, but Birchall sought to lessen that burden by attracting minority shareholders.

Among those potential investors was Michael Kives, a Hollywood agent who had represented the likes of Katy Perry, Bruce Willis, and Arnold Schwarzenegger and founded K5 Global, an investment advisory firm. Musk sometimes stayed at Kives's Beverly Hills mansion when he was in Los Angeles and dined with the agent's celebrity friends. Kives was a connector and had been texting with Musk two days before he announced his bid, suggesting he should meet with Sam Bankman-Fried, the FTX leader who hoped to make improvements to Twitter. (Kives served as one of Bankman-Fried's biggest hype men in Hollywood and earned $125 million after the crypto mogul made a $700 million investment into K5.)

Kives would be one of a handful of people in Musk's ear about Bankman-Fried, who pushed hard to invest in Twitter. Birchall vetted Bankman-Fried before connecting him with Musk. He also diligently worked the phones, hitting up every connection in Silicon Valley.

Finding people who said they wanted to back Musk wasn't hard. Turning around a stagnant company like Twitter was unlike anything he had done before, but his reputation was such that investors were willing to take a risk just to curry favor or be close to him. In their eyes, Musk was the greatest technology mind of his generation and his TED interview was a clarion call that he was open for business. Musk's supreme confidence in himself was what had driven him to success at Tesla and SpaceX, and it was that same confidence that Morgan Stanley bottled and sold in their pitch to potential investors in his Twitter.

Morgan Stanley prepared a pitch deck for Musk's Twitter 2.0 and passed it around to potential investors. By 2028, the document claimed, Twitter's revenue would be $26.4 billion, a staggering increase from the $5 billion it accumulated in 2021. The eye-popping figure was far more aggressive than the trajectory outlined by an optimistic Agrawal at the start of his tenure as CEO, and it seemed improbable that Musk could quintuple Twitter's revenue.

To get there, the bankers told investors that Musk would decrease the company's reliance on advertising, which generated 90 percent of its revenue. Within six years, the company would take in $12 billion from advertising; $1.3 billion from a new payments business; and some $10 billion in subscription fees from Twitter Blue, an existing product that offered improved features for the site's fanatics. The deck also mentioned a mysterious product known as X. While the bankers didn't know exactly what X would be, they still parroted Musk's projections for 104 million subscribers to the unknown service by 2028. Overall, Twitter would have 931 million users by then, the pitch deck said, more than triple the 217 million users that it had at the end of 2021.

The picture painted of Musk's Twitter was one of efficiency. The company would become a well-oiled machine, spinning off $3.2 billion in cash by 2025, and $9.4 billion by 2028, the deck said.

At Twitter, Segal puzzled over the numbers. It was his team's job to make similar financial models, and short of Musk pulling a rabbit out of a hat, his figures did not make sense, particularly as the global economic picture worsened. Twitter's sales executives felt the same way. *Did he not realize that the company depends on advertising and that subscription products have provided little impact on revenue?* they wondered. Some speculated the figures were completely made up.

Still, investors—some considered the savviest in Silicon Valley—wanted to place their faith in Musk. He dreamed of a different reality, one where humans were multiplanetary or put computer chips in their brains or built tunnels to avoid traffic. His appeal went beyond normal logic. He would find a way, investors believed.

—

>>> ON THE MORNING OF APRIL 15, Good Friday, Twitter's board regrouped to decide how to handle its new enemy. The day before, onstage at TED, Musk had revealed more about his plans for Twitter than in his conversations with Agrawal. During their chats, Musk had at times played coy, framing himself as a willing adviser who could easily miss the mark.

But at TED, Musk said he wanted to open-source the company's recommendation algorithm, allowing anyone to pore through the code that drove certain tweets to the top of the timeline and fight back against the "scam armies" of bots on Twitter. He also insisted that most of the site's content moderation rules had to be scrapped.

His tweet about going directly to shareholders with his offer also worried Twitter's directors. They would need to act quickly to block him. Taylor called a vote on implementing the poison pill in order to buy them time. The directors unanimously agreed.

If Musk continued buying shares and accumulated more than 15 percent of Twitter, the pill would trigger, making it financially onerous for him to continue. The stall tactic gave the company a bit more time to reach out to other financiers and find out whether any of them could beat Musk's "best and final" offer. But the problem was finding any one person or organization with the kind of money to outbid the world's richest man.

The firm on everyone's lips was Silver Lake. Egon Durban, the firm's managing partner, already sat on Twitter's board and had been willing to splurge on a rapid bailout two years earlier in order to rescue Dorsey. Perhaps Durban would once again tap into the firm's resources to protect Twitter. Because Silver Lake already owned a significant stake in the company, it would need to spend less than other potential acquirers.

But Durban's position on the board made such an arrangement awkward. When he joined in 2020, Silver Lake signed an agreement to not acquire more than 5 percent of Twitter—part of its truce with Elliott Management. The board would have to lift the cap, a move that Musk would certainly attack as preferential treatment.

The other name in the mix was Thoma Bravo. The Chicago-based firm was one of the largest private equity firms in the world and had a track record of taking major software companies private—perhaps Twitter could pique its interest. But even though Thoma Bravo had about $100 billion under management, challenging Musk was a big ask—and Twitter didn't fit the firm's portfolio, which focused mostly on business software.

The big tech companies like Google and Salesforce, and media giants like Disney, of course, were options. But all of them had taken whiffs of Twitter before and did not like the smell. President Trump had seemed to have little interest in antitrust enforcement, but the Biden administration had made it a top priority to investigate mergers and crack down on corporate giants that snapped up smaller competitors. Any tech company that submitted an offer for Twitter was daring the Justice Department and the Federal Trade Commission to come calling.

>>> IN THE PANTHEON of American business deals, a $44 billion acquisition does not even crack the top ten. America Online made an ill-fated $182 billion acquisition of Time Warner at the turn of the century, the equivalent of $309 billion in 2023's inflation-adjusted dollars. In 2015, Dow Chemical acquired DuPont in a $130 billion merger of chemical giants. Meanwhile, Disney paid $71.3 billion for 21st Century Fox in 2018, and United Technologies paid $121 billion for Raytheon, the defense contractor, in 2019.

Still, $44 billion was objectively a shit ton of money. It was more than had ever been paid for another social media company—more than double the $19 billion paid by Facebook to acquire WhatsApp in 2014. It was also about equal to the annual gross domestic product of Paraguay, and larger than the GDP of Iceland, Jamaica, and Senegal. In April of 2022, it represented about 20 percent of Musk's net worth. (Bill Gates was the first person on *Forbes*'s annual billionaires rankings to amass a net worth greater than $44 billion. The magazine calculated he was worth $51 billion in 1998.)

That one man could accumulate such a fortune, much less be willing to spend it on a company for ideological or personal reasons, was a testament to the world's widening wealth inequality. This simply wasn't what the elite did with their money. To buy and operate a global corporation for one's own pleasure was unheard of. Musk was breaking the rules of what it meant to be rich.

But even for one of history's wealthiest people, it was still an expensive and complicated purchase. The vast majority of his wealth was tied up in his Tesla shares, and offloading $44 billion worth of Tesla stock was inconceivable. If the market perceived his sell-off as a lack of confidence in the automaker, it could tank Tesla's stock and hurt his loyal investors. He also would be decreasing his control over the company. Some tech founders, like Mark Zuckerberg, had guaranteed they would maintain control of their companies by setting up dual-class shares and granting themselves outsize voting power to control their boards, but not Musk. Like Dorsey before him, he could stand to lose control over his company if Tesla's board turned against him—however unlikely that would be—and he needed to hold on to as many shares as possible. He would have to find another way to come up with money.

>>> ONSTAGE AT TED, Musk had been candid about the fact that Twitter's board could rebuff him. Although he didn't share the specifics of how he might force the deal through, taking his bid directly to shareholders in a tender offer would be the only route left for him if his offer was not accepted.

Of course, Musk couldn't keep the secret for long. On Saturday, April 16, as Twitter's employees and board members tried to set their phones aside for the weekend's Passover and Easter celebrations with their families, Musk kept up the pressure. "Love Me Tender," he tweeted that afternoon, surrounding the title of the 1956 Elvis Presley hit with emojis of music notes.

The not-so-veiled reference made his intentions clear to Twitter's directors and executives. Instead of waiting for the board to agree to his terms, Musk could take his $54.20-a-share bid directly to the shareholders, giving them a window of opportunity to sell their stock to him at that price instead of the public market price of about $45. The tender offer would not allow

Musk to circumnavigate the poison pill—Twitter's board could still flood the market with cheap shares.

But if shareholders were eager to take Musk's offer, they might clamor for the board to remove the poison pill and let them sell at his higher price. Musk had often relied on his fans to generate public pressure in support of his positions in the past. With winking tweets, he had raised the price of a meme cryptocurrency called Dogecoin. The Twitter deal would be no different. It was a $44 billion game of chicken.

That weekend in Austin, some of Musk's friends asked him if he was worried about overextending himself. He was already trying to electrify the world's cars and get people to Mars; why would he waste his time on Twitter? Musk said he would need five years to turn Twitter around. And, he told them, running a social media company was nothing like launching rockets or manufacturing cars. To him, it was just a website where people posted whatever they wanted. How hard could it be?

>>> WITH THE THREAT of a tender offer on the horizon, Agrawal needed to figure out how many of his institutional investors might take Musk's bait. Other directors were anxious to know, too. If they voted against Musk without the support of investors, they would quickly find themselves in an even more painful bind.

Lane Fox, who already had some relationships with major investors, volunteered to coordinate the outreach. She began the process of arranging meetings for Taylor and Pichette with firms, including the Vanguard Group, the American investment adviser that held a 9.2 percent stake in the company, and asset manager BlackRock, which held a 6.8 percent stake, to gauge their sentiment. Would they sell if Musk came to them with his offer?

Each video call seemed to expand with new faces. The bankers from Goldman Sachs and JPMorgan were there, along with a phalanx of lawyers from two different white-shoe law firms: Gadde's former shop, Wilson Sonsini, and the litigation specialists from Simpson Thacher & Bartlett LLP. Twitter hired both firms to represent it.

The discussions were dizzying. But one thing was clear to Agrawal: little more than four months into the job, his leadership was on trial. If he intended to reject Musk, he would need to show a compelling case that he could raise Twitter's share price higher than Musk's offer—and do it immediately.

While most of the board was focused on rescuing Twitter, one member seemed to be gleefully cheering Musk on. That weekend, Dorsey tweeted, "As a public company, twitter has always been 'for sale,'" and went on to say the board had "consistently been the dysfunction of the company." The public dig at his colleagues landed during their round-the-clock work and pissed everyone off. Taylor called Dorsey and asked him to lay off his public criticism of the company. Dorsey sullenly agreed. After their discussion, a Twitter user pressed him about his criticism of the board, noting he had been a member for years without trying to turn things around. Dorsey replied, "So much to say . . . but nothing that can be said."

Agrawal tried to tune out his former mentor's subtweets. He knew that Twitter was in a bleak position. The hope of hitting its 2023 goals for expanding revenue and users had faded from a long shot to nearly impossible. To win investors over, he would have to make the case that he could run a leaner Twitter—and that meant accelerating the layoffs he had been considering.

The cuts would have to be much more significant than he had originally planned. News of layoffs usually sent a company's shares rising, since jettisoning employees meant more money on the company's balance sheet. And the bigger the cuts, the bigger the boost. Agrawal set his sights on as much as 25 percent, a drastic reduction that would shatter Twitter's friendly culture and turn him into an enemy of his employees, but possibly—just possibly—save the company from Musk.

Execution would matter, too. Agrawal couldn't just get rid of employees. He planned to cull the executive suite as well, and to drive those who remained to build and launch new products as quickly as possible. He had already pushed to reform the company's annual review and bonus process, which under Dorsey had effectively prevented managers from ranking and incentivizing workers. Twitter's mellow pace, governance by consensus, and days of rest, monthly recharge days introduced under Dorsey, would all have to disappear.

The staff reduction, Project Prism, was the ace up Agrawal's sleeve. If he and the board could woo investors with his plan and show they had a way of creating more shareholder value than what was presented by Musk, they could reject the $44 billion offer.

"No one thinks Elon credible. . . . yet," Durban texted Segal on Easter Sunday, trying to comfort him as he and Agrawal drew up the cuts. While he didn't know it, the investor's language hewed closely to the private texts between Musk and Calacanis. Both sides focused on deep cuts and saw the workers as cannon fodder.

"You and Parag will bring the cattle," Durban added, a crude rancher's reference that compared the layoff plan to livestock being sent to slaughter.

Segal responded by proposing a whopping $1 billion in cuts over the coming year, with $700 million of that savings coming from the workforce reduction. "Winner winner chicken dinner," Durban cheered.

The Silver Lake partner was pleased. "We just drew three of a kind and have a chance for a full house," he said. "Elon has a pair right now." If all the numbers kept lining up right, Twitter would hold the winning hand.

Selling the rest of the board—and Twitter's biggest shareholders—would be another matter, Durban cautioned. The layoffs and strategy for earning more revenue would need to be airtight.

"You're going to be an old school great company," Durban texted.

"I'm learning," Segal replied. "Parag been great. I've done this before but not at this pace under duress."

"Both of you will be assassins," Durban reassured him. "Turnarounds make the career."

Companies under pressure molded great leaders, he continued. The pain would be good for Segal in the end. Iron sharpened iron. Musk's attack would make Segal a stronger executive.

"You guys are scared to get tossed out," Durban noted.

"Scared is good!" Segal replied, with a caveat. "Scared of losing not getting thrown out."

"They're the same thing," Durban wrote.

That afternoon, Durban was late to a call with Twitter's bankers and legal advisers because he was tied up on another phone call with Morgan Stanley's chief executive, James Gorman. Morgan Stanley was working with Musk, and Durban wanted them to know exactly what it was like to orchestrate a deal with the mercurial Tesla executive.

—

>>> MUSK SPENT MOST of his Easter weekend continuing his trolling campaign. He flew back to Austin on his Gulfstream jet on Saturday and readied for a family Easter egg hunt the next day with his dog, a Shiba Inu named Floki. The dog was a recent addition to Musk's entourage, and the billionaire used it as a prop to gain online support from investors in Dogecoin. On Sunday, he dressed Floki up as an Easter bunny next to a basket of golden eggs bearing the Dogecoin logo. The cryptobros ate it up.

Between all the Dogecoin shilling and his winking tweet about a tender offer, Musk used his account to attack Twitter's board members. Besides Dorsey, they collectively owned little stock in Twitter, he announced, a flaw that he believed meant their financial interests were not aligned with those of shareholders. He also criticized them for earning salaries that ranged from $250,000 to $300,000 for what were essentially part-time jobs. Twitter would save a significant sum under his management, Musk said, simply by no longer needing to pay a board of directors.

The board decided it should reach back out to Musk. They worried that total silence would give Musk too much time to stew, and that he might rush ahead with his tender offer if he didn't believe the board was taking him seriously.

Taylor sent him a polite yet stern text. "Elon, I am just checking in to reiterate that the board is seriously reviewing the proposal in your letter. We are working on a formal response as quickly as we can consistent with our fiduciary duties. Feel free to reach out anytime."

Musk brushed off the message, but he was eager to speak with another member of Twitter's board.

He'd heard that Durban, his one-time partner, had been gossiping with Gorman at Morgan Stanley about how he was a nuisance. The Tesla take-private deal that Durban had worked on with Musk in 2018 had collapsed spectacularly, and the litigation that followed ensnared many of Musk's high-profile backers and bankers in years of depositions and legal discovery. It was an endless mess, and didn't suggest that Musk would be an easy man to work with on the biggest take-private deal in the history of Silicon Valley.

The backchanneling infuriated Musk. He hated it when anyone criticized him, especially someone who had once drifted through his inner orbit. He lashed out at Durban in a text message. "You're calling Morgan Stanley to speak poorly of me . . ." Musk wrote.

Durban didn't text him back.

15 > Parag's Last Stand

On the morning of Friday, April 15, Agrawal was in Twitter's San Francisco offices with Gadde, finalizing the plans to enact a poison pill strategy. But amid those discussions, his mind wandered. He pulled Gadde into a conference room and sketched his plan for Project Saturn, the idea he had dreamed up with Jay Sullivan to transform Twitter into a more permissive, freewheeling platform.

The concept began as a product. In more than a decade at Twitter, Agrawal had heard every version of the debate over online speech. It was one area where he believed he could leave his mark as chief executive, radically changing not only the way his company handled the content flowing across its platform but the way the entire social media industry did.

The problem was that Twitter mostly relied on people to review tweets by hand, a bespoke process that didn't fit the explosive pace of the timeline, and in which everyone wanted to second-guess the company's every move. Agrawal wanted to use technology to create an elegant solution.

Twitter spent endlessly on contractors who sifted through reports about harassment, violence, and child exploitation in more than a dozen languages from disgruntled users, trying to figure out which were legitimate complaints. Particularly thorny situations, like breaking news from mass shootings or questions about how to handle an account like @LibsofTikTok, a user who drove harassment against LGBTQ people, were escalated up to Twitter's staff and usually landed on the desk of Yoel Roth, who became Gadde's content moderation deputy after Harvey's departure. Problems with world leaders or political figures needed to be reviewed by Gadde herself before any action was taken. That human process could be slow.

As controversial content flowed through the moderation process, critics found it still sitting in their timelines. Members of Congress howled over election interference, special interest groups highlighted discrimination and hate speech, and misinformation went viral. Everyone had a tweet they hated and wanted it to be removed faster, then turned around and questioned why a tweet they approved of had been taken down. Even employees occasionally rebelled.

Twitter's goal was to grow—but even if Agrawal managed to attract new users in droves, his content moderation problem would grow in proportion. Twitter had never truly gotten a handle on harassment, and users consistently told the company in surveys that they didn't log on as frequently as they did to other social apps because they experienced abuse, or worried they might. Twitter's toxic reputation limited its growth.

Since his whiteboard session with Sullivan months earlier, Project Saturn had been one item among many on Agrawal's to-do list. He hoped to unveil his new idea with a splashy debut that summer. Saturn would slowly begin to make an appearance on the platform, and Agrawal would order tweaks to the system as Twitter tested it out in real time. But Musk's surprise offer wrecked Agrawal's plans. Instead of an iterative public launch, Project Saturn needed to happen at light speed.

Gadde's plate was full with wrangling the horde of lawyers brought in to manage Musk's offer. So she called Roth on Saturday, April 16, yanking her deputy away from Passover celebrations with his family. As Roth listened, Agrawal explained his vision once more.

"I've been thinking about this," Agrawal said. "We were going to have the conversation later in the year, but we're having it now because of Elon."

With the takeover looming, Agrawal wanted to supercharge his plan.

"The role Twitter plays is around what gets attention, not what's hosted," Agrawal said to the two trust and safety executives. "We define the attention economy on social media. Speech on the internet is already free, but reach isn't and it shouldn't be."

Agrawal wanted only tweets in the tightest ring, the ones ranked at 1, to be boosted by Twitter's algorithm. Those would represent the best of what Twitter had to offer. The lower scores would receive less of a boost from Twitter, and users would have to seek them out if they wanted to see them.

Agrawal used Trump's account as an example to explain his vision. Trump had been warned so many times, but went on breaking Twitter's rules, the chief executive said. Those tweets would have landed the former president in an outer circle, probably 3 or 4, and his audience would be diminished. But under Agrawal's new plan, Trump would never have been banned, and he would be able to move back toward the inner circles of Twitter over time if he behaved better.

Roth and Gadde had plenty of questions for Agrawal. His plan required them to drastically overhaul the rulebook they had spent the majority of their careers writing, and they would have to come up with new consequences for misbehavior on Twitter.

Agrawal pressed Roth to move as fast as possible. He wanted to announce Saturn within a week and needed Roth to draft fresh rules by then, and have the product work underway. Agrawal allowed Roth to pull in a small team of fewer than a dozen engineers, designers, and policymakers to make the work go faster.

Roth was intrigued but a little panicked. One week was no time at all, and he was distracted by the constant headlines about Musk's effort to buy Twitter. It felt like whiplash to Roth, flailing as they reacted to Musk's chaotic proclamation that Twitter was failing to protect free speech.

But it was also an opportunity for Roth to make dramatic repairs to the content moderation system he'd grappled with for a decade. The rules were patchwork, and the tools he had to enforce them were weak. Over the course of the pandemic and Trump's final days in office, Roth had become convinced that Twitter needed a more sensible approach to content moderation, and while he had done his best to make the right decisions, he had only an inbox full of death threats to show for it.

>>> HAVING SET ROTH on the path to write up the rules for Project Saturn, Agrawal turned to Segal and began working on Prism, his plan to slash Twitter's budgets.

With the poison pill in place, there were three paths available to Twitter: First, Agrawal could devise a plan to continue to run it as a public company, with institutional investors agreeing to back the chief executive and hold out against a tender offer from Musk. Second, Twitter could find another buyer, ideally one who shared the company's values and would cause less chaos. Finally, the board could agree to Musk's offer and hand over the keys.

On Monday, April 18, Taylor and Pichette met with shareholders. While Agrawal and Segal hammered out their plans to cut costs at Twitter, the two board members locked themselves in meetings with the company's top institutional investors, carefully polling them on whether they wanted to accept Musk's price or back the company against the erratic billionaire, allowing Twitter to reject his offer and remain independent. Whatever their personal views, Taylor and Pichette had to maintain an air of detachment about Twitter's fate as they took the temperature of their shareholders.

Within twenty-four hours, they were concerned by the feedback and called a meeting with Lane Fox and the other board members on the special transaction committee to discuss their limited options. Agrawal, Segal, Gadde, and Twitter's outside counsel called in over video lines, along with Goldman Sachs and JPMorgan bankers. No one had good news to share.

Taylor and Pichette said that mutual fund managers and investment companies like Vanguard and BlackRock had held large positions in the company for years because they thought it had great potential to be the conversation platform for the entire world—but Twitter failed to live up to those expectations. Management, both Dorsey and Costolo before him, had been slow and sloppy, leading Twitter to miss its opportunities.

Maybe that could change under Agrawal, investors agreed, but he was a relative unknown to Wall Street. Could they count on him to force the widespread changes Twitter desperately needed, or would he continue to plod along like Dorsey? What was Agrawal's plan to save Twitter? The investors needed to know. Some were willing to back him, if he could lay out a strategy for flourishing.

Other investors had their heads turned by Musk's advances. *Take his offer seriously*, they told Taylor and Pichette. The price of $54.20 represented a significant payout. Twitter's shares might eventually rise on their own under Agrawal's leadership, but his future success was not guaranteed, they thought.

The bankers couldn't offer a white knight either. Thoma Bravo and several other investment firms had phoned with questions about a potential acquisition of Twitter. But none of them had put together firm offers, and it didn't seem like the private equity shops were serious.

—

>>> AGRAWAL LEFT THE MEETING tense and spent Tuesday afternoon poring over financial models with Segal and ordering managers to finalize their layoff lists. With Musk blabbing away on Twitter, there was no time to waste. Agrawal would need to present the board with Twitter's business plan for moving forward as a public company, so its members could assess whether Twitter had a fighting chance. If Agrawal could convince the board and shareholders that, under his leadership, the company would soon be worth more than what Musk offered, they could reject his offer.

Dorsey's philosophy since the Elliott intervention had simply been to build great products, and the users and revenue would naturally follow. While Dorsey hired more engineers and greenlit more products, the rocket ship never took off.

Agrawal aimed to reverse that strategy, putting finances first. He was ready to promise the board that Twitter could clear at least 35 percent in adjusted profit in 2023—a steep increase from the company's best quarters, where the figure typically hovered around 20 percent. He proposed that from here on out, Twitter's revenue would rise at a faster rate than its expenses.

There would also be severe cuts across the company. The mass layoffs would save Twitter hundreds of millions of dollars, but it wouldn't be enough. Agrawal planned to lay off the revenue and product leads, Falck and Beykpour, and to close some global offices as well as several floors at its San Francisco headquarters, trimming Twitter's real estate footprint. Twitter's lavish marketing budget would also shrink considerably.

The layoffs would save about $257 million in 2022, according to Segal's projections. Those savings would grow in 2023, adding up to the $1 billion he had discussed with Durban. The office closures would save about $10 million in 2023, and slashing the marketing budget would save about $80 million that year.

The next day, April 20, the full board convened to listen to Agrawal's plan. This was his moment to make the case for himself and save his job. Ever since Dorsey had approached him with the idea, he had dreamed of leading the company—but that dream had not included being under direct threat by Musk. He cleared his throat and looked directly into his laptop camera.

"Recent events are a symptom, not a cause, of the problems that have plagued us. We must face and address the root causes with urgency, and set forward an immediate plan to address the root causes," Agrawal said in a script he'd prepared for the call. The root problem, according to the chief executive, was that Twitter had grown aimlessly, adding new employees and expenses without any solid results.

"We need to switch modes as a company," he said. "We need to make hard decisions that we have historically resisted, lead top-down and do so urgently."

As his pitch to save Twitter came to a crescendo, Agrawal was emphatic. He knew that he could transform Twitter if he could just get a little more time. He forced a smile, trying to make the board members feel his optimism through their computer screens. In just five months, he'd already managed to restructure the company and could feel the pace accelerating. He needed them to feel it, too. But as a board member, he couldn't oppose Musk outright. He had to demonstrate his neutrality.

"This plan will change us and define us," Agrawal said. "The future and how we deliver for our shareholders is in our control, and we will succeed or fail based on our focus and leadership through change, and how well we execute now. We can do this."

And how would the board know when Twitter was winning? "When we stop thinking like an underdog and own our role as the most important service in the world," Agrawal said. Twitter wouldn't be a midsize runner-up in the social media conversation—it would dominate the entire discussion.

It was what Durban wanted to hear from Agrawal's impassioned plea. But other members, like David Rosenblatt, appeared to have made up their minds. Rosenblatt, who ran the e-commerce site 1stDibs, was a keen observer of financial markets and knew that Musk was offering a great price. The board needed to understand Agrawal's vision so they could decide what Twitter was worth, and make sure that Musk's offer was indeed as generous as it sounded. But the deal was likely to go through, some board members believed, and once it did, Agrawal's plan would cease to matter.

Then Segal stepped in to walk the directors through Twitter's financial outlook for the rest of 2022 and what they could expect from the business in 2023. The war in Ukraine had driven advertisers off the platform en masse, as brands were concerned about their ads showing up next to violent content. Twitter was $61.5 million short of its expected revenue for the first quarter of the year, Segal said.

The war wasn't the only headwind Twitter was facing, Segal cautioned. The conflict would continue to cost the company money, and the economy was weakening. Inflation was up more than 8 percent that month from a year ago, as the Federal Reserve cautioned it might hike interest rates for a third time and analysts warned of a recession.

Twitter was falling short. The company was on track to miss its target of $7.5 billion in revenue by the end of 2023 by $1.85 billion. It was also going to be 50 million users under its goal of 315 million daily users. If the economy continued to slouch, the situation would get even more dire, Segal explained. If Twitter remained public, he would have to break all this news to investors at the company's regularly scheduled earnings call on April 28.

The directors weren't sure what to do. They instructed Agrawal to give the plans to the bankers at Goldman Sachs and JPMorgan to review against Musk's proposal. The bankers would use it to tally up what Twitter was worth and compare it to what Musk offered. If the board was going to vote in Agrawal's favor, its members needed to know that his vision for Twitter wasn't just wishful thinking. The bankers needed to give it their seal of approval, affirming that Agrawal's Twitter could outperform Musk's offered price of $54.20 by turning him down and continuing as a public company. Only then could the board of directors offer their support without betraying their fiduciary duties to shareholders.

The board agreed to reconvene four days later on Sunday to make their final decision.

\>\>\> A DAY AFTER Agrawal's last stand, Musk surprised the board again by announcing his most coherent plan yet to buy Twitter in documents filed with the SEC. Morgan Stanley and a group of other lenders had offered him $13 billion in debt financing and another $12.5 billion in loans staked against his shares in Tesla, the billionaire revealed. Musk also signed a commitment letter agreeing to pour $21 billion of his own fortune into the deal. For the first time, Musk's offer was more substantial than what he could fit in a tweet. He was saying he had the money to cover the $44 billion sales price plus the additional $2.5 billion in closing costs.

While Musk was lining up his cash, Twitter continued to lose money. That Thursday evening, the board of directors received a note from Segal. The chief financial officer wrote to tell them that, in just twenty-four hours, Twitter's revenue projections had become even weaker.

Segal had met with his team hours after the board meeting on Wednesday, and they told him Twitter was about $10 million short of meeting its revenue goals for the quarter. The swirl caused by Musk's chaotic acquisition offer had caused some advertisers to hesitate, and employees who were supposed to be focused on sales were distracted. It seemed likely that Twitter would be short another $10 million by the end of the following week and, if the chaos continued, could fall $100 million short of the annual revenue Segal had described to the board.

"We recognize flagging these issues in such short order after Wednesday's meeting may cause additional questions; the business environment is very dynamic," Segal wrote.

Fully aware of the revenue shortcomings, the board agreed to move forward with Agrawal's layoff plans while they continued mulling Musk's offer. They voted by email to approve the plan. Even Dorsey agreed to the cuts, responding to the email discussion with a simple thumbs-up emoji. While the $44 billion bid hung over them, the board still needed more time to evaluate their options. Twitter would follow Agrawal's agenda for at least seventy-two more hours.

On Friday, April 22, Taylor and Pichette updated the rest of the board members on their ongoing conversations with institutional shareholders. Most of them were interested in Musk's offer, and it didn't seem like Twitter could count on them to present a show of force against Musk if he launched a tender offer.

The tide was shifting in Musk's favor, slowly but forcefully pulling the board toward his irresistible promise of $54.20 per share. If the shareholders themselves were eager for Musk's money, the board couldn't stand in their way. It was time to scrutinize Musk's financing and make sure he could pay.

The board dispatched its bankers at Goldman and JPMorgan to meet with Musk's team at Morgan Stanley and get more information about his financing. They wanted to go over Musk's terms so they could decide whether or not his offer was sound.

16 > Just Say Yes

Musk's fresh financing brought with it fresh faces. Among those brought in to work with the mercurial billionaire was Michael Grimes. As one of Morgan Stanley's hard-charging rainmakers, the firm's head of global technology investment banking did whatever it took to impress, from knocking on the doors of prospective clients' homes to snooping on his daughter's internet usage to understand teens' online habits. He had even moonlighted as an Uber driver to win the company's trust ahead of its 2019 initial public offering, which Morgan Stanley led. However, Grimes, a sharp-faced man, had a mixed reputation, having overseen Facebook's disastrous 2012 IPO in which the company's valuation plummeted by billions of dollars due to volatile early trading. The fifty-five-year-old had seen it all in the Valley, but had yet to work closely with its richest figure. This was his chance.

At his bankers' insistence, Musk also hired merger and acquisition lawyers from the firm Skadden, Arps, Slate, Meagher & Flom to support the transaction, instead of relying on Birchall and Spiro, his defense lawyer. Founded in 1948, Skadden was a relatively young firm competing with other blue-chip American names, many of which could trace their roots to the nineteenth century.

By the 1980s—a period defined by the excesses of Wall Street and swashbuckling corporate raiders—Skadden had made a name for itself as the go-to international law firm for hostile takeovers. While other white-shoe firms turned their noses up at the idea of taking over another business without its consent, Skadden saw proxy fights and boardroom wars as an opportunity. The firm had overseen historic deals representing billionaire Ronald

Perelman's $2.7 billion takeover of Revlon and media company Capital Cities' $3.5 billion acquisition of ABC.

By the 2010s, however, Skadden had become almost allergic to the hostile takeover work on which it had built its name. Once happy to lay waste to a company board and ram through a deal for a private equity client, the practice was taking more of its cues from the establishment corporations they once battled with glee. It was steady business, and by 2021, Skadden was a top-five law firm by revenue, with annual bookings of more than $3 billion. Still, despite its successes, the firm had never truly cracked the big consumer tech companies of Silicon Valley.

Musk's Twitter deal presented Skadden with an opportunity to buck that trend and do one of the most novel tech deals ever. Musk was a tricky client who was known for his questionable legal demands and churning through lawyers who didn't anticipate his whims. The challenge of representing him in the once-in-a-lifetime deal fell to Mike Ringler, a new partner on the firm's mergers and acquisitions team.

With a Georgetown law degree and a gleaming bullet pate, the deals lawyer had spent twenty-two years of his career at Wilson Sonsini, which represented Apple, LinkedIn, and, of course, Twitter. At Wilson Sonsini, he had worked closely with some of the lawyers now representing Twitter on the opposite side of the acquisition, and together they had closed industry-defining deals, among them Pixar's $8 billion sale to Disney in 2006, YouTube's $1.7 billion sale to Google that same year, and Hewlett Packard's $25 billion acquisition of Compaq in 2002.

Wilson Sonsini's lead partner on the deal, Marty Korman, one of the Valley's biggest dealmakers, had worked alongside Ringler for nearly two decades, giving him unique insight into the man who sat opposite him at the negotiation table. A broad-shouldered lawyer with a buzz cut, Korman knew his one-time colleague had been dealt a bad hand: Ringler had no power and was doing the bidding of one very capricious man, who often didn't listen to legal counsel.

Ringler at times did not even have direct access to Musk. In meetings between the two sides' attorneys, Ringler sometimes disclosed he hadn't had the opportunity to run certain terms by his client. Most of his interactions were with Birchall and Spiro, and Ringler made it known that his directive from Musk was to "get this deal done." Musk anticipated resistance to his offer, and told his lawyers that moving quickly and aggressively would allow him to pluck Twitter out of the sky without a protracted fight.

>>> ON THE AFTERNOON of Saturday, April 23, after more than a week of silence with Twitter's side, Musk contacted Taylor again. "Would it be possible for you and me to talk this weekend?" he texted.

"That would be great," Taylor responded. He suggested a call that evening with himself, Musk, and Sam Britton, a banker who led Goldman Sachs's tech business from its San Francisco offices. Britton and Segal were close from their days at Goldman, and Britton had spent countless hours on video conferences with Taylor and the rest of the Twitter team over the past week, sussing out the strengths and weaknesses of the company's business plan and Musk's financing.

At 4:30 p.m. in San Francisco, Britton and Taylor dialed Musk's cell. After his blowup at Agrawal, when he questioned the chief executive's productivity, and his constant taunting tweets, Musk was more collegial. He told Taylor and Britton he was unwavering about the price he was willing to pay for Twitter.

Taylor had held out hope that there might be a way to squeeze a few more dollars out of the billionaire. Securing an even higher price would demonstrate his ability to tame the erratic billionaire and would satisfy the investors who recalled Twitter's heyday in 2021, when its shares had floated comfortably above $70. Twitter had closed trading that week at $48.93—buoyed by investors expecting that the deal would close at or above Musk's price. But just as Musk had told Taylor previously, he insisted he would pay $54.20 a share. If the board rejected him, he would go directly to Twitter's shareholders with the offer.

Informed by his conversations with institutional investors, Taylor knew it was a nuclear option he couldn't risk. A tender offer would wrest any control of the sale process away from the board, and Twitter's chairman feared that if he sought a higher offer, Musk would blow the whole thing up. Taylor politely told Musk he would be back in touch shortly and ended the conversation.

With his phone pinging constantly as Twitter's bankers and lawyers sent him updates, Taylor could barely get in a few hours' sleep. But he went to bed that evening convinced that Twitter would have to be sold to Musk.

At 11:00 a.m. in San Francisco, Sunday, April 24, Taylor joined the rest of Twitter's directors on a video call to make their final decision. Lane Fox dialed in from her home in London, where she'd spent most of the spring as she struggled with health issues. Rosenblatt and Pichette joined from the East Coast, while Taylor and Agrawal called in from the Bay Area.

Britton walked the board through Goldman's recommendation. Based on everything Twitter's management had said it might accomplish, the company's past performance, and the state of the stock market, Musk was offering a fair price. JPMorgan rendered the same opinion.

Taylor calmly described his conversation with Musk the previous day. The chairman's demeanor reminded some of the directors of a judge presiding over a courtroom, coolly directing his advisers, some of whom appeared sleepless and frazzled by the frantic pace.

Their tormentor was unwilling to budge on price, Taylor said, and an attempt to negotiate would likely end in disaster. Musk was clearly ready to charge ahead with a tender offer if he did not get his way, and he could even offer a lower price to Twitter's shareholders than the $54.20 currently on the table. He could also bail on the entire transaction, leaving the shareholders high and dry.

The best thing to do, Taylor admitted, was to take Musk's offer. It would lock the billionaire in at an acceptable price and give the board some control over how the transaction played out.

Other board members agreed. Twitter's closest comparable public company, Snapchat, had seen its share prices fall about 20 percent that month, while six tech giants—Apple, Amazon, Google, Meta, Microsoft, and Netflix—collectively had lost more than $2 trillion in market value since the start of the year. It was a dicey moment for the wider American economy. Market analysts who thought Musk's offer was a lowball effort just weeks earlier had started to change their tune as well.

Most of the board members felt resigned. Any emotions about the prospect of selling the company to someone who planned to upend it had fallen away, and all that was left was a grim determination to make Musk pay full price. They were focused on their lawyers' advice to secure the best price for Twitter, come what may.

The conversation shifted to ensuring that Musk couldn't weasel out of a deal. If the board did take Musk's offer, how could its members be certain that the Tesla executive would follow through? He had already agreed to join their ranks, only to flit away mere days later. Any agreement with him would have to be airtight.

With the bankers' recommendations in hand, the directors gave the go-ahead: Taylor and the transaction committee should meet with Musk and attempt to thrash out an acquisition agreement that would bind Musk to his price.

While the other directors dropped off the call, Taylor stayed on with Lane Fox, Pichette, Twitter's management team, and the phalanx of bankers and lawyers to start hammering out what the deal should look like. The lawyers would have to draft something from scratch, and quickly began brainstorming shackles to ensure Musk didn't wiggle away.

As they huddled that Sunday afternoon, Britton felt his phone buzz. The Goldman banker was stunned. In all his years of managing complex tech deals, he had never seen anything like the email that was sitting in his inbox. Musk's team had drafted up a seller-friendly agreement on their own and were offering it to Twitter. The billionaire was completely bypassing the ritual dance that preceded massive transactions like this one, in which the seller outlined the deal, the buyer revised it, and the two sides negotiated back and forth, sometimes for months.

But Musk wanted none of it. Not only had he thrown together his own acquisition agreement, he was insisting that it be finalized and signed by the very next morning, so that it could be announced to the public before the stock market opened that Monday.

The draft agreement was accompanied by a letter from Musk himself. "$54.20 has been and will remain my best and final offer, period. This is binary—my offer will either be accepted or I will exit my position," Musk threatened.

Just as the board had during its meeting that morning, Musk noted that the market had continued to decline, making his offer all the more lucrative.

"With your cooperation, we can negotiate changes that you require to be able to announce a transaction before the market opens tomorrow that the shareholders can then vote on. I would respect the outcome of that vote if the shareholders prefer the management plan to my $54.20, and exit my position entirely if that is the outcome of the vote," Musk added.

Britton jumped in, interrupting the meeting to tell everyone about the bizarre missive he'd just received. The conversation quickly shifted from how to approach Musk with an agreement to how to revise the one the billionaire proffered.

>>> AS THE NEGOTIATIONS BEGAN, Gadde, Twitter's head lawyer, found herself in comfortable territory. Before her time overseeing Twitter's legal and policy matters, she had spent ten years as a corporate governance lawyer at Wilson Sonsini, where she had been privy to the ins and outs of boardroom proxy fights and tender offers. She had been part of the team of lawyers—along with Korman and Ringler—who had defended biotech giant Genentech against an unsolicited offer from the Swiss healthcare firm Roche and advised it to hold out for a larger offer in the depths of the 2008 financial crisis. The following year, Roche acquired Genentech for $46.8 billion, wringing out $3 billion more from the original offer.

Gadde knew that Twitter would have to lock Musk in a straitjacket of an agreement, giving him no way to extricate himself once he signed. She trusted Korman, her former mentor, and was confident in Twitter's ability to get what it wanted, telling members of her legal department almost from the outset of negotiations that the deal was going to happen.

The speed at which Musk wanted to close the deal gave Twitter's board an incredible advantage. Acquisitions of this size and complexity could typically take months or more than a year to close, as bankers and lawyers engaged in due diligence, negotiating, and paper pushing, but Musk wanted Twitter and he wanted it now.

So his side sped ahead. Still firm in his belief that he wouldn't allow the Twitter board to muzzle him, Musk refused to sign nondisclosure agreements that would have allowed him and his advisers to review nonpublic information about Twitter's inner operations and finances.

JUST SAY YES

Twitter's deal team was in a state of disbelief. No normal buyer would engage in such an expensive and elaborate transaction without first looking under the hood. Nondisclosures were common in any corporate takeover, and as one person who worked on the deal put it, signing one was as natural as "deciding to wear pants when you walk out into public." But Twitter never sent anything to Musk before the deal was signed that it hadn't already made public to its shareholders. He had prevented that from happening.

Had Musk signed an NDA, he and his bankers would have been able to see Twitter's depressing financial projections. Or he could have asked for figures to do external analyses of Twitter's user numbers or the amount of fake and spam accounts that he professed to worry about. Had he known the full picture, he might have lowered his offer or pulled it altogether. But his bankers proceeded with the deal sight unseen. At Musk's insistence, Ringler continued to aggressively push for the deal to close.

Twitter's lawyers believed Musk was trying to shove the deal forward at breakneck speed to force them to balk. It seemed like a tactic: Musk appeared to want the board to reject him, giving him a reason to launch the hostile tender offer.

So Korman and his counterpart at Simpson Thacher, Alan Klein, came up with a new strategy: "Just say yes."

Wall Street had been accustomed to the "just say no" defense, a tactic among public company boards that arose in the 1980s to deal with corporate raiders. The idea was simple and stemmed from the idea that boards could discourage a hostile takeover by rejecting any offer immediately and simply refusing to negotiate with a bidder. Twitter's lawyers flipped this approach on its head. Because they believed Musk's side wanted them to stall or refuse, they advised the company's board to agree to Musk's terms—which locked in his inflated price. Give no indication Twitter wanted anything other than a quick and easy sale, they said.

Korman and Klein could hardly believe their luck. They knew they were dealing with a slippery character and angled for an agreement that placed all the legal risk on the buyer.

Musk's draft did most of that work for them. By pushing his lawyers to get the deal done immediately, Musk undermined his own position. After his team sent the agreement on Sunday, he was eager to finalize everything by Monday morning.

On a video conference line that evening, Klein, Korman, Ringler, and bankers from both sides hammered out a deal through the night. Ringler insisted that the bankers at Goldman Sachs and JPMorgan had to stop shopping the company around to other buyers in the private equity world. He didn't want Twitter's board to back out at the last minute and recommend another acquirer to shareholders. Korman and Klein sensed they could get a few more concessions and marked up the agreement with changes.

Korman was adamant that Musk had to use his own name on the agreement. He demanded that Musk sign the document—not representatives of one of his companies—an unheard-of move in corporate deals that placed Musk legally on the hook. He and Klein also insisted that Musk be personally liable to come up with all the money, aside from the loans he had secured from his banks, to complete the transaction. In the eyes of the two board lawyers, it had to be the way. If they agreed to the transaction only for Musk to renege, the damage to the company's reputation and stock price would be irreparable.

Each side agreed to a hefty breakup fee to hold them to their word: $1 billion. If Twitter's board went out and found a new buyer, they would have to pay Musk the sum to walk away. Musk would have to fork over the amount only if the deal failed because of the collapse of his $13 billion in debt financing. But the billionaire had no other outs. Twitter's lawyers included a "specific performance" clause in the contract that allowed the company to sue him to force the deal through. Musk couldn't just change his mind and pay the breakup fee to escape.

With little fuss, Skadden agreed to all the amendments. By about 10:00 a.m. on Monday in San Francisco, a few hours past the Musk-imposed deadline of market open, the details of the merger agreement, a contract that laid out the union of Twitter with a holding company controlled by Musk, had been hammered out. Twitter's board gathered on another video call to discuss final steps.

The meeting was a virtual funeral. The eleven directors knew they had done their duty—getting the best possible price for shareholders. Durban, who had cheered Agrawal on as he sketched his plan to keep Twitter independent, had a new fixation: making his one-time client pay.

Others, like Agrawal, had been with the company for more than a decade and seen it through high highs and low lows. He felt pride but also overwhelming angst. He had believed in the potential of Twitter and his ability to help the company achieve it, and the sale seemed like an admission of defeat. He had been chief executive for only 147 days, and it seemed as if half of that time he had been wooing or jousting with Musk. The rest of his tenure as Twitter's leader would be spent transitioning the company to new ownership.

Lane Fox had also enjoyed a long tenure on Twitter's board and held its mission sacred. She mused about the corner she and the directors had been painted into, as they were forced to put a price tag above all else. Was the share price really the only thing that mattered? What about the impact on Twitter's employees, or the likely destruction of the platform?

The lawyers reminded everyone that their role was simple—they were there to represent the interests of the shareholders. The only thing to consider was whether the people who owned Twitter's stock were getting a fair price and the best possible outcome.

Dorsey sat silent. His fellow board members pressed him—not for the first time—on his relationship with Musk. They wanted to know how much planning the two men had done behind the scenes. Dorsey had promised his colleagues several times that he had not agreed to help Musk with the transaction. Although most of the board members owned tiny fractions of Twitter's stock, Dorsey held about a 2 percent stake and stood to make nearly $1 billion from the deal.

But Dorsey didn't need to sell off his shares if he didn't want to. Musk gave him the option to roll his shares into his new ownership, allowing him to remain involved with the company long after the rest of the directors were gone. *Had Dorsey talked to Musk about sticking with Twitter?* the board wanted to know. Dorsey swore he had no plans to roll his stake over to Musk.

No one was sure they could believe him. Dorsey, always a cipher, had become even harder to read after he had resigned as chief executive five months earlier. Even the Twitter executives in the meeting—Agrawal, Gadde, and Segal—who had spent years working alongside Dorsey, had fallen out of touch with him.

But Dorsey's potential involvement in a new Twitter wouldn't change the price that Musk paid. Several board members were livid with him and considered him a snake. After years of backing his leadership, they thought he had bad-mouthed them and gone behind their backs.

With that, the board voted unanimously to approve the terms of the sale and agreed they would recommend the same to shareholders, who had to vote to approve the deal.

As Twitter's board voted, Musk was in Austin at the Tesla Gigafactory with Luhut Binsar Pandjaitan, an Indonesian minister who oversaw the country's investments. They were supposed to discuss mining, as Indonesia sat on reserves of minerals that were crucial for electric car batteries. But the billionaire was distracted by the commotion of the deal.

"I think this is the craziest thing I've ever done," Musk said to Pandjaitan. "I'm going to regret this for the rest of my life."

Just before 11:45 a.m. in San Francisco, Twitter's shares stopped trading on the New York Stock Exchange.

Five minutes later, the company published a press release.

> **Twitter, Inc. (NYSE: TWTR)** today announced that it has entered into a definitive agreement to be acquired by an entity wholly owned by Elon Musk, for $54.20 per share in cash in a transaction valued at approximately $44 billion. Upon completion of the transaction, Twitter will become a privately held company.

The release also included statements from Taylor and Agrawal. The chairman chose to focus on "the best path forward for Twitter's stockholders." The chief executive said he was "deeply proud" of his employees and their work. Neither statement touched on what either man thought was in store for the future of the company. That was left to the new owner.

"Free speech is the bedrock of a functioning democracy, and Twitter is the digital town square where matters vital to the future of humanity are debated," said Musk. "I also want to make Twitter better than ever by enhancing the product with new features, making the algorithms open source to increase trust, defeating the spam bots, and authenticating all humans."

Just after the press release went live, Dorsey reached out to Musk. "Thank you," he texted Musk, including a heart emoji.

"I basically following your advice!" Musk responded, so excited that he didn't check the message for typos before sending it.

"I know and I appreciate you," Dorsey wrote. "This is the right and only path. I'll continue to do whatever it takes to make it work."

—

>>> BY THE TIME the deal was announced, Twitter employees were on tenterhooks. Most found out along with the rest of the public, reading the press release or news stories that were passed around on text threads or posted on Twitter. Outrage and frustration reigned on Slack.

"Oh man it would be awful if someone looked over all my Slack messages and discovered I think Elon Musk is a thin skinned, self aggrandizing, egotistical troll with a authoritarian management style and an understanding of the issues involved in social media roughly rivaling that of a well read 8 year old," one person wrote in a Slack channel called #social-watercooler, where employees gathered to crack jokes and banter.

It was easy enough to find the dissenters, as they mouthed off in Slack, and reignited the criticism of Musk from when the company had first announced his joining the board weeks earlier. Still, there were plenty of employees who welcomed Musk, or at least didn't immediately dismiss him.

That camp had become disillusioned with the lack of growth and innovation at Twitter or seen their ambitions and ideas sidelined by a checked-out Dorsey. And they saw Musk's track record at Tesla and SpaceX as more than enough proof that he could lead a turnaround at Twitter. They found a place to commiserate in Slack channels like #I-Dissent, and their reactions ranged from cautiously optimistic to downright gleeful. Rolling the dice on Musk was better than slogging through the malaise that had plagued the company since Dorsey's tenure.

"I remember the first oneteam in SF where Elon was a guest and everyone loved him," one employee wrote on Blind, an anonymous chat app for workers. "Then the herd found out later that Elon disagrees with them on some issues and they became enraged, threw tantrums on Slack and denounced him from the human race."

Dorsey chimed in that evening on Twitter. He chose to communicate his feelings about the sale of his company with music, sharing a not-so-subtle link to a Radiohead song titled "Everything in Its Right Place."

Just to make sure no one missed his meaning, Dorsey fired off more tweets, extolling Musk and laying out his complicated views on the beast he created.

> **Jack**
> @jack
>
> I love Twitter. Twitter is the closest thing we have to a global consciousness.
>
> 10:03 PM Apr 25, 2022
>
> 1.1K   5.9K   27K

> **Jack**
> @jack
>
> The idea and service is all that matters to me, and I will do whatever it takes to protect both. Twitter as a company has always been my sole issue and my biggest regret. It has been owned by Wall Street and the ad model. Taking it back from Wall Street is the correct first step.
>
> 569   5K   31K

> **Jack**
> @jack
>
> In principle, I don't believe anyone should own or run Twitter. It wants to be a public good at a protocol level, not a company. Solving for the problem of it being a company however, Elon is the singular solution I trust. I trust his mission to extend the light of consciousness.
>
> 4.7K   17K   70K

Twitter's employees couldn't help but think that Dorsey was looking for a scapegoat. He had known the stakes and expectations of running Twitter as a business when he rejoined the company as CEO in 2015, and while it was noble to think that Twitter could be a protocol instead of a corporation, he and the other cofounders had made the decision to head in the opposite direction years earlier. Employees also saw an incredible irony in calling

Twitter a public good and then endorsing its sale to the world's richest man. They mocked Dorsey's phrases "singular solution" and "extending the light of consciousness" in group chats.

Some executives who had also become disillusioned with Dorsey by the end of his tenure began to speculate about their former boss. They wondered if Twitter's creator had actually been speaking to Musk about buying the company earlier than previously known, and began to ask if his August 2021 visit to SpaceX's Starbase in Texas was where Dorsey kick-started a takeover conversation. They never found solid evidence for their theories.

One person who was bothered by Dorsey's message was Agrawal. Dorsey was the reason he was CEO, but Dorsey's wooing of Musk had completely undermined his position. After the acquisition was announced, Agrawal didn't follow Dorsey on a victory lap. Instead, he sent a curt email to his executive team, telling them to put the layoffs on hold. "Given our announcement today, we will not be moving forward with Prism," Agrawal wrote. "While we will continue the work to build the best Twitter and operate efficiently, the Prism plan is no longer the right path forward. Please know that you are still under NDA and should continue to keep Prism strictly confidential. Thank you for all the work you did to get here, and for your leadership now as we move forward." Although they didn't know it, the deal spared Falck and Beykpour from the chopping block.

Executives didn't get any other directives from their leader for the rest of the day, leaving them to fixate on the headlines just like their underlings. It was the first of many times that Agrawal's communication during Musk's disruptions would leave employees wanting more. But he didn't know what else to say. The process had drained him. At an all-hands meeting with employees that day, his message wasn't uplifting. "Once the deal closes, we don't know what direction this company will go in," he told Twitter's employees.

Agrawal would later tell a confidant that Musk was unlike anyone he had ever had to deal with. There was just no way to assess Musk's baseline behavior on any given day. Agrawal could normally get a measure of the abilities and limits of the people he worked with, using that as a gauge to predict how they would act, he told his friend.

"With Elon, he can always take it to another level," he said. Musk could always do something crazier. Working with him was next to impossible.

17 > Golden, Golden

Vijaya Gadde had barely slept by the time the deal closed. The negotiations had stretched Twitter's chief legal officer thin as she stayed in constant communication with the board's lawyers. The comfort of reuniting with familiar faces at Wilson Sonsini had quickly worn off as they sprinted to comply with Musk's demand to sell one of the world's most prominent social media companies in twenty-four hours.

This was her first brush with one of Musk's favorite operational tricks—imposing an impossible deadline and testing people by seeing how quickly they could race to meet it.

On the afternoon of Monday, April 25, she held a call with her legal and policy employees. Gadde was direct, but her strong vocal fry became more pronounced when she spoke in front of large groups, a side effect of her nerves. She projected certainty even as her voice cracked: Twitter has a watertight deal with Musk. That meant that at some point, they should expect Musk to own the company, and with that would come some drastic changes, she said.

Gadde had a reputation for stoicism. But as she recited her teams' accomplishments and impact, she began to tear up. They should take pride in the work, she said, and encouraged them to continue doing so, no matter the outcome of the sale. It was a tribute to all they had accomplished that Twitter was worth $44 billion.

With the deal signed, Dorsey dove in. That Tuesday, he started back-channeling with Musk in earnest, telling the Tesla executive that he would help set priorities for the transition and get Twitter's management in line. In a text message, he downplayed the abilities of Agrawal, having named him only five months prior as his ideal successor.

"He is really great at getting things done when tasked with specific direction," Dorsey wrote, making Agrawal sound like a freshman in need of a homework assignment. "Would it make sense for you me and him to get on a call to discuss next steps and get really clear on what's needed? He'd be able to move fast and clear then."

The trio gathered on a video call at 5:00 p.m. in San Francisco. Dorsey had prepared an agenda that covered Twitter's biggest problems and outlined short- and long-term goals, and Musk tore into it immediately. Following Dorsey's advice, he began directing Agrawal on how he wanted Twitter to be run.

Agrawal tried to explain his own views. There were reasons Twitter had moved slowly, and particular roadblocks it needed time to overcome, he said. Dorsey sat back, watching his successor and his idol debate.

It was clear that Musk was getting frustrated. He fired off a new demand at Agrawal: fire Gadde. He viewed her as the architect of Twitter's censorship regime. Twitter needed to become a more lax, anything-goes environment, Musk believed, and he was unwilling to wait for change.

Agrawal was appalled. Musk didn't seem to understand that, although he had signed a contract to buy Twitter, he didn't own it yet. Until Musk's money transferred, Agrawal was still the CEO, and he would hire and fire whoever he liked. He looked to Dorsey, who had worked in lockstep with Gadde for years, but Dorsey remained impassive.

Agrawal refused. Gadde would stay, he said firmly. Musk was free to do whatever he wanted once he owned Twitter, but until then, Agrawal would call the shots.

Musk was furious. All he heard were excuses for why Agrawal had failed to execute and why he wouldn't try hard enough in the coming months to meet Musk's expectations. Keeping Gadde at the helm of Twitter's content moderation was inexcusable, Musk believed.

"You and I are in complete agreement," Musk wrote to Dorsey after the call ended in a stalemate. "Parag is just moving far too slowly and trying to please people who will not be happy no matter what he does."

Just as Dorsey had waived the opportunity to stand up for Gadde, he threw Agrawal under the bus. "At least it became clear you can't work together," Dorsey responded.

It took two days for news of Gadde's meeting with her staff to leak, but by Wednesday, April 27, of that week, *Politico* had a story: "Twitter's top lawyer reassures staff, cries during meeting about Musk takeover." Saagar Enjeti, a conservative podcast host, tweeted a screenshot of the headline, but put a different spin on it. "Vijaya Gadde, the top censorship advocate at Twitter who famously gaslit the world on Joe Rogan's podcast and censored the Hunter Biden laptop story, is very upset about the @elonmusk takeover," he wrote, tagging Twitter's prospective new owner.

Within two hours, Musk responded. If Agrawal wouldn't fire her, Musk could always turn up the heat. "Suspending the Twitter account of a major news organization for publishing a truthful story was obviously incredibly inappropriate," he wrote.

Twitter's decision to block links to the *New York Post* story about Hunter Biden's laptop was easy to attack since the company itself had already admitted it made the wrong call. But Musk needed to vent, and this was his way of throwing his toys out of the crib when he didn't get what he wanted.

Over the years, Gadde had faced waves of abuse for Twitter's policy and enforcement decisions. She had faced death threats after Twitter banned Trump's account in the wake of the January 6 riot at the Capitol. That experience and others had inured her to the abuse that stemmed from tough policy decisions. But Musk's post ignited something else. His comment was retweeted more than 50,000 times, including by many right-wing sympathizers. Some of them added racist commentary and called her an ungrateful immigrant. As the abuse cascaded, some business leaders, including former Twitter CEO Dick Costolo, slammed Musk for attacking an employee.

Twitter's lawyers had slipped a line into the merger agreement that prohibited Musk from disparaging the company or its executives in his tweets, but Musk had little regard for his contractual obligations. He would say what he wanted.

Employees at Twitter tried to find a silver lining as threats rained down on their top lawyer. Perhaps Musk's tweets violated his agreement not to bad-mouth the company, and Twitter had a way to escape from the transaction, they speculated.

Gadde, however, tried to calm them. Even if Musk's missives technically broke the agreement, she knew that the company would never jeopardize the $44 billion deal over a tweet.

"Letting this cycle pass and focusing on the important work we do everyday is the best path forward," Gadde wrote on Slack.

The cycle did not pass—at least not as expected. As he saw comments about Twitter's chief legal officer flooding in beneath his tweet, Musk became even more fixated on Gadde and what he saw as a great injustice against the *New York Post*.

Agrawal attempted to manage the situation as best he could. Not wanting to upset the deal, he posted a vague tweet a few hours after Musk's that attempted to defend his colleague without further fanning the flames. "I took this job to change Twitter for the better, course correct where we need to, and strengthen the service," he wrote. "Proud of our people who continue to do the work with focus and urgency despite the noise."

On Twitter, the CEO's post was pilloried as spineless. It didn't mention Gadde or Musk by name, and seemed like a sterile message crafted by public relations staffers. Agrawal knew that he was muzzled from saying more, and in the meantime, he took the blame and dealt with more immediate matters. He authorized Twitter's security team to provide around-the-clock detail at Gadde's San Francisco home in case the online attacks translated into real-life ones.

Having learned to disassociate from the platform, Gadde stopped reading the notifications on her Twitter account—but her protective husband, a start-up founder named Ramsey Homsany, could not remove himself. He monitored her replies as the abuse rolled in.

Gadde wanted to get on with her work and life. That night, to mark the signing of the deal, Twitter's executives had organized an outing together. It was game five in a best-of-seven NBA playoff series between the Denver Nuggets and the Golden State Warriors, and Twitter had some of the best seats in the house.

They would be blowing off steam in a "bunker suite" at San Francisco's Chase Center. Twitter had signed a multiyear deal for access to the fully catered, subterranean lounge underneath the stands, which also came with seats near the court and cost the company millions of dollars. The suite was one of Segal's favorite haunts and, with Dorsey's permission, he had signed off on the lavish expense to give executives and salespeople a place to woo important clients. (Dorsey maintained his own suite in the stadium, with seats he didn't have to share with his employees.) But that night, the bunker, fully stocked with wine, top-shelf booze, and charcuterie, would be used for the executives' own pleasure.

Gadde was excited for the game—anything to take her thoughts away from a sleepless and mind-bending ninety-six hours. She was determined not to be affected by Musk's trolling. But the optics of Gadde sitting courtside as Musk railed against her content moderation policies worried Twitter's communications team, and security staffers fretted about keeping her safe. A top communications staffer called Gadde and said it would be awkward for her to attend while so many people were feeling sorry for her. What if a camera captured her there cheering and broadcast it to a national television audience? And what if Musk caught wind of that?

Gadde, who had stayed silent throughout a week of Musk's provocations, felt disappointed. While the rest of the executives who had worked on the deal watched the Warriors win and advance to the next round of the playoffs, she had to follow the game from home.

>>> GADDE TRIED TO distract herself with work. Agrawal called another meeting with her and Roth to discuss Project Saturn. This time he brought Sullivan, his product lead, to meet with the trust and safety team as they hashed out what the future of Twitter would be.

"Our current approach appears to be one where we are swimming upstream. It feels unsustainable," Agrawal said, his voice laced with exhaustion. The chief executive had been pushed to the brink by Musk over the past two weeks and he couldn't recall a time he'd worked longer hours on a more frenetic series of problems.

"No matter how good our policies and enforcement are, without fundamental shifts in product, transparency, and control, we're not solving for the global optimum of people feeling that they can trust us and that they can feel safe speaking on the service," he continued.

Roth walked the executives through his draft for Saturn. They hated it.

Although Agrawal had insisted users ought to know which ring they were in, and how to progress to a better one, he didn't like it once it was in front of him. It seemed creepy and surveillant, he said.

Gadde and Sullivan agreed. The ranking system felt like a social credit score, the kind of surveillance system that an authoritarian regime might use to track its citizens and determine their trustworthiness. In her time at Twitter, Gadde had staked her reputation on fighting surveillance and government overreach. She had fought back against governments who tried to convince Twitter to identify pseudonymous users and resisted calls to remove dissident tweets. She had blocked the CIA from accessing Twitter's firehose and challenged gag orders from the Justice Department that prevented Twitter from speaking publicly about surveillance requests it received from the U.S. government. Gadde didn't want Twitter to institute a content moderation system that resembled a surveillance apparatus.

Sullivan tried to put it gently—the idea was right, but the implementation was wrong.

"With clear published rules and norms, and transparency about why someone is in the state they're in, you've provided sufficient information for fairness and trust to exist," Sullivan interjected. "In that world, you may still not like that one company is making a judgment."

After sprinting through the harrowing week that culminated in the sale of Twitter, Roth was tired and frustrated. He had to go back to the drawing board.

A product manager named Keith Coleman stepped in to help. Coleman, a former Google executive who came to Twitter in 2016 when the company bought his small start-up, had decamped to Hawaii during the pandemic to pursue his passion for surfing. Early in his career at Twitter, Coleman had been responsible for boosting the platform's stalling user growth. But in recent years, he had begun experimenting with content moderation products.

Inspired by Wikipedia, which allowed anonymous internet users to write and edit its corpus of encyclopedia entries, Coleman had dreamed up a similar system for Twitter. He called it Birdwatch, and it allowed Twitter users to fact-check controversial tweets and add notes to them with more context. The notes could debunk misinformation or correct the record as fresh headlines emerged. Wikipedia's army of editors enabled the site to move at light speed, and the site was often updated within minutes after news of a celebrity death or an international conflict broke.

Coleman's idea was that Twitter could do the same, letting users fix misinformation on the platform faster than its content moderators could. In January 2021, he began rolling out Birdwatch, which quickly became popular with Twitter's leadership. Alongside Dorsey's nascent investment in Bluesky, Birdwatch showed that Twitter was opening its gates to users, letting them have more control over their experiences on social media.

Since he'd become a darling of the C-suite based on Birdwatch's success, Coleman seemed like a natural partner for Roth as he redesigned Twitter's content moderation strategy.

While Roth and Coleman tried to reimagine what Saturn would look like, Agrawal and Sullivan focused on rallying the troops. Roth wasn't the only Twitter employee who was burned out and frustrated—employees were quitting Twitter in droves, and those who didn't barely wanted to work.

On May 4, Agrawal held a company-wide meeting to talk to employees about Musk's looming acquisition. While he waited for workers from around the world to log in to the virtual meeting, the beat of Jill Scott's hit "Golden" blared incongruously through the office, defiantly wallpapering over the grim mood. "*I'm living my life like it's golden, golden*," Scott crooned as disgruntled employees filed into Aviary, the event space in Twitter's headquarters where workers gathered in person for all-hands meetings.

Agrawal had come down with COVID and stayed home but told his staff over a video line, "We are shifting our focus back to our work."

The chief executive quickly kicked the conversation over to Sullivan, to the disappointment of some employees who still hoped their leader would speak candidly with them about Musk. Sullivan tried his best to inject life into the room.

"One of the things I'm hearing is, 'Why bother?'" Sullivan said. "We have responsibility to hundreds of millions of people every day, all over the world. They are using our products to communicate and participate in conversations about the most urgent and important things happening on earth.

"We don't know what the future is going to hold, but what we'll know is we left it all on the field for the people who rely on us every day," he said.

Sullivan then walked through the work that lay ahead, including Agrawal's vision for Project Saturn. Although the secret content moderation overhaul wasn't ready to be revealed to all of Twitter's employees, Sullivan hinted at what was coming.

"Social media is in a crisis of confidence right now," he explained. "We've been ahead of our peers and there's something we can do there with product mechanics. People producing more healthy content can get more exposure, and people have more control if they want to see the wild and woolly."

Some employees rolled their eyes at the pep talk, while others were invigorated. But most focused on Agrawal, who sat silently on the screen as his deputies owned the conversation. They wanted to hear directly from him.

>>> WITH THE MERGER agreement signed, Musk's quest for funders went full-steam ahead. The billionaire's bankers had already secured $13 billion in debt financing that would be leveraged against Twitter, leaving Musk needing to come up with $33.5 billion—enough to cover the rest of the $44 billion price tag and $2.5 billion in closing costs.

The cash needed to cover that astounding figure would come from different sources. Musk, of course, already had a 9.6 percent Twitter stake, which amounted to $4 billion. There was the possibility he could sell off shares in Tesla or SpaceX to raise money. And Musk had also secured a $12.5 billion personal loan that had used his shares in Tesla as collateral.

Musk viewed the personal loan as something of a stopgap. It was incredibly risky to layer a loan on top of a volatile asset like Tesla's stock, where Musk's decisions could swing the price and add or shave off billions of dollars in valuation on any given day. Knowing this, the billionaire had already started courting investors in mid-April to raise outside funding. And in those conversations, one person in particular had stood out to his advisers as a potential cash cow.

Along with Jared Birchall and Michael Kives, Michael Grimes pushed to get Sam Bankman-Fried to invest. Wooed by the FTX leader's head spinning accumulation of wealth—he was already reported to be worth more than $22 billion at thirty years old—and his coterie of connections from Tom Brady to Bill Clinton to Tony Blair, Grimes told Musk the crypto billionaire would be willing to put at least $3 billion and up to $10 billion into Twitter.

Grimes called Bankman-Fried an "ultra genius and doer builder" in the mold of his current client.

"Does Sam actually have $3B liquid?" Musk asked the banker, skeptical of the crypto entrepreneur. Musk continued to dodge him.

"Sorry, who is sending this message?" Musk wrote to Bankman-Fried on May 5 after the crypto billionaire texted him, despite having been connected to the FTX leader twice via message.

Bankman-Fried never put into the deal any of the billions he promised to Grimes and others, though Musk did invite him to roll over his existing stake in Twitter—about $100 million—to convert into shares in his private company. (Musk would later claim that Bankman-Fried never did.) Six months later, FTX collapsed and Bankman-Fried landed in jail for defrauding customers. He was later sentenced to twenty-five years in prison.

Musk had privately described the opportunity to invest in his Twitter as "oversubscribed," meaning there were more investors than shares available. It might have been true, or it might have been a bit of salesmanship, but Musk went around with a donation box in hand as if he was a member of the PTA raising money for a new school library. The figures he discussed with his friends were eye-watering, out of reach for all besides the yacht-owning class. He had texted Ellison on April 20 recommending that the Oracle founder invest at least $2 billion or more and selling him on the "very high potential" of his company.

"Since you think I should come in for at least $2B . . . I'm in for $2B 💸," Ellison wrote back four days later, just a day before Twitter announced the merger agreement.

"Haha thankss :)," Musk replied.

While Morgan Stanley had sent around a pitch deck to potential investors, some of Musk's friends backed him on gut instinct alone. They had an unwavering faith in his entrepreneurial ability, with some asking to get into the deal to forge tighter connections to him. Marc Andreessen, the heralded head of venture firm Andreessen Horowitz, texted Musk that his fund was ready to drop in $250 million with "no additional work required."

Musk's conversations were with a who's who of the world's rich and elite. There was Reid Hoffman, a former PayPal associate and LinkedIn co-founder, who told Musk he'd probably be able to put in $2 billion from one of his venture capital funds. There was also Satya Nadella, the CEO of Microsoft, who was connected to Musk by Hoffman; Sean Parker, a cofounder of Napster and first president at Facebook; and John Elkann, heir to the Italian dynastic family the Agnellis. He texted with James Murdoch, the son of News Corp mogul Rupert Murdoch and a Tesla board member, and his wife, Kathryn, who asked if Musk would bring back Dorsey as CEO. ("Jack doesn't want to come back," Musk replied. "He is focused on Bitcoin.")

Not everyone that Musk or his bankers approached, however, forked over checks. Founders Fund, the venture fund founded by Peter Thiel, Musk's PayPal cofounder and frenemy, was among the tech investors who seemed unimpressed by the venture.

Thiel was not a fan of Twitter. "We wanted flying cars, instead we got 140 characters," Thiel famously once said, mocking Twitter's original value proposition. His firm passed when approached by Musk's team. Morgan Stanley also reached out to a diverse group of potential backers including Japanese investment conglomerate SoftBank, former New York City mayor Michael Bloomberg, and Joe Rogan.

Even some of Musk's closest associates couldn't meaningfully contribute. Kimbal declined to invest. Antonio Gracias, the financier who promised his friend that he'd go "to the fucking mattresses no matter what" for his old friend, did not deploy money from his private equity firm, and only made a personal investment. The same went for David Sacks, the former X.com colleague who had helped remove Musk decades earlier. The men had made up since then, with Sacks, a successful entrepreneur in his own right, becoming enamored with the power and opportunity that came with being close to Musk. But when Musk asked if Sacks's fund, Craft Ventures, could invest, the investor demurred.

"I don't have a vehicle for it (Craft is venture)," he texted Musk on April 28. Venture funds, which raise money from outside investors like pension funds and school endowments, typically have narrow areas or focuses in which they agree with their funders that they can invest in. "I'm in personally," Sacks added, noting that his personal investment "would be mice-nuts in relative terms."

One friend who was very enthusiastic about helping Musk was Jason Calacanis, the angel investor and podcaster who continued to provide advice on how to run Twitter and to offer his services. "Board member, advisor, whatever . . . you have my sword," he texted Musk in the days before the billionaire signed on the dotted line. While Calacanis had never been involved in a take-private transaction like this, he pushed the billionaire to allow him to establish a "Special Purpose Vehicle" (SPV), or shell company in which investors could pool their money to invest. Instead of seeking large checks in the hundreds of millions of dollars from individual institutions, an SPV allowed entities that wanted to invest single-digit or low two-digit million-dollar checks to get into the deal. But it also gave Musk less control over who he allowed to invest, and he became frustrated with his swordsman.

"What's going on with you marketing an SPV to randos?" Musk texted Calacanis. "This is not ok."

Calacanis recoiled like a puppy being yelled at by its owner. This was how he invested, he assured Musk, and he had fundraised more than $100 million in commitments. He stared at his phone as a barrage of texts arrived from his lord.

> **JC:** People really want to see you win.
>
> **EM:** Morgan Stanley and Jared think you are using our friendship not in a good way.
>
> **EM:** This makes it seem like I'm desperate.
>
> **EM:** Please stop.

Calacanis's noisy roundup of micro investors made it appear as though Musk couldn't secure the famous executives and blue-chip firms needed to close the deal. Begging for a million here and two million there undercut Musk's claims that he was "oversubscribed." It made Twitter look less like the deal of the century and more like a GoFundMe campaign.

Calacanis would reach out to Birchall to smooth things over before returning to his texts with Musk. This was a deal that captured the world's imagination, he said, massaging the billionaire's ego. Everyone wanted in.

"And you know I'm ride or die brother—I'd jump on a grande for you,"

Calacanis texted, misspelling *grenade*. Would Musk do the same for his loyal friend?

>>> BY MAY 5, Musk had secured commitments from outside investors to invest $7.1 billion into his deal. He filed a document with the SEC to disclose the money and show that he was serious, though behind the scenes, Musk's plans were not all rosy. Despite Ellison's texts a few days earlier about investing $2 billion, the filing showed him contributing only $1 billion.

There were other interested parties, however. The Qatar Investment Authority, which had more than $460 billion in assets, committed $375 million. As a condition for the deal, the Gulf State asked for Musk to attend the 2022 FIFA World Cup Final, which it was hosting later that year. There was also funding from Vy Capital, a Dubai-based venture fund founded by former Goldman bankers. The firm—which counted Birchall's son as an analyst—had already invested hundreds of millions of dollars in Musk's Boring Company and Neuralink, and agreed to contribute $700 million to the Twitter deal.

Sequoia Capital, the famed backer of Apple, agreed to $800 million. Andreessen Horowitz upped its earlier commitment of $250 million to $400 million. Musk had earned fortunes for some of these firms in the past, and they put their faith in his track record.

Musk also left room for crypto money. Even without Bankman-Fried, he saw plenty of opportunity in the FTX head's crypto contemporaries. Morgan Stanley pitched the deal to Brian Armstrong, the chief executive of Coinbase, and to the Web3 Foundation, an organization that sought to develop other decentralized digital currencies beyond Bitcoin. But it was Binance, the world's largest crypto exchange, that pulled the trigger. The company, led by an enigmatic Chinese-born founder named Changpeng Zhao, was flush. It facilitated more than $75 billion in daily trading volume of cryptocurrencies including Ethereum, Dogecoin, and others. It committed $500 million to Musk's cause with the intent of weaving blockchain technology into the social network.

>>> ON THE AFTERNOON of May 6, Musk walked through the glass doors of Twitter's San Francisco headquarters for the first time since his deal was

announced. He was about half an hour late, but he strolled unhurriedly across the lobby.

Although he preferred to surround himself with a handful of his most trusted associates, on this overcast Friday, Musk was accompanied by two people he barely knew: Kristina Salen, a bleach-blond executive who had recently left her job overseeing the finances of World Wrestling Entertainment, and Pat O'Malley, another financial executive with decades of experience in tech.

Salen and O'Malley had been dispatched to audition for Musk by Grimes and his team of bankers at Morgan Stanley, who seemed to be aware that the billionaire would likely require some adult supervision as he remade Twitter to his liking. Either of them would make a fine chief financial officer, Grimes thought.

It was quite clear that Musk did not understand the terms of the $44 billion deal. The day before, he had finally inked a nondisclosure agreement that would give him access to private information about Twitter's finances and technology. He told James Murdoch that he was in San Francisco to do "due diligence" on Twitter.

But the period for due diligence was long gone. If Musk had wanted a peek under the hood to inform his purchase, he would have needed to get it before signing the merger agreement. As Twitter's buyer, he was entitled to information about the company—but whatever he learned couldn't change the terms of his contract.

Salen and O'Malley trailed Musk into a flashy fishbowl of a conference room on the second floor of the company's offices. The room was at the center of a suite decked out with giant hashtags and bird paraphernalia, which Twitter used to entertain statesmen, celebrities, and major advertisers.

Segal shook Musk's hand and invited the billionaire to sit down across from him at the table. The room was packed with about twenty bankers, lawyers, and Twitter staffers, but one person was notably absent: Agrawal. He instead dialed in on a video screen at the head of the table, still fighting off COVID.

Twitter's chief financial officer tried to push past the awkwardness of sitting down with Salen and O'Malley, his potential replacements. It wasn't clear to Twitter's team why the two executives were there, and whether they were advising Musk on the deal or being considered for full-time roles. Segal handed Musk a single sheet of paper that outlined the basics of Twitter's performance over the past several years. There were only about ten lines on the sheet, ticking through the company's key financial data like revenue, profits, losses, and user numbers.

Musk scoffed. "I want to see the most detailed spreadsheet you have," he demanded.

Several members of Segal's staff flinched. The finance department used hundreds of spreadsheets to manage budgets, sales figures, and revenue projections. Twitter's active-user tallies were calculated by engineering teams outside of Segal's direct management chain, and those employees probably used hundreds of their own datasets to arrive at the figures. The most detailed spreadsheet at Twitter was not something anyone had ever considered.

Segal tried to redirect. "Let's just start at the right altitude," he suggested. Twitter could always round up data for Musk later.

Musk was persistent. He didn't want a bird's-eye view of what was going on at Twitter. He wanted to dive into the data, wherever it happened to be. At Tesla and SpaceX, he was known for peppering engineers about minutiae to do with car parts or rocket engine designs, and he expected those who worked for him to know the answers or be able to find them quickly. He commanded Segal to bring him details.

As they went back and forth—Segal trying to keep Musk focused on the one-pager and Musk insisting he needed to see more—some of the CFO's staff believed they were watching their boss lose his job in real time.

Finally, Musk agreed. "Okay," he said. "We'll do what you suggested." Segal steadied himself and began rattling through Twitter's metrics as he had dozens of times for investors over his five years at the company.

Salen and O'Malley, taking cues from Musk, pressed Segal on the details and raised esoteric questions to demonstrate their skill and experience. He politely but stubbornly kept deflecting, trying to keep the meeting on the rails while Agrawal stayed muted, concealing his coughs.

Although the meeting focused on finances, Musk briefly turned his attention to the problem of bots. All he had to do was open his replies and he faced an onslaught of messages that were obviously not from real users. He pulled out his cell phone and demonstrated.

His own experience was Musk's most vivid guide, and he was convinced that Twitter was plagued by far more bots than it had ever admitted to investors. The company included a disclosure in each of its earnings reports that claimed fewer than 5 percent of its users were bots or spam accounts. Musk wasn't buying it. That wasn't his experience, as scammers often created accounts that mimicked his username and profile details to hawk cryptocurrency scams or peddle fake goods.

Agrawal broke his silence to describe the way Twitter arrived at its figure. The company pulled a sampling of accounts at random each quarter and reviewed them to see how many spam accounts had bypassed its filters. He'd also been working on improving Twitter's automated detection systems, Agrawal added, to make it harder for spammers to spin up fake accounts in the first place. And Musk's account was a bad example of a regular user's experience on the platform—he had tens of millions of followers, making him a natural target for scammers who wanted to trick his fans into their crypto schemes.

Musk felt he was being duped. The Twitter people kept telling him: *Just trust us. We have a system. You don't need to look too closely.* He wanted to make his own assessment with raw data, and it seemed like the executives were still stalling. The meeting lasted more than three hours, stretching into the evening as Musk continued to hammer the executives for more details. He exited the office into the San Francisco darkness wholly unsatisfied.

Musk immediately called Alex Spiro, convinced that Twitter's management was lying, and insisted that the lawyer get him what he wanted. Now.

Spiro, used to these kinds of after-hours demands for urgent action, sent a request to Twitter's board lawyers that very night. As everyone who had worked alongside Musk knew, it was pointless to give him a vague summary—he would insist on looking at the details and making up his own mind.

Included in a list of other items Musk wanted to see in preparation for the acquisition, Spiro asked for access to Twitter's so-called firehose, a real-time feed of all the tweets flowing through the platform and the associated engagements, such as likes and retweets. The firehose was akin to a Twitter timeline that followed every single account. Twitter siphoned from the firehose to make its quarterly bot estimates, Agrawal had told Musk during the meeting. So the billionaire wanted to see it.

The firehose was a peculiar piece of Twitter technology. The raw, unfiltered information was public, at least in theory, but no user followed every account on Twitter as it would make their experience of the site unusable. Twitter allowed news organizations to access a version of the firehose so they could detect breaking news before it landed on their reporters' desks, and some media outlets paid to receive alerts as topics bubbled up on Twitter. Researchers could also gain access to the firehose and to study the platform and its handling of hate speech, misinformation, and other hot-button issues.

But the company also protected the firehose as proprietary, and it did keep some portions of the dataset a secret. After Dorsey became enmeshed in the protests surrounding the police killing of Michael Brown, Jr., Twitter cut off the CIA and other intelligence agencies from accessing the firehose in 2016. The company believed that its data could be easily turned against activists and other users, and that intelligence agencies shouldn't be allowed to take advantage.

The nuances about who was allowed to access the firehose, and how much data they could collect, made it the perfect wedge for Musk and Twitter to battle over. The billionaire had no respect for Twitter's deliberative and meandering governance. Twitter executives, perhaps naively, thought that if they just explained the years of reasoning behind their decisions, the billionaire would understand. But few could reason with Musk once he had made up his mind.

18 > 💩

Musk wanted to pump the brakes.

On Sunday, May 8, he texted Michael Grimes about his worries. The following day, Russian president Vladimir Putin was due to give a speech on the status of the war with Ukraine on a holiday meant to commemorate the Soviet Union's victory over Nazi Germany. Musk had been telling those close to him that he had been in communication with Putin and Ukrainian officials since the start of the fighting, and he would later push for a peaceful solution in which Russia would obtain the disputed territory of Crimea. It was a solution that was wholly unacceptable to most Ukrainians, but Musk urged the appeasement of the Russian aggressors to prevent a global war.

"Putin's speech tomorrow is extremely important," he wrote to Grimes late that afternoon. "It won't make sense to buy Twitter if we're headed into WW3."

Quite the swerve. Musk had already signed on the dotted line to buy the company—and he was in the process of asking his friends for investment. Why was he suggesting he may not follow through? Let's wait to see what Putin has to say and "take stock of where things look after that," Grimes texted back.

Musk's mind then pivoted to a completely different topic: the meeting with Segal and the lack of clear answers around bots and spam. The billionaire believed Twitter's management was toying with him, so he wanted to investigate everything. The more he stewed, the angrier he became, and he started chatting with members of his inner circle about suing Twitter executives.

"An extremely fundamental due diligence item is understanding exactly how Twitter confirms that 95% of their daily active users are both real people and not double-counted," Musk texted Grimes. "They couldn't answer that on Friday, which is insane."

The banker didn't write back.

"If that number is more like 50% or lower, which is what I would guess based on my feed, then they have been fundamentally misrepresenting the value of Twitter to advertisers and investors," Musk wrote thirteen minutes later.

No text response.

A little more than an hour after that, Musk had arrived at a eureka moment. If the bots were swarming Twitter in greater numbers than the company had ever admitted, he should be allowed to back out of the acquisition.

"To be super clear, this deal moves forward if it passes due diligence, but obviously not if there are massive gaping issues," he told Grimes over text.

>>> AGRAWAL WAS FIGHTING on multiple fronts, but the battle for his employees' trust was going against him. When he was first named chief executive, some didn't even know he worked at the company.

Twitter's C-suite and board, who had regular contact with the CEO, came to like him. Similar to Dorsey, he could be quiet and cerebral, but they found him to be warm in small groups, and he took time to listen or explain complex engineering concepts. He towered over people like a friendly oak tree.

Twitter's employees had remained remote after the pandemic, however, and most workers had only interacted with Agrawal via video calls where he was reduced to a small box. In those company-wide meetings, he could be robotic, frequently looking off screen as if he were reading from a script. It also didn't help that he could say very little to them about the deal, as employees hungered for any scrap of information and wondered if they would have jobs once Musk took over.

He either didn't have the answers or was hamstrung by the various nondisclosure agreements and legalese around nonpublic information that came with the complex, multibillion-dollar transaction. One misstep could jeopardize the deal, and with Musk seemingly looking for any excuse to pull the rip cord, the novice chief executive erred on the side of caution and relied on workshopped answers and empty platitudes.

What little internal support might have remained for Agrawal evaporated on May 12. That morning, he sent out an email to Twitter's employees notifying them that most new hiring would be frozen while spending would be cut. More important, however, he was firing two key executives who he had planned to lay off a month earlier, before Musk's offer disrupted everything: Kayvon Beykpour, the head of consumer product, and Bruce Falck, the head of revenue-generating products. Employees were blindsided.

Beykpour's termination was particularly harsh. A personable thirty-three-year-old with a strong jaw and easy smile, he had started his career building apps from his Stanford dorm room and was one of the people employees speculated would take over for Dorsey once the time came. He bled Twitter and was married to another Twitter executive. But he had made several missteps with the company's products, and some workers gossiped that Agrawal's move was a power play as the company got ready to transfer ownership to Musk.

In explaining himself, the chief executive did not mince words. "It's critical to have the right leaders at the right time," Agrawal wrote. He said that Jay Sullivan, who joined Twitter six months earlier during the company's growth phase, would replace Beykpour.

In Agrawal's mind, the decision was simple. At the beginning of the pandemic, the company had decided to grow dramatically, spurred by the activism of Elliott Management. Beykpour and Falck were responsible for driving that growth, and their projections became the basis for the company's announcement of its long-term goals to investors. But both executives had come up far short of their ambitions, and "as a company we did not hit intermediate milestones that enable confidence in these goals," Agrawal continued. So he opened the trapdoor and sent both men tumbling down the chute.

The business reasoning may have been sound, but Agrawal's lack of emotional intelligence would hurt him. Completely surprised, Beykpour was on paternity leave, celebrating the birth of his first child, when he learned of his termination. It was a brutal way to conclude his seven years at the company and he took it personally, particularly after seeing Agrawal deal with Musk's antics during his own parental leave a few months earlier.

"The truth is that this isn't how and when I imagined leaving Twitter, and this wasn't my decision," Beykpour tweeted that day. "Parag asked me to leave after letting me know that he wants to take the team in a different direction."

Sullivan seemed to Agrawal like a natural replacement for Beykpour—he had already been filling in for the product leader during his paternity leave, and he had the efficient leadership style of a former Facebook executive. But Sullivan was horrified by the decision and argued over the phone with Agrawal about it the weekend before the axe fell, begging him to reconsider.

Sullivan, a mild-mannered man with a verbal tic, felt loyal to Beykpour. The product executive had recruited Sullivan to Twitter, and their lively conversations were one of Sullivan's primary motivations for joining the company. Agrawal told him that Falck's performance had faltered and Twitter desperately needed to get its revenue business in order, but he was vague about his motives for getting rid of Beykpour.

The firing would have a chilling effect on all the women who work at Twitter, Sullivan argued. If they saw a top executive removed during a family leave, they would be hesitant to use their own parental benefits. And Sullivan himself would become the living embodiment of a decision he believed was deeply wrong.

Agrawal agreed with it all. "I understand the trade-offs," he said. "I know what I'm doing."

Then the chief executive went further. He told Sullivan that he wouldn't just oversee the consumer products that Beykpour had built, but would also take on the revenue products that were Falck's purview. Sullivan had a bit of experience building features for advertisers, but it wasn't his strength, and he said so. "You're the right leader for the right time," Agrawal said simply.

Sullivan hung up and called Beykpour. An apology tumbled out of him. "This is the worst," he kept repeating.

To some employees, the firings made little sense. It was like starting a full bathroom remodel on a house someone else had already agreed to buy. With Musk looming, they wondered why Agrawal decided to create more waves and toss a lieutenant, particularly while he was spending time away with his family. Twitter was supposed to be a compassionate workplace—at least for now. Some tweeps felt that the company should abide by those standards until the keys were handed over.

Sensing the tide turning against him, Agrawal took to his Twitter account to lay out the reasoning for his recent decisions. Unlike Musk or Dorsey, the chief executive was not a prolific tweeter. He started his procession of tweets by acknowledging his role as a "lame duck" CEO. While he said he expected the deal to close, the company needed "to be prepared for all scenarios," he wrote. He wasn't just working to keep the lights on.

> **Parag Agrawal**
> @paraga
>
> People have also asked: why manage costs now vs after close? Our industry is in a very challenging macro environment – right now. I won't use the deal as an excuse to avoid making important decisions for the health of the company, nor will any leader at Twitter.
>
> 3:10 PM May 13, 2022
>
> ♡ 33   ↻ 134   ♡ 2.7K

Once again, his tweets made no direct mention of the billionaire who had upended his tenure and made his life hell. But there was a veiled reference. Agrawal said he wasn't going to be the "loudest sound bite" and promised to focus on "the ongoing, continuous, and challenging work our teams are doing to improve the public conversation on Twitter."

The thread didn't resonate with the rank and file, and raised more questions than it answered. It felt like the same public relations drivel they had been force-fed since the Musk circus began. And while Agrawal said he believed the deal would close, the qualification about preparing for "all scenarios" led employees—and the public—to once again wonder if Musk was serious. If employees were listening in hope of a voice of comfort and trust, it would not come from their chief executive.

At his first meeting with Beykpour's former product team, Sullivan seemed dejected. "I don't defend that. I don't know how to defend that," he told them about the sudden firing. Sullivan's candid approach made him seem like the more humane executive.

—

>>> THE SHIFTING ECONOMIC landscape spared no company, not even Tesla. While the company's shares—and subsequently Musk's net worth—had entered 2022 nearing all-time highs, they began to plummet by early April, affected by the macroeconomic environment and its chief's own actions.

The day after Musk first announced his deal for Twitter, Tesla's shares fell more than 12 percent as investors worried he'd be selling off stock in the automobile company to fund his purchase. On May 4, the company's shares were down nearly 17 percent from a month earlier, closing the trading day at $317.54.

Tesla's value was of crucial importance to Musk. Not only was much of his personal fortune tied up in his roughly 20 percent ownership stake in Tesla, but he had also pledged some of those shares as collateral, including for a $12.5 billion personal loan to fund the Twitter purchase. (That personal loan was later reduced to $6.25 billion in early May after Musk was able to raise outside funding.)

Tesla's shareholders also worried that Twitter would divert Musk's focus. Adding Twitter to the billionaire's constellation of companies would only further divide his limited time, they thought, and the electric automaker needed to continue to deliver new cars and products. Moreover, analysts, shareholders, and Tesla's own management worried about the potential financial knock-on effects of paying for the social network. In a 2022 financial filing, the company had previously outlined the risks of the personal loans made to its chief executive, "which are partially secured by pledges of a portion of the Tesla common stock currently owned by Mr. Musk.

"If the price of our common stock were to decline substantially, Mr. Musk may be forced by one or more of the banking institutions to sell shares of Tesla common stock to satisfy his loan obligations if he could not do so through other means," the company wrote. "Any such sales could cause the price of our common stock to decline further."

In essence, investors and the company itself were fearful of a domino effect. If Musk failed to pay back his loans, he would have to sell his Tesla shares. And if he sold off his Tesla shares, the company's investors could be spooked and sell off their own shares. That would drive down the share price further, leading Musk to have to sell off more shares at an even greater discount to create the necessary cash to pay off the loans. And so on and so forth.

In the week after May 4, Tesla's stock dropped another 23 percent, reducing the company's valuation by $226 billion. One of Musk's greatest powers was his ability to instill confidence. But the Twitter deal sowed doubts, and, in turn, Musk began to doubt himself. He would tell those close to him he was unsure if he should push through with his acquisition. (By the end of May, Musk completely did away with the remaining $6.25 billion of the personal margin loan by raising more outside funding, though the damage to Tesla's stock had already been done.)

As Tesla's shares plummeted, Musk pushed harder on the bot issue, his doubts about spam and fake accounts evolving into a full-blooded conspiracy theory. Twitter's executives must be hiding something from him, and Musk pushed his lawyers and bankers to do better due diligence.

On May 9, lawyers from Skadden had started messaging Twitter's teams for access to the firehose. The reams and reams of data, they believed, would allow them to do their own research and potentially prove out their boss's bot hypothesis.

On Twitter's side, the requests seemed preposterous. The executives had given Musk's advisers the chance to sit down with their engineers for a live demo so they could walk through their calculations. Moreover, providing firehose access wouldn't allow Musk's team to determine the difference between a real or fake user, which required private account information that didn't come from the enormous dataset that included only public tweets. Twitter combed through personal information including user locations, emails, IP addresses, and other data that simply couldn't be shared because of global data privacy laws.

It also defied logic. Twitter had given Musk the chance to do due diligence on its financials and user numbers before he had agreed to sign the deal, and the billionaire had balked at the idea. Twitter had nearly ripped his hand off shaking on the deal.

Musk, perhaps riled up by his own conspiracy or angered by the tricky predicament where he now found himself, began to uncork in the only way he knew how, unleashing a steady stream of criticism about Twitter on Twitter. He railed against Twitter's algorithmic recommendations and, on May 12, tweeted that he believed "Trump should be restored to Twitter." Then, after an evening of posting about spam, the culture war over paper straws, and supposed media hoaxes, he fired off something that caught everyone by surprise.

"Twitter deal temporarily on hold pending details supporting calculation that spam/fake accounts do indeed represent less than 5% of users," Musk tweeted at 4:44 a.m. Austin time on May 13.

This was the public's first real hint he was having doubts. Reporters scrambled to see if it was even possible for Musk to halt the deal after signing the contract.

To Musk's inner circle, the move was seen as self-sabotage. Even if Musk believed he had a way out of the deal, what was the use in announcing it to the world and tipping his hand? His Twitter habit had once again put him in a corner, and Spiro and Birchall frantically texted and called their boss to fix the situation.

It took two hours and six minutes for them to convince Musk to act. "Still committed to acquisition," he tweeted at 6:50 a.m., just as much of the American financial press was beginning to stir. It was an incredible, confusing set of back-to-back tweets. Few knew what to think. Twitter's stock dropped nearly 10 percent. Tesla's shares rose 5 percent. Musk stopped tweeting for thirteen hours.

>>> BY THE EVENING OF SUNDAY, MAY 15, Agrawal had become fed up with Musk's posturing about the so-called bot problem. To Agrawal and Twitter's board, Musk's logic was particularly shoddy. Correcting bots and spam were one of the particular reasons the billionaire gave for buying the company, after all.

"If our twitter bid succeeds, we will defeat the spam bots or die trying," Musk had tweeted on April 21. He then repeated the line four days later in a press release announcing his Twitter bid. Three weeks later, what had changed?

Agrawal had had enough. The Twitter peanut gallery had pilloried him for appearing weak as Musk led a harassment campaign against Gadde, even though he had resisted Musk's private demand to fire her. He had remained silent as Musk's twitchy fingers used the platform he oversaw to start the board member merry-go-round and push a de facto hostile takeover. After spreading his wings on Twitter a few days earlier to explain the firings of Beykpour and Falck, he felt encouraged. The Tesla chief was just taking his crazy to another level, this time to undermine the work of his employees.

That night he drafted a fifteen-tweet thread, vetting it with the transaction committee and lawyers. Many of those who saw it told Agrawal not to send it. The tweets would only antagonize Musk, they said, and there was no point in trying to reason with someone so erratic by using facts and logic. Agrawal could also risk jeopardizing the deal, giving Musk more fodder to wiggle out of his contract.

But Agrawal believed in the solidity of the merger agreement and he wouldn't be mentioning Twitter's potential buyer by name or handle in any of his tweets. In Twitter parlance, it was a "subtweet," a post that clearly bashes another user without directly naming them. In this case, it was abundantly clear who Agrawal was talking about, and he hit the tweet button at 9:26 on Monday morning in San Francisco: "Let's talk about spam. And let's do so with the benefit of data, facts, and context . . ."

The spam problem, he said, was complex and dynamic, an ever-evolving problem that led the company to suspend half a million accounts a day. Using sophisticated techniques and data science, Twitter estimated that spam accounts were less than 5 percent of daily active users. And because of that, it was impossible for any outside party to accurately assess the bot problem, because they couldn't access the full wealth of Twitter's data.

Musk didn't care. He responded initially with a single emoji:

Elon Musk
@elonmusk

@paraga 💩

1:03 PM May 16, 2022

💬 3.7k 🔁 8.7K ♡ 50K

a troll move that garnered 51,000 likes, three times more than Agrawal's lead tweet in the thread. It was childish, but effective—a grade-schooler winning an argument against a PhD student by calling them a playground insult.

Fourteen minutes later, he used actual words.

"So how do advertisers know what they're getting for their money?" Musk wrote. "This is fundamental to the financial health of Twitter."

Never mind that Twitter disclosed these figures in public financial reports. And never mind that Musk had already agreed to buy the company. He once again won the battle in the court of Twitter opinion. Agrawal's account went silent for a week.

19 > Bots and Horses

Until Agrawal's thread, Musk had expressed most of his ire at the chief executive in private. He sometimes seemed to even enjoy his discussions with Agrawal as they bonded over engineering.

Before he registered his interest in buying Twitter, Musk had poked a little fun at Agrawal, tweeting a meme in December 2021, shortly after Agrawal had stepped into Dorsey's shoes, which showed Agrawal's face superimposed on the body of former Soviet Union leader Joseph Stalin, as if he was executing Dorsey. While Musk had needled the board, Gadde, and the company more generally during the hostile negotiations, Agrawal had been spared direct attacks online. His strongly worded messages, however, changed that. If you hit Musk, he would hit back.

Throughout May, the billionaire continued to slam Agrawal over the bot issue and tagged him in tweets with various user complaints, as if he were a nosy boss directing complaints to a regional manager. In one, Musk asked why some users were unable to load his tweets: "What's going on @paraga?" In another, he challenged Agrawal to take action on a 2016 tweet by Hillary Clinton and called it "a Clinton campaign hoax," insisting it violated Twitter's policies against political disinformation. Political observers pointed out that Musk was simply echoing a Trump talking point. Agrawal never replied.

Twitter's rank and file saw the lack of public reaction as an extension of Agrawal's meek personality during company-wide meetings. Musk was openly denigrating their work—on the platform they built—and their leader said nothing. And while Musk posted like a twelve-year-old off his Ritalin, that approach still seemed better than Agrawal's communication blackout.

Employees began openly poking fun at their subdued chief executive. After suffering a series of embarrassing hacks over the years, Twitter had upped its security and forced top executives to use YubiKeys, hardware devices that generated random passwords, to shield Twitter's digital empire. Once, Agrawal unexpectedly piped up in a Slack group that was open to all employees. But rather than a comforting missive, his message was a randomly generated collection of letters and numbers, apparently copied from his YubiKey. Employees mocked him by posting gobbledygook passwords of their own.

As Agrawal's trust among the rank and file eroded, he also lost one of his earliest supporters in the boardroom. On May 25, Dorsey unceremoniously left Twitter's board at the end of his term as a director. He had made no secret of his disdain for Twitter's board, and his backchanneling with Musk had made him a liability during the negotiations. The man who had helped create Twitter, who had fought so hard to be named its CEO, had nothing left connecting him to the company besides his multimillion-dollar stake.

>>> ON MAY 27, Musk's private jet took off from Los Angeles bound for Saint-Tropez and the French Riviera. It was meant to be a quick trip so that he and Natasha Bassett could attend the wedding of Ari Emanuel, the acid-tongued Hollywood mogul. Emanuel was the CEO of Endeavor, the entertainment conglomerate that owned a talent agency, World Wrestling Entertainment, and the Ultimate Fighting Championship, and had become a friend and confidant of Musk's. Emanuel's nuptials to fashion designer Sarah Staudinger was a star-studded affair, graced by the likes of Larry David, Sean "Diddy" Combs, and Mark Wahlberg. The event, on a private hilltop villa overlooking the coast, was also attended by Emanuel's business associates, among them Egon Durban.

Throughout the deal, Durban had avoided interacting with Musk when possible. While he counseled Agrawal and Segal—and backchanneled criticism of Musk to others in the financial world—he tried not to lock horns. The small affair, however, made their interaction inevitable. On the side of the event, Musk, dressed in a blue suit with no tie, sparred verbally with Durban as guests milled about, unaware that one of the world's most consequential deals was being litigated within feet of them. Neither man got what they wanted out of the kerfuffle and Musk flew home after less than two days in France.

—

>>> BY JUNE 6, Musk was still trying to get his hands on Twitter's firehose data and losing patience with the company's refusal to provide it without guardrails governing how it might be used. Musk's threat of building a competitor was still ringing in the board's ears, and they didn't want to give him open-ended use of all the tweets that made up Twitter.

Ringler delivered a scathing letter to Gadde that day. "Mr. Musk has made it clear that he does not believe the company's lax testing methodologies are adequate so he must conduct his own analysis," Ringler insisted. If Twitter didn't hand over the data immediately, Musk was considering backing out of the deal, the lawyer added.

His message set off alarms in the boardroom. The board had waltzed Musk into what they saw as an inescapable contract, but his threat of backing out was still worrisome. If Musk followed through with it, he would cause even more disruption among Twitter's employees and investors, and Twitter would have to file a messy, public lawsuit to force Musk to buy the company.

Twitter's board and chief executive saw it as their utmost responsibility to get the acquisition over the line. Everything—employee concerns, their pride, and even their own jobs—came second to making Musk pay the $44 billion. And with the bottom having fallen out of the stock market, they knew there was no way the public would value Twitter anywhere near $54.20 a share.

The board proposed a compromise for the firehose. They would give him limited access to one datastream going back to January 2021, and thirty days of historical information from the firehose. It could assuage Musk's concerns that the company was stonewalling, without giving up the entire kingdom.

Agrawal also tried to bring Musk back into the fold, inviting him to meet Twitter's employees. He hadn't led the company for very long, and knew that a closed deal would likely be the end of his short tenure. But there weren't a lot of options. If Musk's bid failed, Agrawal would have the near impossible task of rebuilding a company with a damaged reputation. With a sale, he could pocket tens of millions of dollars—the result of the acceleration of his chief executive compensation package—for less than a year of work, wipe his hands clean, and start anew, having faced off against Musk and won.

Internally, Twitter's workers continued to pressure their chief executive for more information about what the hell was going on. He gave them very little, but in Musk, he saw a potential solution to the communication gap. Maybe he could allow workers to present their questions to their future boss directly.

On Monday, June 13, Agrawal sent out an email to the company, saying that Musk would be joining them on Thursday for an all-hands meeting. Twitter's CEO, however, was not the one who would be moderating. Instead, the responsibilities fell to Leslie Berland, Twitter's chief marketing officer.

>>> BERLAND HAD NEVER interacted with Musk directly, so to prepare, she scheduled two calls with him. She had been around large male business egos for most of her career and had developed a way of not only navigating but harnessing them for her own or the company's wider benefit. As the "Jack whisperer," she was a sounding board for Twitter's former chief executive. Berland couldn't handle everything Dorsey did—sometimes he would make decisions or tweet major company news without vetting it with her— but she did try to channel his energy and innate disregard for authority. Musk was an amped-up version of that.

The man on the other side of the line, however, was a different one from the person who tormented Twitter's board and executives with his tweets. Musk was personable, making small talk about his family. He was engaged, asking what the employees at the company thought about the deal, and interspersing the conversation with small jokes that he'd laugh at himself. It was a pattern that would play out in various conversations with Musk and Twitter's executives over the coming weeks and months. On some days, he could be completely pleasant, normal, even. On others, he was on the warpath for blood. It depended on what else had happened that day, what he had seen on Twitter, and who surrounded him.

Berland drew a good day. Ending the small talk, she prepared him for questions they believed employees would ask, with Musk sharing ideas on how he'd answer them. She told him to expect questions about why he wanted to buy the company and predicted there would be questions about his views on working from home and diversity and inclusion initiatives. The conversation meandered, with the CMO trying to keep Musk on track. He spoke of his love for the platform and his views of it as the global town square.

Having led dozens of these meetings for Twitter employees in the past, Berland understood there couldn't just be easy questions. Twitter's internal culture was one focused on transparency, and its employees often used the opportunity to pose uncomfortable, if not biting, inquiries to management. It was a practice that had once been employed at Facebook and Google, allowing executives to gauge what was important to their workforces and giving employees a modicum of belief that they could hold their leaders accountable. It was her job to prepare him for that.

So Berland presented some banana peels. She asked how Musk would respond if employees asked about a looming class-action lawsuit from Black workers at one of Tesla's factories who alleged rampant racism. She asked how he would answer questions about layoffs. And then she asked about his political tweets, including those about transgender people.

In 2020, Musk had tweeted that "pronouns suck," joining the growing backlash against people whose gender identities didn't conform to a binary structure. Twitter employees feared what would happen to their trans colleagues once Musk took over.

Later that month, Musk's oldest child, one of a pair of twins he had with his first wife, changed her name to Vivian Jenna Wilson, removing her father's last name. News outlets would later report on the name change, which was the first indication that Musk had a trans daughter. On a legal document, Vivian Wilson wrote that the reason for the change was "Gender Identity and the fact that I no longer live with or wish to be related to my biological father in any way, shape, or form."

>>> THE MORNING OF the all-hands meeting, Thursday, June 16, Korman and Klein wrote to Ringler, notifying him that the board had decided to hand over some of Twitter's firehose data. Ringler relayed the message to his handlers as Musk readied himself for the call.

Some Twitter employees were eager to hear from Musk. There was indeed a large contingent of workers who had already made up their minds on their new prospective owner, but others remained ambivalent. The last time they had heard directly from Musk was at the January 2020 OneTeam celebration. These were very different circumstances and some wondered if this was the moment Musk would finally reveal his grand vision for the company.

But Musk was late. He couldn't get his laptop to work with the Google video conferencing software and eventually logged in from his iPhone. Before Musk's video started, a Google account avatar briefly flashed across some employees' screens. It showed two hands shaped into a 69, a number associated with a sex act.

After a brief, halting introduction from Agrawal that was certainly read from a script, Berland took over. Musk shifted his camera to frame his head as if he were video chatting with a friend. An unknown man stood in the background of his shot, fiddling with a laptop. It was a bizarre start and somewhat disappointing for Twitter's employees.

"Why do you love Twitter and also why did you and why do you want to buy Twitter?" Berland asked, kicking off the conversation.

It was a softball, but Musk seemed oddly unprepared. He put his hands to his chin in a thinking pose and stared into space for eight seconds, before delivering an unpolished, rambling answer that spanned his own personal enjoyment of the platform, his need to express himself, his disdain for the media outlets that got his messages wrong, and the need for a place to communicate freely. He segued into why Twitter needed to be more like WeChat, the popular app in China.

"Why aren't more people using Twitter and why do people click away from Twitter?" Musk said. "Think of WeChat, in China, which is a great app. But there is no WeChat equivalent outside of China and I think there's a real opportunity to create that. You basically live on WeChat in China because it's so useful and so helpful to your daily life."

It was a strange statement to some employees. For years, American tech leaders, including Zuckerberg and Snapchat CEO Evan Spiegel had dreamed about making the West's WeChat—a single, universal app where people could chat with friends, hail taxis, order food, or make payments—but they had little to show for it. It was true that WeChat had more than 1.3 billion users, but that was because the app was used in a country whose government closely policed the internet and picked which apps and businesses could be successful and used by its citizens. China certainly didn't allow free speech either.

Berland shifted the conversation to some more immediate employee concerns. They worried about Musk's ending of remote work policies and stock compensation in a private company. They fretted about layoffs as news outlets reported on impending cuts. Musk gave no definitive answers but stressed that he preferred in-person work over remote, feeling that it contributed to the "esprit de corps" of his workers. Then he laid out his very basic principles. "If somebody is getting stuff done, then great, I love them," he said. "And if they are not, then I don't like them. I do not love them."

"Anyone who is a significant contributor should have nothing to worry about," Musk later added. "I do not take actions which are destructive to a company."

The talk lasted for about forty-five minutes, as Musk wandered into politics ("My political views, I think, are moderate," he said, adding that he just voted Republican), advertising ("I would probably want to talk to the advertisers and say, 'Hey let's make the ads as entertaining as possible'"), and alien life ("To be clear, I've seen no actual evidence for aliens"). For all the bluster about bots over the past few weeks, Musk had little to say beyond echoing earlier comments that he sought to make the platform better by making it more difficult for fake and spam accounts to proliferate. Berland also skipped the trans question.

"I want Twitter to contribute to a better, long-lasting civilization where we better understand the nature of reality," Musk said.

The meeting ended with most employees having no clearer sense of what Musk would do when he took over. Those hoping for some grander plan were disillusioned; Musk didn't have one. But at least they knew where he stood on extraterrestrial life.

>>> AS MUSK TALKED with his future employees, he was also working to put down a small rebellion at SpaceX. The previous month, *Business Insider* had dropped a bombshell story, reporting on the existence of a settlement between Musk's rocket company and a former employee, a flight attendant who accused Musk of sexually propositioning her on a 2016 flight to London. On his private jets, the billionaire often hired flight attendants who were trained masseuses.

On the flight, the billionaire reportedly rubbed his flight attendant's leg without her consent and exposed his erect penis. According to her account, he knew that the flight attendant was an avid horse rider and offered to buy her a horse if she engaged in a sexual act. She refused.

The flight attendant worked for Musk for two more years, before being convinced that her opportunities on the job had become limited. She lodged a complaint with SpaceX in late 2018, and by that November they had agreed to a $250,000 settlement.

In response to the story, Musk tweeted that the accusations were "utterly untrue." He told the outlet: "If I were inclined to engage in sexual harassment, this is unlikely to be the first time in my entire 30-year career that it comes to light." When the *Insider* story broke, it gave some Twitter employees brief hope it could tank the deal. On Slack, reactions ranged from disgust to dark humor, as they mocked Musk with equine jokes and horse emojis. At SpaceX, some workers wondered if the story was just the latest piece of evidence of a toxic, male-dominated culture, led by a chief executive who tweeted about sex acts and starting a university with the acronym TITS. Months earlier, female SpaceX employees had already spoken up about stories of sexual harassment and discrimination, which the company brushed aside. Even after the *Insider* story published, Musk continued to deny it, claiming that he doesn't use a flight attendant and demanding that his accuser describe any "scars" or "tattoos" on his body.

For some SpaceX employees, the story was too much. They had signed up to work at the company because they believed in furthering mankind's exploration of space, but they hadn't anticipated their CEO's heel turn. As Musk prepared to speak to employees at Twitter at their all-hands meeting on June 16, they circulated an open letter within SpaceX calling their chief executive's behavior "a frequent source of distraction and embarrassment for us."

"Elon is seen as the face of SpaceX—every Tweet that Elon sends is a de facto public statement by the company," read the letter, which was signed by more than four hundred people. "It is critical to make clear to our teams and to our potential talent pool that his messaging does not reflect our work, our mission, or our values."

By the following day, five of the letter's organizers were rounded up and fired for what SpaceX president and COO Gwynne Shotwell labeled "overreaching activism." Others were brought into meetings with managers, who had clearly been told to discipline their subordinates. In one such meeting, Jon Edwards, a vice president who oversaw launch vehicles, called the letter an extremist act. One employee replied whether Musk had the right to sexually harass workers without repercussions. Edwards didn't respond directly but offered an excuse. "SpaceX is Elon and Elon is SpaceX," he said.

Some at Twitter saw the firings as what lay ahead in their future. Musk, the supposed defender of "free speech," would never be one to tolerate criticism or dissent. And if employees did, or raised concerns, he could use his Twitter account to undermine facts. He had plenty of well-paid, powerful executives who had tied their success to his and were willing to protect him. Despite the revelation of the settlement, Twitter's board remained focused on selling the company to Musk. Nothing could get in the way.

>>> EVEN AS MUSK met with Twitter's employees, his lawyers continued to lob grenades in the background. Twitter had continued to deny their requests for full firehose access, worried about him siphoning data for the competing social platform he had threatened to build.

Musk's arguments that Twitter engineers were lax in detecting bots were ridiculous to Twitter's leaders, who began to wonder if Musk planned to do something else with the data. After rounds of venomous letters between Ringler and Korman, the Skadden lawyer sent a clear shot across the bow on June 17. "Mr. Musk made clear that Twitter's continuing failure to provide him with the data and information he has been requesting since May 9, 2022 constitutes a breach of his merger agreement with Twitter," he wrote. "That breach remains uncured."

Korman and Twitter's legal team were surprised by the statement but viewed it as just another attempt by Musk's advisers to give him an out. They replied three days later, with a strongly worded letter of their own. "Mr. Musk insisted on a rapid negotiation of a merger agreement on a take it or leave it price basis," they wrote, and the sale agreement gave him no right to "unfettered access to vast amounts of proprietary and competitively sensitive data." Then came the hammer.

"It was only after the equity markets entered a steep decline following the execution of the merger agreement that Mr. Musk began making statements questioning Twitter's SEC disclosures, and asserting Twitter had breached the merger agreement, all in an apparent attempt to create optionality should he change his mind about the deal," Korman wrote.

Still, the company would agree to grant Musk's side access to the firehose sets of data anyway and prepared to provide credentials for his engineers to access the information. Twitter's engineers set up the information on a tool that allowed them to track how Musk's side ran queries of the data in order to audit what they were looking at. Early on, they realized the billionaire's team wasn't doing much of anything with the data or running particularly meaningful searches. The request, in their view, had been a farce.

20 > Sun Valley

On June 17, Ringler sent a new letter, demanding Twitter hand over even more information. He was no longer after just tweets but, rather, demanded correspondence between the Twitter board and the company's top executives about its spam calculations, including personal emails and text messages.

Musk's lawyer also asked for fresh mock-ups of Twitter's financial state. He wanted to see the company's updated budgets and projections for 2022, along with the financial models Goldman Sachs had drawn up when they recommended that the Twitter board accept Musk's offer.

Segal tried to rein Musk in, asking the billionaire to sit down with Agrawal, who could personally walk him through the bot estimation process. One of the company's best and longest-tenured engineers, Agrawal would be able to show Musk exactly how it worked in perfect detail, Segal said on a call.

Musk agreed to an in-person meeting on June 21—but then backed out, claiming he had a conflict. Meanwhile, his team had started playing around with Twitter's massive amount of data. Within two weeks, they quickly hit the limit Twitter had imposed, preventing more than 100,000 queries per month on the dataset, and demanded leeway to make more. Twitter raised the cap to 10 million.

Klein and Korman, the lawyers who had steered the board through the transaction, were worried. The request for personal messages was completely out of bounds and suggested that Musk did not plan to go through with his purchase. It read more like a discovery request, the kind of probing a company would receive as part of a lawsuit. They also puzzled over Musk's sudden interest in Twitter's finances. At TED, the billionaire had proudly declared that he was unconcerned about the economics of the deal. Why the about-face?

That month, the stock market had plunged to historic lows. Gas prices had skyrocketed. A recession seemed inevitable. Musk's offer had been generous in April but was outlandish by June. The billionaire was on the hook to overpay wildly.

>>> IF MUSK WAS STILL ANGRY about the bots or the data-sharing information, he didn't tweet about it. On June 22, the day after he was supposed to meet with Agrawal, he began a nine-day period of silence on the app. Some Twitter executives, who monitored his tweets as if they were the weather, wondered if he was all right. Musk also cut back in private communications with his bankers and lawyers, leaving his team hesitant to make decisions in his absence. Within a week, news outlets, which typically reported breathlessly on Musk's posts, began publishing stories about his absence from his favorite platform.

Segal had a clue as to his whereabouts. An intriguing rumor had reached his ears: Musk wasn't available to iron out aspects of the deal because the billionaire had had another child.

When Segal relayed the information to Twitter, some executives and board members wondered if it were a trick. Some information about the negotiations had leaked, and Musk's side might have been using a story about a new baby as a test balloon to see if it would make it into the press or onto the *New York Post*'s gossipy Page Six. So they kept the information secret—not wanting to upset the deal—but considered sending a gift. They never did.

Board members speculated that Natasha Bassett, Musk's most recent girlfriend, was the mother of a new baby. But the child, a boy named Techno Mechanicus Musk, had in fact been born by a surrogate for Boucher and Musk. It was their third child together, following a daughter born the previous December. In both instances, news of the birth was kept out of the press, with Musk keeping uncharacteristically quiet about his eleventh known child. Prior to his sudden silence, Musk had tweeted steadily through June about his fear of human population decline and the falling birth rates in countries around the world. "Past two years have been a demographic disaster," Musk tweeted in mid-June. "I mean, I'm doing my part haha."

When it came to humanity's impending doom through population decline, Musk's hero complex was in full force. He and his first wife, Justine, had six children together, though their first had died tragically of sudden infant death syndrome at ten weeks, traumatizing both parents and contributing to their divorce in 2008.

After two stints of marriage to actress Talulah Riley, neither of which led to children, Musk and Boucher had X Æ A-12 Musk in 2020. Then a second child, a daughter named Exa Dark Sideræl Musk, came via surrogate two years later.

Exa was not the child's planned name. Boucher had hoped to name her daughter Valkyrie but was unaware that Musk was also planning on having twins with one of his employees, Shivon Zilis. A Canadian with piercing blue eyes, Zilis was a venture capitalist and had met Musk while serving on the board of OpenAI, the artificial intelligence nonprofit he had cofounded. In 2017, he recruited her to work for Tesla and Neuralink as director of operations and special projects.

In 2021, while carrying twins, she remained mum even to coworkers about the identity of the father, saying only that she had used in vitro to conceive. In November, she gave birth to Valkyrie Alice Zilis, a girl, and Strider Dax Zilis, a boy, petitioning the following year to add Musk's last name, according to court filings obtained by *Business Insider* reporter Julia Black.

Boucher, whose daughter was born a few months after Zilis's twins, was furious to learn that Musk had not only fathered other children at almost the same time but also used a name, Valkyrie, that she had been saving for her future daughter. She raged at Musk, one of many episodes in their up-and-down relationship, and would tell those close to her that his secrecy would damage her attempts to build and maintain a family. Boucher wrote a poem, titled "Letter to S," which referenced her pain at Musk's betrayal and mentioned the name Valkyrie, and posted it to Twitter a few months later:

But one thing I cannot abide:
The insult to my daughter's pride.

Zilis and Musk would take to calling their daughter Azure, perhaps in an attempt to subside the drama. But it continued. Boucher oscillated between various states of love and animosity toward Musk. She was kept away from the twins and in turn sparred with Musk over access to her own son, X, whom the billionaire often kept by his side. Like his business dealings, Musk's personal life was chaotic.

—

>>> BY JUNE 28, Roth and his small team had written a fresh white paper for Agrawal that sanded down some of the creepy edges of Project Saturn and included product mock-ups to show what users would see when they investigated which ring they were in and how to move to a more popular ring.

"At the core of recent debates about content moderation and free speech are a set of opposing preferences: On one hand, many of the people who use Twitter want us to moderate more, and take more significant steps to curb the spread of potentially harmful content online," Roth wrote. "On the other, we hear the feedback loud and clear that some people feel we moderate too much, or unfairly, and haven't left enough space for people to express themselves and connect with the communities they care about, on their own terms."

Agrawal liked the new version and insisted that Twitter publish it as soon as possible, even with Musk's acquisition seeming more precarious than ever. He left a set of directives as a comment on Roth's draft: "1. take responsibility 2. enable people to have control 3. no one size fits all solutions."

Project Saturn was nearly ready to go. Agrawal was excited to show the world that he was a visionary leader who could rival Dorsey when it came to inventing new technologies. His fresh approach to content moderation could give Musk what he wanted—a Twitter that kept nearly all speech online—while still adhering to the company's principle of keeping the conversation healthy. The entire social media industry, which struggled with the same burdens, might see what Twitter had done and follow suit.

Musk was still cold turkey on Twitter that day, staying silent on June 28, his fifty-first birthday. Still, the backbiting on the deal continued. As Twitter's board grew more and more concerned that Musk was going to walk away from the deal, they directed their lawyers to reach out to his team on his birthday and inquire about the state of his financing so they could get a better sense of whether he had the money.

Klein and Korman reminded Musk that he was on the hook to buy the company and that he needed to line up the financing for it as soon as possible. "You are not presently devoting sufficient resources and attention to these obligations," they wrote.

The formal request for information about his money seemed to anger Musk, who accused Segal and Agrawal of using their lawyers to "cause trouble."

"That needs to stop," he texted Segal and Agrawal the next day.

Segal made a last-ditch effort to reengage Musk. He called Musk directly on June 30 and asked him what Twitter could do to allay his concerns about bots. Would he please take the sit-down with Agrawal? Segal probed, trying to figure out what Musk was really after.

Segal later told colleagues he pleaded with Musk to take another look at the sampling process Twitter had sent him in May, which outlined its strategy for tallying bots.

Musk, sounding bored, admitted he had never read the document. Instead, Twitter was spending way too much, he asserted, and asked if Segal had considered staff cuts. He also said one of his longtime confidants, Antonio Gracias, would be taking over financial matters related to the deal.

Gracias was a handsome son of immigrants who grew up in Grand Rapids, Michigan, and made his name by hitching his star to Musk's rocket ship. While attending the University of Chicago in the 1990s, he met the entrepreneur through a law school friend and kept in touch with Musk as he built Zip2.

While Musk lit fires in Silicon Valley, Gracias had finished his law degree, done a stint at Goldman Sachs, and embarked on a career in private equity. He started two of his own firms, and by 2001 he was running Valor Equity Partners out of Chicago. He reconnected with Musk, who had by then taken over Tesla and was in dire need of capital. Gracias obliged, funding both Tesla and a years-long friendship, at one point providing his buddy with a $1 million personal loan. Musk eventually rewarded him with the second-ever Roadster, Tesla's first car, and board seats at the automaker and

SpaceX. His stature grew with Musk's, as did the size of his private equity fund, which eventually came to count the Church of Jesus Christ of Latter-day Saints as one of its largest investors.

Musk told Segal to wait to hear from Gracias. But in the weeks that followed, Gracias never contacted Twitter's CFO.

>>> EVEN WITH THE launch of Project Saturn imminent, Gadde was skeptical. She was the primary architect of Twitter's content moderation policies and had seen firsthand how much work was required to implement a functional rulebook. Every day, Twitter users found creative ways to test the boundaries of her policies, and it took significant human judgment to determine which tweets should be moderated. It wasn't that technology couldn't help—she was sure that Agrawal's plan could chip away at the problem.

But Gadde focused on the tool she knew best: litigation. For years, she had been deeply worried about the rising demands for censorship coming from India. Although Prime Minister Narendra Modi ostensibly ran a democracy, he had become increasingly autocratic about critical online posts. Twitter found itself at the epicenter of Modi's crackdown. As he struggled to put down protests by farmers in 2021, Modi's government passed new social media legislation that required companies to remove posts that criticized the government and placed criminal penalties on their employees if they refused.

Gadde had several policy workers in India, and they often received visits from government officials. She feared they would be arrested if Twitter continued to resist calls from the government to remove content from opposition politicians, journalists, and nonprofit groups. On July 5, she directed Twitter to sue the Indian government, arguing that Modi had interpreted the law too broadly. The best way to fight for freedom of expression, at least in Gadde's view, was in court.

>>> BEFORE SUNRISE ON FRIDAY, JULY 8, Ringler sent a letter to Gadde, confirming the fears of the Twitter board.

"Mr. Musk is terminating the Merger Agreement because Twitter is in material breach of multiple provisions of that Agreement," he wrote. The letter claimed the company had broken several of its agreements with Musk and misled him—and the public—in its quarterly earnings reports, which he had relied on when agreeing to the acquisition.

His lawyers argued that the company's refusal to give him its firehose data was good enough reason for him to abandon the deal. Musk also cited Agrawal's decision to fire Beykpour and Falck, and Gadde's to pursue legal action against the Indian government's censorship demands. His agreement to buy the company stipulated that its management had to keep things running as usual until the deal could be completed—a routine requirement that prevented resentful or reluctant sellers from recklessly joyriding the company for months before handing it over in disrepair. By firing the executives, Musk argued, Agrawal had gone beyond normal operating procedure and damaged Twitter.

For some on the board, Musk's declaration was a relief. For two months, they'd been tiptoeing around his temper tantrums. They'd appeased him, plying him with reams of data like a mother offering different pacifiers to a fussy toddler. Finally, they could go to war with him instead.

One of the directors who relished the moment was Durban. "We're going to make him pay every fucking penny," he said during a meeting.

Taylor was more circumspect. "The Twitter Board is committed to closing the transaction on the price and terms agreed upon with Mr. Musk and plans to pursue legal action to enforce the merger agreement," the chairman tweeted that afternoon. "We are confident we will prevail." It was the only thing Taylor ever said publicly about his battle with Musk.

Since he helped draw up the merger agreement in April, it had been obvious to Korman that Musk was going to try to back out. That foresight had driven him to insist that Musk sign it personally. But the validation of his instincts was little comfort.

Korman wasn't the only attorney who could plainly read Musk's intentions. Musk's antics had also caught the attention of several partners at Wachtell, Lipton, Rosen & Katz, one of the top corporate law firms in the United States, which in the 1980s had invented the poison pill tactic that Twitter deployed. They were well-known in takeover defense, often fending off hostile acquirers like Musk, and trial specialists, unlike Korman's and Klein's firms.

When Musk began his complaints in early June about Twitter not sharing its data, Benjamin Roth, one of Wachtell's partners, seized the opportunity. "I've been following with interest the news about your pending transaction with Elon Musk," Roth wrote in a June 7 email to Gadde, Segal, and Edgett, Twitter's general counsel. "If what I am reading is accurate, it seems there is a meaningful risk of litigation to enforce the terms of your merger agreement. We would be extremely interested in representing Twitter in preparing for this possibility and in the unfortunate event that it comes to pass."

Roth highlighted the firm's deep ties to the Delaware court where any litigation with Musk would inevitably land. Although Twitter's headquarters sat in San Francisco, the company—along with most American corporations—claimed as its home the corporate haven of Delaware, which offers favorable taxes and limited transparency. But one of its most appealing features was its court system specifically dedicated to corporate disputes. Delaware Court of Chancery, which was founded in 1792 but traced its roots to feudal England, presided over botched mergers and fiduciary duty failures, and did so on expedited timelines so that companies did not have to wait years for the plodding federal court system. Their fast track was presided over by experts in corporate governance law—one chancellor, six vice chancellors, and three magistrates—and offered a curtailed appeals process, which ensured decisions were firm and actionable. If a merger was ordered by the court, money would soon be changing hands.

In his email, Roth pointed out that one of the most reliable experts on the unique court "sits about 25 feet down the hall from me," referring to Leo Strine, the former Chancery Court vice chancellor who was among Wachtell's partners. Strine had made a name for himself on the court in 2001 when he forced the poultry producer Tyson to buy IBP, a beef packer that it had bid $3.2 billion to acquire before attempting to back out of the deal. The ruling proved that Delaware was willing to force a merger, even though significant bad blood had accumulated between the parties, rather than permitting a breakup fee.

"That experience has been taught and handed down to the next generation like my colleague Bill Savitt, who is the cochair of our litigation prac-

SUN VALLEY

tice and I believe is the leading Delaware litigator," Roth wrote. He copied both Strine and Savitt on the pitch to Twitter's top brass.

In response, Twitter quickly began negotiations with Wachtell. While Korman and Klein continued to handle the day-to-day correspondence with the billionaire's team, Twitter quietly worked in the background with Savitt to begin prepping for the eventuality that the company would need to sue to force Musk to complete the transaction.

Savitt had an angular face and dark, arched eyebrows that gave him a slightly perplexed look, but he rarely lacked poise when it came to boardroom showdowns. After college, he spent the late 1980s in New York City, jamming on guitar in rock bands and driving a taxicab to make ends meet. Although he landed a few gigs at the famed punk venue CBGB, he eventually went straight, enrolling in law school at Columbia. By 1998, he found himself clerking for Supreme Court Justice Ruth Bader Ginsburg, and two years after that he joined the ranks at Wachtell, where he clung to his rocker roots by displaying a robin's-egg-blue Fender Stratocaster in his office.

His quirky background and cool demeanor made him an easy fit at Twitter, where he seemed to some almost a Dorsey-like figure, as if the company's founder had gone to law school instead of hacking together dispatch services. Twitter officially retained Wachtell on June 21, as the letters between Twitter and Ringler increased in acidity.

Savitt's past had another twist that made him intriguing to his new clients: a previous run-in with Musk. Tesla had hired Wachtell to defend it in a shareholder lawsuit stemming from its 2016 acquisition of SolarCity, but when the firm lost a motion to dismiss the case, Tesla fired Savitt and his team, showing the lawyer how mercurial Musk could be.

Tesla went on to win the lawsuit, leaving Savitt frustrated about losing the client. Suddenly, he found himself readying for a showdown with the billionaire.

>>> THE DAY AFTER Musk called off the deal, July 9, Agrawal, Segal, and Taylor met in the lobby of the Sun Valley Lodge, a grand, wood-paneled resort tucked in the mountains of central Idaho. In the winter, the lodge played

host to ultrawealthy vacationers and off-duty Olympians who flocked to Idaho for some of America's best snow and most private skiing. In the summer, the skiers gave way to equestrians and golfers from similar tax brackets who paced the halls of the resort, retracing the steps of Ernest Hemingway, who booked a suite at the lodge to finish his 1940 novel, *For Whom the Bell Tolls*.

Agrawal, in a black T-shirt with an ultrasoft cashmere hoodie over top to ward off the mountain air, and his wife, Vineeta, had arrived the previous Tuesday. The first few days of the conference had been a relatively peaceful mix of business meetings. Vineeta, who led the biotech and health investment portfolio at venture firm Andreessen Horowitz, had her own work to do; she greeted the other Twitter executives and set off in search of coffee.

That Saturday morning was different. Out of the corner of his eye, Agrawal could see heads turning. Even after Dorsey had named him chief executive, he had remained relatively anonymous, able to wander through San Francisco with ease. But here in Idaho, at the annual Allen & Company conference, known as a summer playground for those with three-comma fortunes, Agrawal was suddenly a celebrity by way of his highly public clash with Musk. It was his first time attending the conference; while Taylor and Segal were regulars at the event, the chief technology officer of Twitter had never merited an invitation.

It didn't help that Musk had arrived in Sun Valley late that Thursday evening. On Friday morning, as the world woke up to the news that he was abandoning the Twitter deal, he was breathing the same mountain air as Agrawal, Taylor, and Segal. The odds of a chance encounter on the tree-lined paths of the resort kept everyone staring at the Twitter team, not wanting to miss a serendipitous confrontation.

Taylor didn't flinch, launching into the plan for the day. Together, the men would attend the interview Musk was scheduled to give later that morning. They would stay cool and calm, even if Musk insulted them to their faces. And after, they would regroup for a call with their legal team.

Musk was among a who's-who list of tech and media tycoons scheduled to attend that week, including Shari Redstone, the chairman of Paramount Global, and Disney's Bob Iger. Outside the resort hotel, stages peppered the lawn to host speeches. But because the place was crawling with press and nosy onlookers, the attendees were given headsets to listen to the talks so the speeches weren't broadcast to the masses over a PA system.

The Twitter team filed outside and took their seats, greeting the other executives around them before slipping on their headsets. Musk took the stage with a grin on his face, well aware of the question on everyone's mind.

Almost immediately, he launched into a tirade. Twitter was completely overrun with spam, Musk claimed. Then he turned to the audience. "Does anyone really believe that the bots are fewer than five percent?" he asked.

Maybe it was the fact that they all were wearing headsets, but the crowd seemed ill-prepared for an audience participation segment. No one was sure whether to answer aloud or to raise their hands, or to just sit in silence as Musk continued his complaint. The billionaires and multimillionaires looked around uneasily at each other.

Musk was unperturbed, barreling along as though the audience had cheered in agreement. "See, I told you," he said. Taylor sat unruffled, his hands folded lazily in his lap.

Musk brushed off further questions about the deal, citing the impending lawsuit that Twitter had threatened to file. But he didn't shy away from press reports that he had recently fathered a secret set of twins with one of his employees, Zilis. The world was suffering from a severe underpopulation crisis, Musk told the audience, and the only way to start a colony on Mars would be to have more children.

>>> ON SUNDAY, as Segal sat on the metal benches inside a quaint Idaho airport near the resort, his phone buzzed. Jay Sullivan, Twitter's general manager and product head, was on the line. The men were set to discuss Twitter's developer platform, a portal for independent coders who wanted to build new features for Twitter. But Sullivan had other ideas. He had been stewing over the deal and couldn't bite his tongue any longer.

Sullivan pleaded with Segal to reconsider the lawsuit. Musk was clearly going to be a terrible owner for Twitter, Sullivan said. And now that he didn't even want it, couldn't they just let the billionaire walk away?

"I don't want to do business with someone who acts this way," Sullivan told Segal forcefully, trying to hammer home what an awful fit Musk was for Twitter's mission.

"Try to keep an open mind," Segal suggested. The board believed selling Twitter to Musk was the best outcome, he reminded the product lead. It was the right price for shareholders, and they all had an obligation to see it through, even if it meant suing to force the deal ahead.

Sullivan raged at the idea. "I've read my Milton Friedman, okay?" he snapped, citing the famed free-market economist. He understood that the board had a fiduciary obligation to sell to the highest bidder and didn't need it explained to him by the CFO. "There's some higher calling here."

Sullivan pressed on, asking Segal why he hadn't been called in to present to the board when it was weighing Musk's offer. Surely, as the head of product, he should have been given the opportunity to present his vision to the board and make the case that Twitter could be run independently. It was hard for anyone, other than Musk, to claim to be the best man to serve as the steward of the preeminent online speech platform, but Sullivan at least wanted to make his case. He didn't know that Agrawal had made the pitch already, as part of Prism—the only part of the plan that had been shared with him were the layoffs.

"I just care so much about the service," Sullivan sighed. Even though he knew there was no backing out of the deal, he was venting, demanding that Segal acknowledge his argument.

Segal tried to comfort him, saying that Elon might come in peace. There was the bombast he'd seen onstage, ranting about population decline, and online, tweeting poop emojis at his CEO. But Segal still thought there might be a more reasonable Musk somewhere behind all the showmanship, a man who genuinely wanted to do the right thing and just needed to learn how Twitter worked. Everything up until now, even the lawsuit, might be histrionics. Besides, his hands were tied. His personal feelings about Musk didn't matter—he was obligated to complete the transaction.

21 > Court of Chancery

On July 10, Savitt introduced himself to Musk's lawyers with a curt, three-paragraph letter. "The agreement is not terminated," he wrote to Ringler. Musk's excuses were just that, Savitt added. Meanwhile, Twitter's engineers had kept a close eye on the firehose data, monitoring how Musk directed his lawyers and engineers to query it. His behavior raised a glaring red flag.

Although Musk had purportedly called off the deal, his team continued to run thousands of queries against its data. The ongoing analysis was extremely troubling to Twitter employees. If the deal was indeed off, then Musk should have no longer been interested in the company's data and user information. Instead, he was doubling down. After his April musings to Agrawal about perhaps starting a competitor, it seemed to Twitter that Musk might be intending to use the company's proprietary data to do so.

Twitter's lawyers, however, were less conspiratorial. To them, the searches implied that, even though Musk was claiming the bot problem was out of control and using it to wiggle out of the acquisition, he hadn't found a smoking gun yet. He was still trying to justify his decision—which meant he probably knew he didn't have much of a case.

Savitt unleashed Twitter's lawsuit in Delaware's Court of Chancery on July 12. The document was excoriating, venting the pent-up frustrations of a board and company that had been muzzled for months.

"Musk refuses to honor his obligations to Twitter and its stockholders because the deal he signed no longer serves his personal interests," read the suit. "Musk apparently believes that he—unlike every other party subject to

Delaware contract law—is free to change his mind, trash the company, disrupt its operations, destroy stockholder value, and walk away."

The sixty-two-page document laid out what Twitter's board and legal advisers believed was Musk's true motivation to escape—financial losses inflicted by Tesla's falling stock price.

Savitt tore into Musk's bot obsession, calling it "a model of hypocrisy."

"Musk said he needed to take the company private because, according to him, purging spam would otherwise be commercially impractical," he wrote. "But when the market declined and the fixed-price deal became less attractive, Musk shifted the narrative, suddenly demanding 'verification' that spam was *not* a serious problem on Twitter's platform, and claiming a burning need to conduct 'diligence' he had expressly forsworn."

Musk played it off as though getting sued had always been part of his master plan. He tweeted a meme that showed him laughing. "They said I couldn't buy Twitter. Then they wouldn't disclose bot info," the caption read. "Now they want to force me to buy Twitter in court. Now they have to disclose bot info in court."

His fans lapped it up. Of course, their hero was playing four-dimensional chess, luring Twitter into suing him and then trapping them into discovery. But Musk was serious about getting out of the deal, and he saw the lawsuit as defiance by the disobedient Twitter executives.

Musk brushed the whole thing off and deputized the fight to Alex Spiro, his bulldog litigator. But in Chancery Court, the rules were different from the federal courthouses where Spiro had secured victories for Musk, and everything in Delaware moved on a fast timeline.

Spiro brought in his own team of Delaware experts to back him up. He used Andrew Rossman, a lawyer from his own firm who had experience in the state's business court, and Ed Micheletti, a Chancery Court veteran from Skadden who had carried other high-profile busted deal cases. Micheletti had represented Toys "R" Us during shareholder lawsuits over its 2005 leveraged buyout, arguing before Strine when he was a vice chancellor at the court.

The case advanced by Micheletti was clear: Twitter was run by complete idiots. They had no idea how to count their users properly, which made their estimates about the number of bots on the platform completely unreliable.

And worse, they were malicious—they knew their numbers were wrong and, instead of fixing the problem, had gone to great lengths to hide it from Musk for as long as possible.

"Musk was flabbergasted to learn just how meager Twitter's process was," Micheletti wrote in a response to Twitter's lawsuit on July 15. "Human reviewers randomly sampled 100 accounts per day (less than 0.00005% of daily users) and applied unidentified standards to somehow conclude every quarter for nearly three years that fewer than 5% of Twitter users were false or spam. That's it. No automation, no AI, no machine learning."

The papers painted Twitter as a tech company in name only. According to Micheletti, Twitter just pulled up 100 random accounts every day, an oafish, gum-chewing reviewer stared at them and then rubber-stamped them all as real users. The lack of rigor or technical prowess cut at the heart of Musk's business plan, the lawyer argued. Musk wanted to get rid of spammers, verifying the identities of users as they signed up, and then sell them subscriptions so that he could lessen Twitter's dependence on advertisers for revenue. The business plan worked only if the users were real—bots didn't pay subscriptions.

Musk's constant railing against Twitter's executives wasn't harming the company but instead was furthering its economic interests, Micheletti posited.

"With the sense of humor of a bot, Twitter claims that Musk is damaging the company," he wrote. "Twitter ignores that Musk is its second largest shareholder with a far greater economic stake than the entire Twitter board."

Aside from launching barbs at the board, Micheletti had another objective. He asked the judge to delay the trial until February. That, he claimed, would give Musk enough time to get to the bottom of the bot problem. Twitter was pushing for trial as soon as possible, citing the deadline for the deal to close: October 24. Both sides knew that Musk's only way out of the acquisition agreement was if his loans fell through, which would allow him to pay the $1 billion breakup fee and walk away. The loans were set to expire in April 2023. By asking for a February trial, Musk was running down the clock—something Savitt couldn't allow.

But Musk was in for a lesson in how Chancery Court worked.

Chancellor Kathaleen McCormick had been expecting Twitter's lawsuit.

The forty-three-year-old, who wore her thick brown hair cropped plainly below her shoulders, had followed the headlines about Musk's disputes with the company and knew the fight would eventually end up in her courtroom. She even kept a lurker Twitter account under a pseudonym, which she occasionally used to scroll through the news. When Twitter filed its case, her secretary printed it out and plopped it on a ledge that held stacks of incoming papers in McCormick's office. "Here it is," the secretary said, not needing to explain which case she was referring to.

McCormick, the first woman to lead the court since its founding, assigned herself to oversee the Twitter dispute, following a court tradition in which the lead chancellor took cases under the most scrutiny and left less troublesome litigants to the six vice chancellors. By that summer, the small Delaware court had already been stretched to its breaking point with new cases, but Twitter demanded urgency. Despite coming down with COVID on July 17, McCormick barreled ahead with a virtual hearing, and declared the trial would begin in October.

The date became an event horizon for everyone involved. For the lawyers, it was the deadline to uncover a smoking gun, and they began lobbing subpoenas at anyone who had ever worked at Twitter or had associated with Musk. For Twitter employees, it was the outer limit of their suffering—at least by then, they would know the outcome. Some started interviewing for other jobs, realizing they wanted off the ship. Others scheduled leaves of absence for the fall, hoping to avoid the takeover altogether.

As they had when the board instituted the poison pill to block Musk's acquisition in April, employees turned to Twitter itself to learn about the obscure process of chancery trials in Delaware, uncovering accounts of legal scholars and court observers. There would be no jury and no witnesses—just a sumo match between lawyers, with McCormick deciding Twitter's fate.

22 > "Sorry I didn't get your special"

While Agrawal left the legal filings to the lawyers, he remained intimately involved in the litigation strategy. He had no legal experience; everything he knew about mergers and acquisitions had come from overseeing deals that Twitter did to buy start-ups. But public company buyouts were foreign to him, and aggressive negotiations that ended up in court seemed like something that happened on a different planet.

Still, Agrawal dug in and learned. He had an obsessive personality and loved to game out every scenario, needing to know all the possible moves available to him on the chess board. He read case law stemming from previous hostile takeovers and sat with lawyers in their strategy discussions, privately telling associates that he thought he had absorbed more on the intricacies of these complicated deals than Spiro. If this was going to be his legacy as Twitter's CEO, Agrawal wanted to be a decision-maker rather than a passive player who rolled over for the richest man in the world.

Yet, as he gained proficiency in those aspects of running a business, he was floundering in others, unable to communicate the details that employees wanted to hear about the future of their workplace. The trust deficiency between the chief executive and the rank and file was growing. Most employees only ever saw or interacted with Agrawal on video calls, so his staff proposed a tour to different Twitter offices where he could personally introduce himself to employees.

He started in June at the company's London office, which had a coffee bar for employees. While professional baristas typically staffed the kiosk, Twitter's comms staffers had Agrawal and other execs work there for about an hour, serving coffee and talking with anyone who stopped by. They wore khaki aprons and served from a menu that included a coffee drink called a "Parag Special" and "Ned's Cookies"—named after the Twitter CFO's recipe for chocolate chip cookies sprinkled with sea salt.

The U.K. office's head sales exec hosted an onstage discussion with Agrawal, before presenting him with a framed emoji-styled portrait, featuring the chief executive's distinct glasses and five-o'clock shadow. Agrawal sheepishly accepted the gift and continued his employee meet-and-greet at the company's Dublin office later that week.

In late July, Agrawal visited Twitter's New York office in Manhattan's posh Chelsea neighborhood. It was going smoothly, with the CEO greeting each worker as they stepped up to place their order. Then one junior engineer approached the counter and tried to make an impression on his boss's boss's boss with a joke.

"Can I just have a regular cappuccino?" he said to Agrawal with a smile. "Sorry I didn't get your special. I hope I don't get fired or something!"

Twitter's chief looked up from the ordering system with a confused stare.

"Why . . . why would I fire you?" Agrawal said.

He locked eyes with the engineer, who started to panic after realizing that the chief wasn't reading his social cues. Rumors of layoffs had swirled among employees. Surely, Agrawal knew everyone was anxious?

The employee stammered under his breath and quickly moved down the line to wait for his drink.

—

>>> AS TWITTER WAGED a paper war in Delaware, trying to bury Musk's legal team in endless filings, Project Saturn was forced to the back burner. Because the Saturn team included some of Twitter's brightest stars in product and policy, most of them were dragged into endless briefings with lawyers to explain how they detected bots on the platform.

Another one of Agrawal's game-changing projects stalled, too. Jay Graber and her handful of engineers at Bluesky had been patiently waiting to hear from Agrawal about how Musk's bid would disrupt their nascent effort to build a decentralized social media ecosystem that could someday become the backbone for Twitter, Facebook, and any new platform that happened to emerge.

Agrawal had reassured Graber that she had nothing to worry about. He would make sure the checks kept clearing and that she would have the latitude to continue building the protocol.

Bluesky's corporate documents set aside a board seat for a Twitter representative, guaranteeing that Twitter would be able to supervise Graber's process and work to incorporate what she built back into the social media giant. With Dorsey gone, the seat was Agrawal's.

It was a natural fit. Dorsey had always cared about the "what"—building a decentralized protocol for social media—but it was Agrawal who was interested in the "how" and was eager to dive into technical discussions with Graber. But Agrawal, perhaps as a gesture of respect to his mentor, signed over the board seat to Dorsey.

Graber was becoming increasingly impatient with the entire ordeal. She couldn't get a straight answer from Agrawal about when their work would resume, and so she began hiring a small team to build Bluesky its own app. They could always integrate with Twitter later, Graber figured, but if people could start using Bluesky right away, then there would be no need to wait for the resolution of the lawsuit against Musk.

>>> BY AUGUST, Twitter's legal battle was well underway, and its leaders felt increasingly confident that they would be able to force their erratic opponent to buy the company. On Monday, August 1, Agrawal charged into his San Francisco office with a renewed sense of purpose. He wanted to push Project Saturn forward, making his mark on Twitter before he had to hand the platform off to Musk.

"What's not changing: Twitter taking accountability for impact on the world," he wrote to colleagues who were working on the initiative. "What's changing: minimizing centralized control and increase control and transparency for people." He told Gadde and Sullivan that Saturn would be revealed to the public by the fall and instructed them to write it into their plans for the year.

The chief executive also broke the news to his communications staff. They would need to begin editing blog posts and briefing reporters for the launch. But one flack balked. She had worked under Gadde for years, bearing the brunt of criticism from the press for Twitter's controversial decisions about the *New York Post*, COVID, and Trump, and she was disgusted at the idea of allowing Trump and other banned users to return to the platform. It seemed like a ridiculous regression to her and she refused to work on it.

It was a hint to Agrawal that his grand vision might not be welcomed by many people who had come to rely on Twitter's rules. Still, he pressed on.

Agrawal also informed Savitt about what he was planning. Savitt, too, had serious concerns. Each time Agrawal made a decision that he thought would improve Twitter, another letter from Musk's lawyers landed in the litigator's inbox, claiming that Agrawal had made material changes to Twitter's business that opened another loophole for Musk to escape from the acquisition.

"Are you operating in the normal course of business?" Savitt asked Agrawal. Upending Twitter's entire content moderation strategy—which was one of Musk's stated goals when he offered to buy the company—seemed risky. But Agrawal had been planning Project Saturn before Musk signed the merger agreement, he argued. Wouldn't it be more normal to continue with his plans as though nothing had happened?

Savitt asked that his lawyers be allowed to review all the documents related to the initiative, but ultimately agreed to the launch.

In the early morning hours of August 23, Agrawal's phone started buzzing frantically, piercing his sleep. He grabbed it and started scrolling, and the headlines he saw about a former Twitter executive named Peiter Zatko opened a pit in his stomach. Saturn would have to wait—again.

23 > Mudge

Hired by Dorsey in November 2020 to become its head of security, Peiter Zatko was better known by his hacker moniker, Mudge, and had earned renown in the tight-knit security community in the 1990s for research into a type of security vulnerability known as a buffer overflow. While many hackers shared Dorsey's anarchistic skepticism of the government, Zatko realized that he could advocate for broader security reforms by working with the feds than by peppering them with attacks from the shadows. He took a job at the Defense Department's research wing, DARPA, and worked with Bill Clinton's administration to mitigate cyberattacks.

When Dorsey hired him, there was just one problem—Twitter had, just weeks earlier, hired another executive for essentially the same role. Her name was Rinki Sethi, and while she didn't have Zatko's industry prestige, she had experience running security organizations. Sethi was told to report to Zatko.

A power struggle quickly unfolded. Both were charged with patching the company back together after a bruising hack in July 2020, when a group of teenagers commandeered the site. To her new employees, Sethi seemed sharp-elbowed and suspicious of her staffers, often micromanaging them. She quickly drew up a list of people she wanted to fire, alienating longtime employees. Zatko had prestige, but his employees suspected he had checked out from security at the height of his fame in the '90s and didn't have a working grasp of modern security systems.

Both Zatko and Sethi embarked on land grabs, trying to wrest parts of the security organization for themselves. Sometimes they assigned conflicting work, and employees didn't know which executive they were supposed to be following—the one the CEO preferred or the one who actually managed them. Some Twitter security workers were particularly suspicious of Zatko's deep ties to the government, which Twitter's staff largely viewed as another adversary to be prevented from spying on users. The former DARPA official appeared on video calls with a folded American flag and military trophies displayed in the background.

It was no secret that Zatko and Sethi were at odds, and the executives quickly launched competing complaints against each other with the human resources department questioning each other's ethics. Within a year, it was obvious the arrangement wasn't working.

Zatko, at least, seemed to realize that the tide was turning against him. He appealed to Agrawal, asking the new chief executive to come visit him at home in New Jersey during Agrawal's holiday trip to the East Coast. Agrawal initially agreed to the sit-down but then put the meeting off, saying he was unable to make time during his trip. But Zatko used his access to Twitter's systems to check the travel arrangements for Agrawal's security detail and found no last-minute adjustments. He believed Agrawal was lying to him.

By January 2022, Agrawal was ready to act. In one of his first moves as chief executive, he decided to fire both Zatko and Sethi and start fresh.

In a last-ditch effort to save his job, Zatko appealed to the board of directors, claiming that Agrawal had blocked him from briefing the board about the extent of Twitter's security problems. Thousands of employees had broad access to Twitter and could take down the site if their accounts were compromised, he warned. And such a compromise was likely, Zatko continued, because the employees weren't installing regularly scheduled security updates that would protect their devices from being hacked. Zatko argued that the gaping holes ran counter to what Twitter had been telling the Federal Trade Commission for years—that it had a robust security program and could be trusted with its users' personal data.

"I joined because I felt an 'attachment to mission,'" Zatko wrote in a plaintive email to Pichette that February, several weeks after his dismissal. "I'm different from a lot of the world that way."

Zatko extracted a $7 million settlement from Twitter in June, after his lawyers argued that his firing was unjustified. But he was wounded and watched from the sidelines as Musk buffeted the company. He believed that Agrawal was toying with Musk as the engineer had done with him. He hadn't overseen Twitter's bot tallies, but Zatko had asked around about it while at Twitter. He didn't think Agrawal was being honest about the scale of the problem.

The former security executive signed on with Whistleblower Aid, a nonprofit legal organization, and the group agreed to help him submit a complaint to Congress, the FTC, and the SEC. The securities regulator paid awards to whistleblowers whose claims led to enforcement actions, so Zatko stood to make a tidy sum on top of his Twitter settlement. Whistleblower Aid ensured that they would help his claims go far and wide. On August 23, the group did just that, coordinating news stories of Zatko's whistleblower report in *The Washington Post* and on CNN.

Twitter's security program was a complete disaster, Zatko claimed. The company had always struggled to secure user data. At first, it grew too quickly, and its databases were held together with virtual bits of tape and string. Then, hackers recognized its influence over global discourse. It became too enticing a target to pass up. The FTC intervened in 2011 and had kept watch over the company ever since.

The agency's ongoing oversight of Twitter gave regulatory significance to Zatko's complaint. The security problems meant that Twitter was still violating its 2011 agreement with the agency to clean up its act, Zatko argued.

The report was a boon for Musk and his lawyers. His case had been floundering, as round after round of discovery failed to turn up proof that Twitter had manufactured its spam estimates. But here, finally, was someone who was willing to confirm Musk's theories. Better yet, Twitter had already paid him off handsomely in exchange for his signature on a nondisclosure agreement. It seemed as if the company had known about its bot problem all along and covered it up.

Spiro vowed to subpoena Zatko immediately. Agrawal, meanwhile, was frustrated. As far as he was concerned, Zatko was yet another problem inherited from Dorsey. He had tried to move decisively, as he always did, and clean it up quickly. But here Zatko was, using a loophole to escape his nondisclosure and splash his whistleblower complaint across the internet.

Agrawal launched a strident defense of the company to his employees, many of whom wanted to know if Zatko's claims were true. On August 24, the day after the whistleblower complaint was published, Twitter's chief executive summoned his workers to a company-wide meeting.

"There are accusations in there without any evidence," Agrawal said. Twitter's staff could hear the frustration in his voice, and to some it seemed that Agrawal was finally relatable. "Honestly, some of it just doesn't make any sense. The narrative that has been created is false."

Sean Edgett, Twitter's general counsel, took the mic next. "We have never made a material misrepresentation to a regulator, to our board, to all of you," he said. "We are in full compliance with our FTC consent decree."

Twitter's top security and privacy executives, Damien Kieran and Lea Kissner, debunked Zatko's allegations one by one. Twitter had more work to do on its security, but things were not so desperate, they said. By the end of the meeting, most employees were reassured that the former executive wasn't credible—after all, it had been his job to fix Twitter's security problems, and instead he'd taken a massive payout and walked away.

But Spiro and Musk's team of lawyers wasted no time. On August 29, a week after Zatko's claims were published, they wrote to Twitter to break off the deal again. His suggestion that Twitter had misled the FTC was reason enough to end the acquisition, even if Musk's concerns about bots turned out to be nonsense, Musk's legal team argued. "These violations would have material, if not existential, consequences to Twitter's business," Ringler wrote in a fresh breakup letter.

Spiro also worked Zatko's claims into the lawsuit, demanding that Musk be allowed to walk away from the deal because the damning whistleblower report revealed Twitter had misled the public about its user numbers, which the company referred to as monetizable daily active users, or mDAUs.

"Stunning events over the last week, however, have revealed that the misrepresentations regarding mDAU were only one component of a broader conspiracy among Twitter executives to deceive the public, its investors, and the government about the dysfunction at the heart of the company," Musk's lawyers wrote to the court in mid-September. "The Musk Parties and Twitter's many other investors were sold a different company than the Twitter that actually exists—one that was more valuable, more popular, more secure, and more compliant with governing law."

Chancellor McCormick agreed that Zatko's claims should be incorporated into the trial but seemed somewhat skeptical of Musk's entire approach. In a hearing about the whistleblower report, as Musk's side argued that there would have been no way for them to find out about Zatko if he had not come forward publicly, McCormick gently offered a reality check.

"We'll never know, will we?" she said lightly. "There was no due diligence."

As the October trial date approached, McCormick was becoming increasingly concerned about the media firestorm and the security risks it could bring to her court. Her staff arranged for a secret entrance on a loading dock, where Musk could come and go without being seen. They also worried about mysterious graffiti that appeared outside the courthouse and around Wilmington, Delaware, that made cryptic reference to the case. "Dievest," a tag in bright Twitter-brand blue paint read, splashed on the wall of the parking garage used by McCormick and other staffers. "Twitr paid." Was it a backing message for Musk? A threat? A signal of support for Twitter? McCormick threatened to come into work with a power washer and remove the graffiti herself before the owner of the garage buffed out the strange message.

With the trial date looming, Twitter's leadership became more and more optimistic that they would win. Musk's excuses for vacating the deal were too thin to hold up in court. Throughout the litigation, Savitt had braced himself for surprises, knowing Musk would find more cards to play. While the Zatko report was a challenge, Savitt thought it would take more than a disgruntled former employee to crack the merger agreement.

Musk's best path to escape owning Twitter was for the board itself to recommend that shareholders vote against the deal. But on September 13, the board took the deal to Twitter's shareholders, who voted overwhelmingly—98.6 percent—in favor of it. Twitter's stock was puttering along at about $41 per share, and Musk's offer of $54.20 was too good to refuse. All that was left was the looming trial.

24 > An Accelerant to X

Even as Musk's lawyers pleaded with Chancellor McCormick to air Zatko's allegations, they recognized that their client's path to victory was narrow. Zatko had been long gone from Twitter by the time Musk made his bid and hadn't been directly involved with the FTC's oversight of the company. Some of Musk's lawyers believed McCormick was skeptical of the billionaire, as she awarded Twitter wins on motions, and thought their client wouldn't get a fair trial.

So Musk began to test the waters. Emanuel, Musk's friend, made an overture to Durban to settle the dispute. Musk's representatives later asked for a 30 percent discount, which would have valued the company at about $31 billion. The board dismissed the lowball.

Musk was scheduled to be deposed by Twitter's lawyers on September 29, but some of his lawyers thought it best to prevent Musk from talking under oath. Musk could barely contain himself from tweeting things that jeopardized his case. Who knew what he might say in person?

Spiro and the Twitter board began discussing a new offer—a 10 percent discount. The price would value the social media company at roughly $39.6 billion, a slim premium above its $39.2 billion market capitalization. Entertaining discussions of a discount was risky—shareholders had already voted to approve Musk's offer, and undercutting the vote would certainly expose them to lawsuits. The board's goal was to get Musk's money into its shareholders' hands, and even though they were confident they would win at trial, nothing was certain. A courtroom loss would leave shareholders with nothing, and a small discount could be worthwhile if it guaranteed Musk would pay.

The smaller discount left Musk squeamish. Since he had agreed to buy the company in April, interest rates had increased and a change in purchase price would mean he would have to renegotiate the terms of his debt financing. The higher interest rates would erase a significant chunk of his potential savings.

Still, Musk's side warned that if the Twitter board didn't agree to settle the case, Musk would pursue them to the ends of the earth with lawsuits. *You don't want Musk as an enemy,* the members were told. The billionaire chafed at the idea of paying the Twitter executives' multimillion-dollar severance packages, known as "golden parachutes," and he wanted to wage a scorched earth campaign against everyone he thought had wronged him over the course of the deal. Twitter's board was worried about Musk's threats. Some members already had security outside their homes after receiving sinister online messages, and they sought assurances that Musk wouldn't sue them personally if they agreed to a discount. Spiro thought the demand was outrageous and shady. It seemed to him that the board wanted additional legal protections after a bruising summer of litigation, another sign they had something to hide.

Musk's September 29 deposition came and went without the billionaire showing up. He would reschedule, his lawyers claimed.

Inside Twitter, Roth was losing faith that Project Saturn would ever see the light of day. It seemed obvious that Musk would lose the lawsuit, and that whatever work he accomplished would be immediately reversed by Twitter's new owner.

Sullivan had his doubts, too. He was eager to build the technology that he and Agrawal had brainstormed all those months ago. But Project Saturn had been stripped down from a grand product vision to a couple of blog posts while everyone waited to see what would happen with the deal. It didn't make sense to Sullivan to do a splashy announcement about a product that didn't exist yet—especially with Musk about to take over and upend everything.

He went to Gadde with his concerns. The lawyer, who had harbored doubts about Saturn for months, agreed. There was a lot of flash and no substance yet, and the two executives tried to coax Agrawal off the ledge. Even though he was committed to leaving a legacy, there was no point in doing it unless Saturn would stick, they told him.

In early October, Agrawal sailed through his deposition without being baited into saying anything stupid, giving Twitter's lawyers more confidence in the strength of their case. Hours after Agrawal finished testifying,

Musk's team suggested the billionaire might finally be ready to pay full price for Twitter. Musk's desire to control Twitter may have won out, or he may simply have been painted into a corner by Twitter's legal team. He told his lawyers that, even if he won the lawsuit, he would lose in the long run—because he wouldn't end up owning Twitter.

"It seemed inevitable to me that he was always going to acquire the company. There's nothing on earth that guy loves more than he loves Twitter," said one lawyer who worked for Musk's side. "It may be that he never really intended to go to trial, he never thought he could win, but was using it as a tactic to renegotiate the deal."

>>> GADDE STILL WASN'T sure what to expect when she strode into the dull gray lobby of One Market Plaza on October 3 and rode the elevator up to Wilson Sonsini's offices on San Francisco's Embarcadero. She was one of the few Twitter executives and board members who had not yet given a deposition to Musk's legal team, but she was finally due to face them that Monday morning.

Despite Musk's offer to buy Twitter at the original price, Gadde didn't trust him to follow through and showed up ready to testify. She felt confident that she could handle Spiro's bombastic approach. But Spiro was late.

As the minutes ticked by, Gadde became confused. Was this a tactic to frustrate her and bait her into an argument? It seemed like an amateur move, but she'd come to expect bizarre antics from Musk's prized lawyer.

After a half hour, Gadde realized: Spiro wasn't coming. She chatted with her lawyers about what his absence might mean. The teams that oversaw spam and platform manipulation reported to Gadde—surely Musk's lawyers wouldn't miss the opportunity to grill her about Twitter's problems with bots.

His team said they'd cancelled the deposition and Gadde somehow hadn't gotten the memo. Finally, that evening, a letter from Musk arrived confirming the reason Gadde had been stood up. The deal was officially back on. Musk would buy Twitter for $44 billion after all, put Twitter's lawsuit against him to bed, and avoid ever sitting down to talk under oath.

The following day, it was as if he'd never tried to bust up the deal in the first place. "Buying Twitter is an accelerant to creating X, the everything app," he tweeted.

25 > "Not owned by a fucking moron"

'm trying to think what Twitter would get by NOT accepting the exact same offer," a Twitter engineering manager wrote on Slack, in response to the sudden news. He shared his musings in a 2,000-person channel called #stonks, where employees traded advice about how to manage their Twitter stock options and debated the company's financial performance.

Bitter responses flooded the channel.

"A lot lower stock," one worker responded.

"It'd get to not be owned by Elon Musk," another said.

"Not owned by a fucking moron?" an engineer chimed in.

Despite their disgust, the chattering employees understood that his $44 billion offer overvalued their company. Twitter's stock rocketed upward with confirmation that a deal was all but done. Trading of Twitter's shares was temporarily halted, and once trading resumed, Twitter closed out Tuesday up more than 22 percent at about $52 a share. The deal, as one employee in #stonks put it, would "put bread on the tables of shareholders" and that was all that mattered to Wall Street.

Edgett sent a terse note thanking employees for their patience. "We received the letter from the Musk parties which they have filed with the SEC," he wrote. "Our intention is to close the transaction at $54.20 per share."

The email seemed clownishly redundant. Of course Twitter intended to close the deal—it had been the refrain that executives drummed into employees all summer. But Twitter's board members and legions of lawyers didn't know what to make of Musk's sudden pronouncement. He had ripped the rug out from under them, and they had no reason to trust anything he said. This could be a stall tactic, a way to wiggle out of his deposition or to delay the lawsuit until his loans expired the following April. Still, his pronouncement heightened the grim sense of duty in the boardroom. The members knew they must sell, but any optimism that Musk might be a good steward of the platform had long been erased.

In court, Twitter's lawyers were fighting to get their hands on even more of Musk's text messages that might reveal his true motives for tanking the deal—the billionaire had turned over no messages between May 24 and May 30, when Musk was arranging funds to replace his $6.25 billion margin loan, nor between June 1 and June 7, when Musk was whipped into a fervor over the supposed bot problem and directed his lawyers to tell Twitter he might back out of the deal. Savitt hadn't found concrete proof that Musk's bot complaints were a smokescreen, but it was possible.

Whatever Musk's intentions, Twitter should press on until they saw $44 billion hit their accounts, Savitt insisted. He still wanted to proceed to trial in Delaware.

But Musk's lawyers had other plans. They told Chancellor McCormick that they would close the deal by October 28 and that a trial was no longer necessary. On the evening of October 6, the chancellor agreed to give Musk what he wanted. If the deal wasn't done by the end of the month, she told Twitter it could have a November trial date instead.

—

>>> THE DEADLINE TO BUY Twitter set off a frenzy in the Musk camp. After spending the summer insulting Twitter's top executives online and claiming he wouldn't buy the company, Musk suddenly had to corral the funds all his friends had promised him back in April. No matter that some of them might have had doubts about the value of Twitter in the meantime, Musk was coming to collect.

From the outset, Morgan Stanley had pitched the Twitter deal as an opportunity to back a generational entrepreneur. But rising interest rates meant investors were more cautious than they had been. Some of Musk's potential funders, like Reid Hoffman, who had once considered putting in $2 billion from his venture fund, pulled out completely.

Other investors needed cajoling. Dorsey had wavered in his commitment to Musk. While he had been one of the biggest cheerleaders of Musk's takeover in April, he had recently waffled about rolling his nearly $1 billion Twitter holding into Musk's private version of the company. Dorsey had two options: he could either let Musk buy him out at $54.20, just like every other shareholder, or he could take a small ownership stake in whatever Twitter would become, perhaps increasing his fortune if Musk later re-listed the company on a public stock exchange.

Since early May, the two billionaires had discussed how to run the social network, but Dorsey seemed unconvinced by the man he once called the "singular solution I trust." Dorsey was worth $6 billion, but most of that was tied up in equity in his company, Block, the new name of Square, and Bitcoin, and he didn't have much cash on hand to invest. He considered taking the cash at the $54.20 price and leaving the site he helped create in the rearview mirror.

This left Musk in limbo, so he cut his friend a sweetheart deal. Not only did he promise not to damage Twitter, but he also swore that, if Dorsey ever wanted out, he'd pay Dorsey the full price of $54.20 per share for his stock. Unlike other investors, the Twitter cofounder had a guaranteed floor price if the company's value decreased under Musk. That meant no inherent risk for Dorsey—other than relying on Musk's word—and he agreed. For Musk, it seemed like a necessary compromise to ensure he had another $1 billion to put toward his acquisition. Musk believed there was no way Twitter could lose value under his management.

Another major Twitter stakeholder who experienced doubts was Al Waleed bin Talal Al Saud, a sixty-seven-year-old Saudi Arabian prince. The grandson of the first ruler of the Kingdom of Saudi Arabia, Al Waleed was a flashy figure who had studied at American colleges before becoming an investment manager of the royal family's money.

In 2011, Kingdom Holding had put $300 million in Twitter, taking a more than 3 percent stake in a company that was valued at the time at $8 billion. Al Waleed's company held on to the investment for more than a decade and he became an avid user, tweeting pictures from his palace compounds, his meetings with world leaders, and his private jet.

Kingdom Holding still retained its Twitter stake when Musk came knocking, but it had become less clear who actually controlled Al Waleed's company. In December 2017, the Saudi billionaire was one of 320 royals and government officials who were rounded up and detained on the orders of Crown Prince Mohammed bin Salman, Al Waleed's cousin. The prince, who came to be known simply by his initials, MBS, locked his family members and bureaucrats in Riyadh's Ritz-Carlton for months, extracting concessions, payments, and pledges of allegiance to his cause to become the country's next leader. *The Wall Street Journal* reported that MBS sought at least a $6 billion ransom payment from his cousin, while Al Waleed sought to obtain his freedom by offering MBS an ownership stake in Kingdom Holding. He was released in January 2018.

By 2022, Al Waleed had become an MBS loyalist. On paper, he remained Kingdom Holding's chairman and was not a fan of Musk's offer to buy Twitter. Like other longtime shareholders, he fondly recalled when Twitter's stock soared above $70 per share in 2021 and tweeted that he was opposed to Musk's offer, arguing that $54.20 didn't come close to the "intrinsic value" of the company.

The insult stung Musk. He had held a grudge against Saudi Arabia after its sovereign wealth fund had declined to take Tesla private and told those close to him that Saudi money in the U.S. tech industry was everywhere, representing a threat to national security. Although he had his own hand out, he wasn't wrong. The Saudi government had invested billions of dollars in companies across Silicon Valley. He fired back at Al Waleed's post with a pointed tweet that referenced the execution of *Washington Post* columnist Jamal Khashoggi, a killing carried out at MBS's command.

"Interesting. Just two questions, if I may," Musk wrote on April 14. "How much of Twitter does the Kingdom own, directly & indirectly? What are the Kingdom's views on journalistic freedom of speech?"

By October, however, both men had changed their tune. Al Waleed perhaps saw an advantage in remaining close to the world's richest man, and Musk, whatever his morals, needed money. Kingdom Holding rolled over the entirety of its stake, which was valued at $1.9 billion, and was set to become one of the largest outside shareholders in Musk's Twitter.

—

>>> WITH THE SALE of Twitter all but certain, Agrawal began having private conversations with his top executives about whether they would work

for Musk. He asked them to set aside the way he'd been treated by the billionaire.

True to his engineering background, he encouraged them to gather information about the new owner before making up their minds. "Just because this is the way it's gone for me, doesn't mean it will go this way for you," Agrawal said. "Get the data to make the right decision."

That data would, of course, include personal calculations about money. Many of Twitter's top executives, including Agrawal, had contracts with change-of-control clauses. That meant if there was an acquisition or new leadership brought in to run Twitter, they would receive lucrative exit packages that cashed out grants of stock they would otherwise have earned out over years of service. The payout was intended to erase biases against a buyer, keeping executives open-minded about an acquisition without fear of losing their salaries. If Musk wanted them to stay, he would need to extend new job offers to them. But the exit packages gave many of the leaders, who figured Musk would fire them, a comfortable financial cushion. Agrawal would earn his salary and unvested stock options, giving him a golden parachute of more than $57 million. Segal would earn more than $44 million and Gadde around $20 million.

But for workers outside the executive suite, a soft landing wasn't guaranteed. Some decided they would work for such an erratic owner only if Musk offered improved compensation. Others, realizing their bosses would likely vanish, dreamed of climbing the ladder. Workers panicked about the impending deal, fearing layoffs or cuts to Twitter's cushy benefits. Executives tried to reassure them by pointing out Musk's contract required him to leave pay and benefits unchanged for a year after buying the company, but the promises rang hollow. Employees knew Musk didn't always honor his agreements.

The executives' exit packages, which were publicly disclosed in the company's financial filings, drove a wedge between the company's leadership and its rank and file. The money was more than any regular worker dreamed of making in a lifetime. They couldn't believe Agrawal would be paid so much for spending less than a year in the top role. Twitter had supposedly been a place where executives cared about what happened to their employees. Now they were on their own.

>>> WITH THE DEAL BACK ON, Twitter executives were called in to a series of whirlwind meetings to provide some semblance of a transition before October 28, the court-mandated closing date. Most acquisitions included a coordinated transition period, with both parties hammering out the details over months. But this was unlike any takeover. In mid-October, Musk and Gracias met virtually with two finance leaders: Julianna Hayes, a vice president on Twitter's finance team, and Robert Kaiden, Twitter's chief accounting officer.

Kaiden was not a typical Twitter employee. He had spent almost twenty-seven years as an accountant at Deloitte. He joined Twitter in 2015, and stood out among the hoodied crowd with his khakis and blue oxford shirts. Kaiden had developed a reputation as a hard-ass who scrutinized the company's financials to the point of driving his counterparts in other departments mad.

Kaiden and Hayes were meant to give Musk an overview of the company's finances, but Twitter's lawyers still believed that Musk might abandon the deal and insisted no one give him insider information. It quickly became apparent to the execs that Musk lacked basic knowledge of the merger agreement he had signed.

Hayes and Kaiden brought up the fact that on November 1—after Musk was supposed to take over—many employees were set to receive their vest, a bonus grant of company stock. But because Twitter would no longer be a public company, the workers would be entitled to cash bonuses with each share they held equivalent to a $54.20 payout. It would cost Twitter about $200 million.

Musk, however, was confused and asked why he couldn't reissue new stock in his soon-to-be private company. That simply wasn't in the agreement, Kaiden told him, much to the annoyance of the billionaire who was now on the hook for a massive sum.

"No, you agreed to that," Gracias noted.

"Interesting," Musk responded.

>>> ON WEDNESDAY, OCTOBER 26, Twitter headquarters was buzzing. The Friday court-ordered deadline for the deal to close was looming, and employees still knew little about what to expect.

One cluster of employees huddled around a table in the cafeteria at lunchtime, swapping rumors about Musk and what he might do to their beloved company. While they openly debated which executives might be on the chopping block, one woman's face dropped. "*It's Ned,*" she hissed at her coworkers.

The clean-cut chief financial officer had walked up behind them and was quietly eavesdropping on the discussion. He flashed a bright smile, trying to put the workers at ease. Despite the gloom in the building, Segal appeared buoyant. After months of anxiously waiting, it was time for him to get to work on the transaction and hammer out all the financial details.

"Ned, what's going to happen?" one of the workers asked.

"I think we're going to be fine," Segal said. "We're in good hands."

The assembled group of employees was taken aback at Segal's apparent confidence. Surely he couldn't be referring to Musk? But the finance chief seemed so assured and relaxed. *He must know something we don't*, the employees thought. It was one of the few moments when they felt like maybe, just maybe, Musk wouldn't be so bad.

Segal chatted for a few more minutes, cracking jokes with his staff. They had held together through months of uncertainty and pain, continuing to put the company first while other coworkers resigned or threw up their hands, and Segal seemed eager to comfort them. Then he excused himself. Musk would arrive within hours. He needed to prepare.

26 > Let That Sink In

Musk arrived at Twitter on the afternoon of Wednesday, October 26, to tour his coming conquest with a made-for-Twitter stunt.

Grinning widely, Musk walked through the glass doors of one of Twitter's lobbies, carrying a white porcelain sink as if it were a stack of pizza boxes. He let out a guttural chuckle as he walked by the empty front desk.

"You can't help but let that sink in," he said to no one in particular.

He later tweeted a video of his staged entrance with the same corny phrase: Let That Sink In! The groan-worthy pun was a symbol of the changing of the guard. Agrawal had used his Twitter account sparingly. But Musk was *of* Twitter. He thrived on the dopamine rushes he got when his tweets went viral and he knew the attention-grabbing phrases that would pump up a post.

The overuse of the phrase online had generated a meme of a sink waiting at the doorstep of a house, but "Let that sink in" had become an especially hackneyed refrain on Twitter, used by angry liberals during Trump's presidency. Slapping "let that sink in" at the end of a tweet about Trump's latest outrage was a gimmick to get retweets.

Musk himself had joked about the Twitterism earlier that year. "I'm dressing as a sink on Halloween, as they will have no choice but to let me in," he tweeted the previous June, attracting 150,000 likes for the post.

In a tight black T-shirt and jeans, with a silver chain dangling around his neck, Musk seemed to be striving for an extremely online meme lord cool factor. Some employees recognized the Twitter-speak and saw it as confirmation that Musk viewed the acquisition as an expensive joke. Others grew anxious. Musk was finally in the building. The deal was real.

Musk's sink gag played out in the lobby of one of Twitter's smaller office buildings, a structure erected next to the company's primary Art Deco high-rise in the 1970s that employees referred to by its address, "1 Tenth." Agrawal had planned to welcome Musk to 1 Tenth to keep his arrival from causing too much commotion among employees, and to give the billionaire a celebrity welcome. The side building held a suite of gleaming, freshly remodeled conference rooms decked out with giant hashtags, selfie walls, and bird statues.

The Twitter chief waited patiently while Musk filmed his grand entrance, then escorted him up the elevator to the second floor.

Earlier that day, Leslie Berland had flown to San Francisco to welcome Musk personally. On her way to the office, she typed out a quick email to employees.

"As you'll soon see or hear, Elon is in the SF office this week meeting with folks, walking the halls, and continuing to dive in on the important work you all do," Berland wrote. "If you're in SF and see him around, say hi! For everyone else, this is just the beginning of many meetings and conversations with Elon and you'll all hear directly from him on Friday." She had been texting with Musk and was attempting to once again play the CEO-whisperer role she had held with Dorsey, translating the boss's edicts to his deputies.

But Musk's plans seemed to change by the hour. Even though he had an assistant, a doe-eyed woman from the Boring Company named Jehn Balajadia, the billionaire controlled his own calendar. Twitter's leaders couldn't discern if he had any plans for his first day at Twitter, beyond filming his elaborate inside joke.

Musk allowed Berland to ferry him through Twitter's primary building on a tour. Over the years, plenty of celebrities had visited Twitter. Employees usually tried to keep a respectful distance, occasionally slinking forward to ask for a selfie. But Musk attracted unprecedented gawking. Streams of anxious employees trailed him from a distance like a school of curious fish in the wake of a shark. Some held their phones in the air to take photos of their prospective new owner as if he were a rock star at a concert. No one came up to greet him.

Some in leadership believed Berland's presence might make Musk's arrival more palatable for employees who were clearly unnerved by the uncertainty. But the workers didn't know how to act. Musk remained calm and personable as Berland led him into Perch, a coffee shop on the tenth floor where people often congregated over free espressos or cold brew served by baristas. Musk ordered a hot coffee and fidgeted with his hands as the chief marketing officer tried to motion people over, with little success. Most people were content to stay away, using their phones as shields through which they could peer at their new overlord.

Esther Crawford saw Musk's jaunt through the office as an opportunity. She had sold Squad, her group video chat start-up, to Twitter in 2020 and was focused on wooing influencers and creators to Twitter. While apps like Snapchat, Instagram, and TikTok had worked hard to attract influencers, sometimes paying them to post, Twitter had lagged behind. Its users were politicos, comedians, and journalists, who didn't necessarily convince their fans to spend money. Crawford, a product manager who oversaw some of the company's forays into new revenue streams, like payments, tried to change that.

She watched as Musk went up to the counter to get his coffee and told a colleague that it was strange no one was going up to him. "Should we go say hi?" Crawford asked her friend. "Or will it look too thirsty?"

They decided to approach. Crawford introduced herself and started listing off the products she oversaw at Twitter. When she mentioned payments, Musk lit up. Payments were essential to his vision, he told her. He had laid out a road map at X.com to make it easy to exchange money online, and with the purchase of Twitter, he hoped to finally complete it.

"Email me tonight," he told Crawford. "I want to meet with you tomorrow."

After their interaction, a white-haired man with a genteel New Orleans accent approached Crawford. He introduced himself as Walter Isaacson, the biographer of Leonardo da Vinci, Benjamin Franklin, and Henry Kissinger. In the tech world, he was best known for his 656-page tome on Steve Jobs, which affirmed him as an author who sought to explain and sometimes lionize the changemakers of the human race. Isaacson was shadowing Musk for his next project, an authorized biography of the genius behind Tesla and SpaceX.

Isaacson asked Crawford for her contact information and shot her a toothy grin. "I can tell you'll be important," he said.

Crawford eventually walked away, but her interaction broke the ice for everyone else. Dozens of employees soon formed a semicircle around Musk, taking selfies and asking questions as he leaned against a counter. He rubbed his coffee mug as he spoke openly with his soon-to-be employees, riffing on his expectations for Twitter. They listened intently, looking for ways to impress him.

"Often I open Twitter and people will be saying things, and it makes me feel sad," he said. "And I don't want to feel sad. People shouldn't feel sad when they come here and I feel sad."

The employees nodded along. One female worker mustered up some courage to ask what was really on people's minds. "Sorry I have one more question," she said. "I know we're all super excited to meet you, but I think the real question everyone is thinking is: Are you going to fire 75 percent of us?"

The question elicited a fit of nervous laughter. The employees had all read a *Washington Post* report from a week earlier that said Musk had told prospective investors that he was planning to cut three-fourths of the company. The billionaire raised his left eyebrow and gripped his mug tightly.

"You know, I'm not sure where that number actually came from because, uh, no," Musk replied, staring directly at the worker. Some in the crowd smiled as if a weight had been lifted off their shoulders.

"Spread the word," Berland said, nodding. "Spread the word."

—

>>> BY THE AFTERNOON, Musk's lieutenants, among them Birchall and Gracias, had set up camp in the 1 Tenth conference rooms. In the coming days, they would be joined by a small core of advisers and members of his inner circle, including Balajadia and Steve Davis of the Boring Company; the investors David Sacks and Jason Calacanis; Sriram Krishnan, a venture capitalist with Andreessen Horowitz who had once worked at Twitter; and even Musk's infant son, X, who was followed around by his nanny and own security detail.

During the tense summer of litigation, there had been virtually no contact between the two sides besides the back-and-forth between the lawyers. But Musk's posse had demands. They wanted to know which employees were in charge of what operations, and who excelled at their jobs. It was clear the men were preparing for job cuts, but Twitter executives were wary of giving them sensitive employee and financial information in case the deal collapsed at the eleventh hour.

On the Twitter side, executives were also talking about cuts. Twitter's transition team was holed up in a suite of offices reserved for Agrawal's staff. Many figured they wouldn't stay for long but wanted to see the transition through and attempt to protect their teams from the coming layoffs and preserve the platform they had come to love.

Others wanted out. Roth went to Gadde and begged for a separation agreement that would allow him to quit with severance when Musk took over.

"I won't get the exit you'll have, and I'm worried," Roth confided in his boss. Musk would surely get rid of him anyway, he said, and he wanted to avoid the massive wave of online harassment that he assumed would follow when Musk sent him to the guillotine. Gadde refused, saying that it would be unfair to other employees if Roth was given an escape hatch.

Kathleen Pacini, one of Twitter's human resources executives, agreed to take on the task of keeping track in her head of who would leave and who would pick up their work when they were gone. She was not to write down the succession plan, in case Musk got ahold of it. But she was the go-to person for anyone who wanted to know who they would report to after their boss vanished.

As Musk froze out much of Twitter's C-suite from meetings or any of his plans, he kept Berland close, using her input to decide which employees he should meet. He was immediately interested in Twitter Blue, a small part of the company's business that allowed die-hard users to pay $4.99 a month for premium features, like the ability to edit tweets or change the appearance of the Twitter icon on their phones. To Musk, this represented an untapped financial opportunity. If people relied on the site as much as he did, then of course they would pay for it. Berland set up a meeting for him with Tony Haile, a senior product director who oversaw Blue.

Haile was exasperated and stressed. He had prepared a presentation for

Musk that wrapped up all of Twitter's experimental bets to find new revenue, including its newsletter service, content creator initiatives, and Twitter Blue. He'd been told to brief Sacks on it, but then one of his employees, Crawford, jumped the gun and went straight to Musk. Haile was stuck in a tough position: instead of meeting with Sacks, he now had to try to walk Musk back from Crawford's pitch and get him to see the bigger picture.

Berland also suggested that Musk meet with Gadde, an idea that initially seemed preposterous to the billionaire. She was the enemy, the leader of a legal team that had ensnared him in the unbreakable deal to buy the company. While he had given up on escaping the deal, Musk still held on to his conviction that Twitter's executives were cheats who had wronged him and undermined his legend as an entrepreneurial genius. He also seemed to grasp that he had made Gadde's life hell by targeting her on Twitter and he couldn't understand why she might engage with him.

"She wants to meet with me?" Musk asked incredulously.

Of course, Berland replied. She told him that these people loved Twitter and wanted to see it succeed.

Musk agreed, and his assistant penciled in a thirty-minute meeting toward the end of his day.

Gadde steeled herself before walking into the conference room. It was her first time meeting Musk in the flesh. After his tweets had directed a hateful mob against her six months ago, she remained wary.

Musk, who preferred working late into the night, had asked Gadde to meet him at 6:00 p.m. But as other meetings about the transition dragged on, he kept postponing the sit-down with his nemesis. Finally, after 8:00 p.m., he was ready.

She strode past baby X, who was puttering about the hallway with his nanny in tow, and took a seat at the long wooden table, knowing she wouldn't have long to transfer more than a decade's worth of legal lessons and concerns to Musk.

The man sitting across from her did not seem like the troll behind @ElonMusk. Musk was reserved, his voice sometimes barely fluttering to an audible level. He was exhausted from his speed run through the office.

"I'm fried," Musk admitted apologetically. "It's been a long day."

With little time for pleasantries, Gadde launched into her prepared agenda.

Twitter faced an abundance of tense legal challenges, she explained. The FTC remained on high alert about the company's privacy practices after negotiating its settlement and tighter restrictions on Twitter in May. The European Union was on the brink of implementing its Digital Services Act, a landmark piece of legislation that imposed new moderation responsibilities on Twitter and other major internet platforms. The overhaul Twitter had undertaken in order to comply with the new European legislation had not been completed, and it would be Musk's obligation to see it through if he did not want to face major financial penalties.

Musk listened, but it was unclear if what she said registered with him. She knew he was a person who didn't seem particularly moved by laws and regulations, but still she was shocked at his nonchalance. Failure to comply with the laws could potentially even lead to the European bloc banning Twitter on the continent.

As the conversation moved on, Gadde continued to focus on issues outside the U.S. She wanted Musk to understand that Twitter was a global service. While news outlets tended to focus on content moderation issues in the United States, there were far greater and immediate concerns about authoritarian governments abroad, and Musk's business relationships presented a number of potential conflicts. He ran a car business that relied on manufacturing and sales in China, she pointed out. *What would happen if the Chinese government put pressure on you to censor something it didn't like on the platform and threatened Tesla?* She gamed out a similar scenario in India, a market that Tesla had courted while Gadde spearheaded a lawsuit against its government.

Musk nodded. "Interesting, interesting," he said. "I haven't really figured that out yet."

To Gadde, it was a stunning answer for a man who had planted his flag on the protection of free speech. How could he not have thought about the implications of his business relationships and the tough global content moderation calls that the company regularly had to make?

Sensing they weren't getting anywhere, Gadde decided to play to his ego. She wanted to see Twitter continue to succeed, and she had seen the company's growth stall. Perhaps, she suggested, Musk could use Starlink to

Jack Dorsey (right) and fellow Twitter cofounder Biz Stone on the rooftop of their company's San Francisco office in 2008.

Elon Musk at age twenty-nine, before he was usurped as the chief executive of X.com, the precursor to PayPal.

Jack Dorsey starts a video call with Elon Musk during Twitter's 2020 #OneTeam event. Later, Twitter's chief executive asks him if he has any suggestions for the company.

Vijaya Gadde, who served as Twitter's chief legal officer before Musk's takeover, testifies before a U.S. House of Representatives committee on February 8, 2023.

Leslie Berland, Twitter's chief marketing officer, helped translate Dorsey's whims to his employees.

Yoel Roth once led Twitter's trust and safety team but resigned following Musk's takeover.

Twitter's chief financial officer Ned Segal, who oversaw its earnings and relationships with shareholders, poses in the company café.

Gadde, Segal, and Dorsey participate in a bike race at the company's Boulder office.

Dorsey addresses Twitter employees from an undisclosed tropical locale to announce his resignation as chief executive and the appointment of Parag Agrawal in November 2021.

Twitter's chief executive Parag Agrawal walks into the Allen & Company Sun Valley Conference with his wife, Vineeta, on July 7, 2022.

Twitter chairman Bret Taylor negotiated the sale of the company to Elon Musk.

Twitter board member and Silver Lake managing partner Egon Durban invested in Twitter in 2020, rescuing Dorsey from an activist investor who wanted to push him out.

Musk and then-girlfriend Claire Boucher, better known as the singer Grimes, pose together during a 2018 event at SpaceX headquarters in Hawthorne, California.

Elon Musk and Steve Davis, the chief executive of the Boring Company, at an event for the tunneling start-up in May 2018.

Jared Birchall, a former senior vice president at Morgan Stanley and the head of Elon Musk's family office, is seen as the billionaire's personal fixer.

Alex Spiro, an attorney for Musk, arrives at a San Francisco courthouse in January 2023 to defend his client in a Tesla shareholder lawsuit.

Michael Grimes, head of global technology investment banking at Morgan Stanley, arranged financing for Musk's Twitter acquisition.

Musk carries a sink into Twitter headquarters on October 26, 2022, the day before he completed his acquisition of the company.

Esther Crawford, a director of product management, rose quickly in the aftermath of Musk's takeover.

After pulling long shifts in Twitter's San Francisco office trying to relaunch Twitter Blue, Crawford had another employee stage a photo of her in a sleeping bag. The image became an early, divisive symbol of Musk's new ownership of the company.

Musk and his mother, Maye Musk, attend Heidi Klum's Halloween party in New York City on October 31, 2022.

Musk and his son X Æ A-12 meet a Formula 1 driver during an event in Miami in May 2023. The boy, known better as X, became a regular fixture by his father's side in the days following the Twitter takeover.

Musk and Davis set up beds inside Twitter's San Francisco headquarters to encourage employees to work around the clock. The eighth-floor setup became known as the "Twitter Hotel" and prompted an investigation by city building inspectors.

Following his takeover, Musk displayed a collage celebrating important free speech moments in human history near the Twitter cafeteria. Those moments included the publication of John Milton's *Areopagitica*, a treatise on free expression; the writing of the U.S. Bill of Rights; the 1964 Free Speech Movement at the University of California, Berkeley; and Musk carrying a sink into Twitter's headquarters.

In July 2023, as Twitter was transformed into X, the company's new logo was projected in a largely empty cafeteria that still had a giant bird logo.

Musk erected a large X sign on top of the company's San Francisco headquarters for a few days in July 2023.

connect places with little internet coverage—like parts of India—to bring more potential Twitter users online. True free speech meant democratizing Twitter's platform not just in the U.S. but around the world. The billionaire nodded disinterestedly.

As the conversation moved on, Musk became laser-focused on one of Gadde's decisions in particular: banning Trump. *Why did she do it? And how was the ban implemented?*

The questions caught Gadde off guard. The Trump decision had been aired out over and over again during the past year, starting with Dorsey's demand that Twitter publish its rationale for removing a sitting president. The matter had been relitigated extensively in congressional hearings and by the January 6th Committee, which subpoenaed several former employees to speak about the risks that Trump's tweets presented. If Musk was curious about it, all he had to do was a quick Google search.

But he insisted that she run through it again.

"We were very public about this decision," Gadde reminded him. "Jack was very involved and ultimately approved everything."

It wasn't easy for anyone, Gadde continued. Removing a world leader from the global discussion platform wasn't something she had taken lightly. Violence was always the red line, Gadde told Musk. If a world leader used the platform to incite violence, Twitter would treat them like any other user and ban them.

Musk brushed off her explanation, moving on to a more recent content moderation decision that he had fixated on for months: Why did Twitter ban the Babylon Bee?

To Gadde, reviewing individual content decisions—particularly ones that had been picked over in the press—was a waste of time. She wanted Musk to see the bigger picture: the responsibilities and power that he would now wield as the owner of Twitter, and the pressures and pitfalls that came with running it. But he seemed more interested in the American culture wars.

Gadde mentioned Project Saturn, Agrawal's pet project. It could give Musk a framework to allow the kinds of content he was interested in, she noted. There was a structured, sensible way to accomplish his goals.

The meeting ended after nearly thirty minutes and Gadde left the Twitter office—never to return again.

It was clear. Musk had not bought Twitter to be a responsible steward and guide one of the world's most heavily used websites and forums for human communication. He had bought it as an object of personal obsession and was going to shape it to his whims. Musk had come to love Twitter, and he believed that the people who had run it had led it astray.

He was going to make them pay.

27 > Trick or Tweet

At 5:11 a.m. on Thursday, October 27, Bret Taylor's phone buzzed. In his inbox was a letter from Alex Spiro, who instructed that "in anticipation of the imminent completion of the merger" that day, Taylor and Twitter's executives should immediately freeze payments to outside vendors and contractors. Having loaded Twitter with debt, Musk was already attempting to manage his costs.

Taylor sighed, and forwarded the message on to Gadde, Edgett, and Savitt four minutes later. The breakneck speed of the closing had worn him down, and he still needed to freshen up before his final meeting with the Twitter board at 7:00 a.m.

Spiro's message continued to reverberate in the inboxes of Twitter's leadership. Edgett passed it on to Segal and Kaiden with an addendum. "FYI. The latter half of this letter asks us to revoke all payment authority internally as of the closing. Let's discuss how best to make sure that happens without anything breaking," Edgett wrote.

Again, Musk appeared to be overreaching. Until he owned the company, Twitter's employees were free to ignore Musk. He clearly believed differently.

At 7:00 a.m., Taylor joined a video call with several of Twitter's other directors to close out their final bits of business. Once the call ended, the directors planned to submit resignation letters—they would no longer have a role at the private company. But several of the members had already checked out and didn't attend.

In a call days earlier, Edgett had briefed Taylor and Pichette on a proposal from Savitt, in which the outside lawyer asked that Twitter consider paying his firm $95 million for their work. White-shoe law firms like Wachtell sometimes charged what they called "success fees," lavish honorariums they were awarded only if they were victorious in their cases. The success fees vastly exceeded their already robust hourly fees and could enrich the partners for years to come. By securing the $54.20 price per share from Musk, Wachtell argued it had increased the value of Twitter by billions of dollars. After discussing the request with Edgett, Taylor and Pichette had settled on bonuses that took the total billing for Wachtell's few months of work to $90 million. In average human terms the figure was an extravagant amount, but as a percentage of the deal, it was around 0.2 percent of Twitter's price tag.

At 7:29 a.m., as the board ticked through its agenda items, Edgett forwarded an email to the directors that included a chart of the proposed payment.

Lane Fox responded to Edgett almost instantaneously:

O

My

Freaking

God

She broke her sentence into a haiku-like format for extra emphasis. Lane Fox never dreamed the total would be so extravagant.

But the board knew there was no time to waste. If they didn't pay Wachtell before the deal closed, it was unlikely that Musk would pay them at all. And didn't the lawyers deserve a reward for trapping the rule-flouting executive into the transaction?

—

>>> SHEEN AUSTIN WAS at his South Bay home sleeping when his phone buzzed just after 4:00 a.m. that Thursday morning. He had worked at Tesla for eight years as an infrastructure engineer, maintaining the servers and services that the electric automaker depended on, and had become accustomed to receiving messages from work at all hours. He was, after all, in service of Tesla—and therefore, Musk—and their mission of electrifying the

world's automobiles and slowing climate change. It was a mission that was bigger than him or any other individual employee, and these workers threw their convictions behind whatever Musk asked of them.

This request, however, was unlike any he had seen before. The message said that he and some of his colleagues were expected to report to Twitter's San Francisco headquarters that morning.

The Tesla employees who received the pings didn't know whether to take them seriously. They knew their boss planned to buy Twitter, as his exploits had saturated the news for months, but many felt that it didn't concern them. They worked for Tesla.

But it was unavoidable. The Tesla workers messaged one another, asking if they'd received similar commands, and grumbled as they stepped out of bed to ready themselves for an unfamiliar commute. Austin made the hour-long drive north along the 101 freeway up the San Francisco Peninsula and toward the fog of the city.

Born in India, Austin, a round-faced man who often flashed a smile from behind a thick beard, had been living in Toronto in 2013 when one of Tesla's recruiters called to ask if he'd be interested in relocating to California to work at an up-and-coming automaker. Tesla had just launched its Model S sedan the year before. Austin jumped at the opportunity.

Initially, Austin was confused by the command. He was bright and well educated, but as a rule, he and other Tesla loyalists avoided most news about the boss. The press was slanted against Musk, they believed, and the day-to-day media drumbeat about him was simply noise that got in the way of the mission. Inside Tesla, workers often joked that their next marching orders or deadlines could come from one of the billionaire's tweets. Some of them checked Musk's account for clues and found a message posted at 6:08 a.m. that began "Dear Twitter Advertisers."

Sarah Personette, Twitter's chief customer officer, had asked Musk to say something. As the head of the company's ad sales organization, Personette knew that major brands were deeply worried about the takeover and yearning for reassurances that Musk would protect them from illicit content. Advertisers, the company's main source of revenue, were jittery at the prospects of his ownership, fearing a decrease in content moderation and the harboring of hate and abuse, Personette warned him. Instead of his usual flippant jokes, Musk had posted a grown-up message. Twitter would not be a "free-for-all hellscape" but rather a place where "a wide range of beliefs can be debated," Musk tweeted.

"That is why I bought Twitter," he wrote. "I didn't do it because it would be easy. I didn't do it to make more money. I did it to try to help humanity, whom I love. And I do so with humility, recognizing failure in pursuing this goal, despite our best efforts, is a very real possibility."

Austin was among the first to arrive that morning and was soon joined by a few dozen of Tesla and SpaceX's most trusted engineers. Among them was Ross Nordeen, a young supercomputing engineer, and Musk's two cousins: James Musk, a Tesla autopilot engineer, and his ginger younger brother, Andrew, who worked at Neuralink.

Within an hour of arriving at Twitter, the engineers began to realize there wasn't much of anything for them to do. They had not been informed of any clear transition plan, and it was unclear if Musk even had one. While most of them milled about in Twitter's side building at 1 Tenth, some of the engineers wandered both offices, marveling at the bird-themed decor, free snacks, and preparations for a company-wide Halloween party that would be unthinkable at the spartan offices of Tesla and SpaceX.

The invaders were easy to spot by leery-eyed tweeps, who sized them up and down. They didn't have the blue badges carried by employees, and instead held red visitor or green vendor keycards. Most of them were men. Some of them sported Tesla- or SpaceX-stamped apparel. To the Twitter employees, these outsiders could not be trusted, and they came up with a nickname for the intruders: "the goons."

—

>>> AT AROUND 9:00 A.M., Agrawal summoned his leadership team into one of the large glass-doored conference rooms that lined the suite of offices on the seventh floor of Twitter's San Francisco headquarters. After months of tension and worry, there was a grim clarity in the air—Musk was finally completing the acquisition.

Twitter's top-ranking employees crammed into the room. Agrawal's deputies were there, as well as vice presidents from finance, product, human resources, and sales. Even more executives dialed in on video conference from New York and around the globe, their faces tiling the screen at the end of the room.

It was clear to everyone there that it was Agrawal's last meeting. He sat at the conference room table, Segal by his side. The mood was somber—everyone in the room understood that many among them might soon be swept away by Musk's tsunami.

No one was more likely to be fired than Agrawal. For months, Musk had made clear his disdain for Twitter's chief executive in barbed tweets, curt text messages, and explosive video calls. Agrawal had taken most of Musk's outbursts quietly, advised by Twitter's battalion of lawyers not to argue with the billionaire or speak about the deal to employees—or even executives—because anything he said might leak to the media.

After months of near-silence to the wider group, Agrawal spoke, remaining calm and analytical. "We might close today," he announced. The court-imposed deadline for Musk to complete the transaction was the next day, Friday, but it seemed he could get it done a day early. Agrawal told the executives he was proud of what they'd accomplished.

There was no agenda, he told everyone, and opened the discussion. "What's going to happen now?" one executive in attendance asked. Segal tried to explain how the closing would work but said candidly that no one could be sure. After all, the man on the other side of the transaction was unpredictable.

There was plenty of work left to do to finalize the deal, but Agrawal allowed Twitter's leaders to riff, share, and ask anything they wanted. They had never had a meeting quite like it before. Sales executives wanted to know what they should tell advertisers. HR leaders wanted to know what they could tell employees, and when they were allowed to share any information.

Then one of the employees in the room broached the question that everyone was thinking but no one dared say: "What's going to happen to you guys?"

Segal repeated the same line he'd told employees before. "I haven't talked to him," he said. "I'll remain open until I do." Agrawal nodded along.

"Each of you needs to make your own decision," Agrawal said.

The executives had endless questions, but their leaders had few answers.

Segal could sense their frustration and, after months of facing unanswerable questions, he cracked. Fighting to keep his composure, he told them he didn't know what was coming next. "People remember how you handle yourself when it's hard, not when it's easy," he said, his voice choking with emotion. He tried to express the weight of the responsibility all of them had—to the company and to each other—to see the sale through.

Several of the executives in the room were startled to see Segal, normally polished, perky, and on message, get emotional. As the meeting ended, some of them embraced each other, while others hung back to say their goodbyes to their bosses.

—

>>> GRACIAS, Musk's de facto finance shepherd, had told Twitter's team on Wednesday that he had all the money in place to close the deal. It was a pleasant surprise to Segal, who, upon learning that Gracias had the funds, nudged the board. They should move up the close, the chief financial officer suggested. Finishing the transaction early would leave Musk one less day to back out. The board's bellwether all along had been deal certainty, and moving quickly to collect Musk's cash before another unforeseen crisis derailed the whole thing was just another way Twitter could deliver certainty to its shareholders. While Twitter's leadership had no idea where some of Musk's money was coming from—new, undisclosed investors had joined Musk's take-private effort—they were more than willing to take his $44 billion.

Members of Twitter's finance teams had adopted a gallows humor approach to the deal and made a joke of trying to track Musk's money. When he sold new tranches of Tesla stock and filed the required public disclosures of the transactions, they tallied up his funds, trying to figure out if Musk had enough cash on hand to buy their company. At one point, Musk's lawyers also accidentally sent Twitter's finance team a full spreadsheet of all the people and investment firms from which they solicited money. That screwup was immediately followed by a legal threat to the Twitter recipients to delete the email and its contents.

Of course, there was no way of knowing where the billionaire kept all his money or how he planned to use it. Twitter employees debated whether

Musk was sitting on a secret stash of cryptocurrency or had obtained fresh margin loans using his private SpaceX shares as collateral. *The Wall Street Journal* later reported that Musk borrowed $1 billion from SpaceX that October, paying the money back with interest the following month.

To Twitter, it didn't really matter where Musk's money came from—so long as he paid. But given how many agreements Musk had already tried to break, nothing was certain. There was a world where the richest man on earth, they believed, could test the court-appointed deadline by saying he simply did not have the available funds to do the deal.

In a normal transaction, the buyer would be transparent with the seller about where his funds were flowing from. But Musk, in what Twitter executives believed was an effort to protect his investors from scrutiny, had dumped all the funds into a single account so that Twitter couldn't trace their origins.

On the call with Segal and Twitter's finance executives and lawyers, Gracias changed his tune. His boss was actually short, Gracias explained. Musk was missing more than $400 million, and Gracias demanded that Twitter wire money from its own coffers to Musk so that the deal could close. Segal was dumbfounded. Kaiden and the half dozen other people who listened in to the conversation couldn't believe what they were hearing.

Gracias knew that Twitter had more than $2.5 billion in cash on its balance sheet. And with Musk still short on financing—some of the expected funds had not arrived at that moment—the private equity buff tried to pull a mafioso move.

"You need to wire us the money," he growled. It wasn't clear to the listening Twitter executives what had changed since Wednesday, when Gracias said Musk had his funds ready. But Twitter had the cash on its balance sheet, and it would be Musk's soon enough anyway once he owned Twitter, Gracias reasoned. Why not simply use Twitter's money to shore up the remainder of the financing and make this easier for everyone?

But the Twitter executives didn't trust Gracias, given Musk's attempts to back out of the deal. What if they did wire the sum and Musk once again tried to call the whole thing off? "No one on this call has the authority to do that," Segal replied.

Since Musk barged in with his offer in April, Twitter had stuck to its "just say yes" defense. Segal knew better than to refuse Gracias's request, potentially handing Musk a reason to blow up the deal during its final hours. He wouldn't say no, but he couldn't agree either. So he kept his answer open-ended but truthful, focusing on the fact that only the board had the authority to consider and authorize a last-minute transfer of funds.

"You're not going to wire me the fucking money?" Gracias said, becoming more impatient. "Are you saying no to Elon Musk?"

"I'm just saying that no one on this call can make the decision to send the money," Segal responded. He was the highest-ranking Twitter executive in attendance, and no board members were dialed in. If there was any possibility of Segal staying to work for Musk, it evaporated at that moment. At one point, Korman, who was also on the call, asked the Skadden lawyers to keep Gracias in check.

As the call ended, Twitter's finance executives gossiped amongst themselves about the demand, which they found inappropriate. Musk had agreed to find the funds, independent of Twitter, at the price he set himself. They couldn't move Twitter's cash around in some corporate shell game just to appease him. "This feels a bit fucked up," one member of Twitter's C-suite told Hayes, the finance vice president. Others thought it was potentially fraudulent and criminal.

An hour or so later, Gracias called back. He'd found the money from somewhere else, he said. Some Twitter executives speculated that they had gone back to Qatar's sovereign wealth fund, but could never confirm it. The deal could proceed.

>>> THE FINAL SCRAMBLE to close and the strange backroom calls never trickled out to Twitter's rank and file. But throughout the building, the atmosphere was tense. Some employees trickled into the office early on Thursday, rubbernecking to catch a glimpse of the billionaire and witness history. Others had already planned to be in the office that day for a Halloween party, which had long been scheduled for that afternoon.

Twitter and its employees took the holiday seriously, dubbing the party, which took place in offices around the world, "Trick or Tweet." In London, staffers on the corporate events team had blanketed the office in piles of decorative pumpkins and black paper cutouts of bats, while the New York office featured bales of hay and fake cobwebs that draped over giant hashtags and @-sign statues. In Mexico City, workers prepped for a spooky art session, in which employees would get individual canvases to paint jack-o'-lanterns. At noon in San Francisco, workers were adding the finishing touches, stringing up lights in the ninth-floor common area, organizing group costumes, and setting up tables that employees' children would visit to trick-or-treat. Everyone was encouraged to invite loved ones to the festivities.

But no number of pumpkin-flavored cocktails or pieces of candy could cut the tension. Workers hovered in communal areas, ignoring work, which seemed pointless considering the looming ownership change. People gossiped and compared notes as the San Francisco office became one large, rumor-swirling game of telephone. *What did you see yesterday? Who did you meet? What have you heard?*

Others tried to maintain normality. Across Market Street from Twitter's headquarters, some employees opened the doors to NeighborNest, a community center the company had funded. NeighborNest had been in the works since before the pandemic, when Twitter struck a deal with the local government to keep its offices in San Francisco in exchange for a tax break on payroll. Instead of decamping to the peninsula and building offices alongside Google and Facebook, Twitter agreed to stay on the ramshackle Market Street and contribute to its revitalization. As part of the plan to give back, the community center would offer technology courses to the neighbors, helping them catch up to the wave of techie newcomers sweeping into the city and driving up rents. But the entire endeavor had stalled during the COVID years.

On Thursday, NeighborNest finally reopened to the public. Twitter had bought up hundreds of laptops and partnered with local groups to bring in a crowd of recent immigrants from Central and South America for technology training and a laptop giveaway. A young mother of three cried with excitement when she received her computer, and the Twitter employees shelled the laptops out as quickly as possible, nervous that Musk would barge in and shut the whole thing down. They were painfully aware that NeighborNest's first event was also likely to be its last.

After the training was over, the employees drifted reluctantly back across the street to join the Halloween party. Their excitement at doing something good for the community quickly gave way to fear.

In the executive suite, some of Twitter's leadership wondered if they should include Musk and his henchmen to give them a taste of what the company was about. They mused about getting X, Musk's young son, a last-minute costume so he could be like one of the other kids and join in the festivities. Unsure how their future boss would react, they eventually decided against it. In late September, when Musk agreed to close the deal, Twitter had already ordered him a welcome box of swag that included a custom letterman jacket and other Twitter-branded goodies, running up a tab of $6,397. Getting presents for X, too, might overdo the welcome.

—

>>> MUSK'S GOONS CAME without costumes. They meandered through the halls as they awaited instructions, the glares of Twitter employees burning into the back of their necks. Some of them felt awkward invading Twitter, but they knew they served the whims of one man. In a brief meeting with the billionaire that morning, Austin and some of the other engineers received a clear directive from his boss.

"Make sure the site doesn't go down," the billionaire said. "Make sure no one does anything."

As the closing of the deal crept closer, Musk became increasingly paranoid, just as he had done in times of crisis at SpaceX and Tesla. All the Twitter employees hated him, he believed, and he'd seen some of them tweeting openly about their opposition to his ownership. Musk didn't expect a warm welcome from his new workers and had spun up an imaginary scenario in which a vigilante deleted some of Twitter's code or unleashed a cyberattack that took down the site and humiliated him.

Austin and several other Musk employees rounded up the Twitter executives they thought would be able to prevent a disgruntled employee from going rogue, including Lea Kissner, the chief information security officer; Carrie Fernandez, vice president of engineering; and Damien Kieran, chief privacy officer, to relay Musk's concerns. They demanded that Twitter implement a "code freeze," preventing any changes to Twitter's site or apps, effectively grinding half the company's work to a halt.

The Twitter executives pushed back. Even though his people were on the ground and in the Twitter office, Musk had not yet closed the deal. He was not the owner and had no right to order them around. Twitter's employees had been instructed not to follow any commands from Musk's team unless they received approval from Segal or Edgett.

"Elon's your boss," Kissner told the outsiders. "But he's not ours."

Besides, Kissner and Kieran had other work to do. They were responsible for submitting quarterly audits to the Federal Trade Commission that documented Twitter's compliance with the agency's ongoing oversight of the company's privacy program. The audits were grueling and deeply detailed, documenting each task Twitter had to do to preserve users' privacy and the specific employee who was responsible for making sure that task got done. The next report was due in two weeks, and Kissner and Kieran were both legally liable for its accuracy—if anything went wrong, the executives could face criminal charges.

Kissner called Roth and asked him to manage Musk's paranoia, summoning him to the second floor of 1 Tenth where Musk and his aides were gathered. Roth opened his laptop and showed Musk the @ElonMusk account in Twitter's moderation dashboard. Roth was one of the only employees who had back-end access to high-profile accounts like Musk's, he explained to the billionaire, a security measure that ensured that few employees could tamper with it. Musk seemed reassured.

Roth seized the opportunity to remind Musk that the Brazilian presidential runoff election was coming that Sunday, and the company would need to be on high alert for misinformation. Musk nodded. "Very risky," he said.

—

>>> AFTER HER LATE sit-down with Musk the night before, Gadde decided to work from home and avoid any possible mess that might await her at the office. She expected the deal might close that day, and Segal had confirmed her suspicions.

Most of the paperwork had been signed and cleared by lawyers, but

Twitter still waited for all the money to arrive. For a massive deal brokered by the world's leading banks and law firms, the process was extremely haphazard, Gadde told her associates. The sale of Twitter—one of the biggest moments in Silicon Valley history—had been reduced to a stochastic series of wire transfers.

Twitter's bankers at Goldman Sachs sat refreshing the screens of their web browsers. They were logged in to view a third-party administered account, which held Musk's payments in escrow until the total amount—$44 billion plus $2.5 billion in closing costs—was gathered in full. Gadde constantly called her bankers that afternoon, like a child on a family road trip. *Are we there yet?*

While she waited for Musk's payment, Gadde also scrambled to get cash out the door for Savitt and his attorneys at Wachtell. After the board signed off on the $90 million invoice that morning, it was her duty to make sure the lawyers actually got paid before Twitter changed hands. Just after noon, Twitter's accounting department approved the eye-popping wire transfer from the company's Citibank account. At 3:50 p.m., just ten minutes before the deal closed, the transfer was posted.

Once Musk's money arrived, the final step in the sale was Gadde's. She had already signed her name to the merger certificate and nodded her approval for it to be sent off, officially relinquishing control of Twitter to Musk. Then her lawyers shipped the freshly signed document to Delaware's Division of Corporations, the government agency that oversees the more than a million companies that claim the tiny state as their home.

Gadde sat back in her chair and let relief and grief wash over her in alternating waves. Her home office, lined with white bookshelves and artfully arranged plants, was strangely quiet after the frenetic sounds of her constantly buzzing phone died down.

She had sold Twitter.

—

>>> AS GADDE PREPARED to send off the merger certificate bearing her signature, the final step in closing the deal, Segal walked into Agrawal's office. It was time.

Agrawal had never gotten a chance to finish decorating the space, and a few pieces of unhung art leaned against the walls. The chief executive's standing desk overlooked several tall, narrow windows with a view of Market Street.

Segal held his cell phone in one hand, a video conference with Gadde and Twitter's external lawyers still playing in the background. The people on the call watched as he gestured for Agrawal to come with him. The deal was imminent. It was best that they leave the premises on their own terms before Musk made them do so. Agrawal nodded solemnly and hit send on an email explaining that the change-of-control terms in his contract had been met, and that he stood ready to discuss the future of the company with Musk. Then he gathered his things and walked out of his office for the last time.

Segal, still on the video call, framed his face with his iPhone as he walked through the office. He didn't stop to warn the employees who were still working about what was coming, although he had told a few trusted senior executives that he planned to go. He shoved open the security door and walked into the stairwell, Agrawal in tow. They descended from the ninth floor down into the parking garage, avoiding the television crews that were waiting on the street with cameras. Segal's Wi-Fi connection faltered and his video flickered in and out, giving the others on the call the impression that the CFO was about to disappear. He said an abrupt goodbye to Agrawal before jumping in his car, and, with the video still rolling, pulled out of the garage. Agrawal got into his car, too, and drove out into the bright October sunlight of San Francisco, just another software engineer heading for home.

28 > "The bird is freed"

Segal had exited the building with little fanfare. Among those on Segal's team who were unaware of his departure was Jon Chen, a vice president of corporate development.

A nine-year Twitter veteran, he lived in Los Angeles and had come north to headquarters only a few times in recent years. Typically the place had been empty, but with the Halloween party and the rumblings of the takeover, the office was busier than ever. People had come to witness history or drink their worries away.

Chen had arrived at San Francisco International Airport that abnormally clear morning on a one-way ticket, at Segal's request. He had no hotel reservation and didn't know how long he'd stay in the city, but the previous day he had logged in to a video call from his home to find Gracias, Birchall, Davis, and Sacks staring back at him.

Musk had tasked his friends, who became a de facto transition team, with scouting out Twitter employees best suited to bring about his vision. He believed that Twitter was a bloated company with too many managers. But he also thought there'd be a few motivated individuals outside of the untrustworthy C-suite who would leap at the opportunity to work for him. It was his transition team's job to locate them.

Chen, a gregarious former investment banker at Morgan Stanley, was one of the people on their list. Unlike some of his colleagues, he wasn't morally opposed to the idea of working for Musk. But he remained hesitant. To Chen and the other select Twitter employees who were invited to meet with Musk's goons, these conversations felt like auditions, in which they tap-danced for the outsiders in a desperate attempt to keep their jobs.

Chen was immersed in the financial discussions during the topsy-turvy negotiations with Musk and had steeled himself for antagonism from Musk's friends. But when he joined the call with them on Wednesday, they weren't adversarial at all. The men peppered him with questions about Twitter's strategy for mergers and acquisitions, and Chen found common ground with them. Gracias and Sacks were investors themselves, and Chen played up his investment banking background. He ended the hour-long call with a sigh of relief. *That felt . . . normal*, he thought to himself.

The meeting had gone well, Segal told him on a call later that Wednesday night. Musk's people had asked to meet Chen in person. He took the hour flight up to San Francisco on Thursday morning with little more than his laptop and a few changes of clothes. Without an assigned desk in the office, he found himself milling around with the rest of the Twitter and Tesla employees as he waited to be summoned to Musk's war room.

—

>>> SHORTLY AFTER SEGAL and Agrawal fled the building, Chen finally got the call and walked over to the conference rooms where Musk held court. Chen waited inside, as people filed in and out of Musk's room, and spotted some of the men who had interviewed him over video the day before. They shook hands and launched into more of the same discussion.

Chen sensed he had won them over. Davis and Birchall spoke about what Musk wanted as if they were his disciples, reverently laying out his plans to transform Twitter. Sensing opportunity, Chen riffed off their proclamations with his own hopes for what the billionaire could build. As a finance pro, he was not involved in product or engineering at Twitter and he knew his suggestions weren't exactly novel ideas. The way Musk's transition team hung on his words made him wonder how much actual thought they had given to how they would run the company.

"We should obviously be taking advantage of payments on the platform," he said. The goons' eyes lit up as if he had said the magic words.

"Maybe we could build for less appreciated communities that are already on Twitter," he later suggested.

"Like what?" Davis asked.

"Well, like gamers," Chen said matter-of-factly. "These are highly monetizable communities. They spend all types of money on in-app purchases. Why aren't we focused on that type of stuff?" Chen said. The men's faces melted.

"You have to tell Elon what you've been telling us," Davis said.

After the lengthy discussion, Musk strode to the room and sat down opposite Chen, much larger in real life than what the finance exec had expected. At first, Chen felt he was interviewing for his job, but he was now confident he'd be sticking around. Why else would the world's richest man be asking for his ideas?

Musk listened as Chen shared his vision for the "everything app," but the conversation was halting. Lawyers constantly bustled in and out with stacks of paperwork, which Chen strained to read. He had expected the closing to run up to the Friday deadline, but the flurry of lawyers, bankers, and documents led him to realize: *Oh shit, the deal is closing right now.*

Chen was the only Twitter employee in the room as Musk bought Twitter. Musk was nonchalant, barely glancing at the papers he autographed, as if he was signing a check for dinner at the end of the night. Watching the casualness with which Musk closed one of the most momentous deals of all time was jarring.

And then someone opened the door: "We are closed!" A shock wave of excitement rushed through the room and a smile crept across the face of Twitter's new owner.

The bankers high-fived and Musk laughed, his months-long, on-again, off-again pursuit of the world's town square finally completed with a few signatures and wire transfers. Chen sat across from his new boss, stunned.

"Do you guys need me to step out?" the Twitter vice president asked the room.

"No, no, no, you're good," Musk replied. Then he slammed his fist on the table and let out what could only be described as a battle cry.

"Fuck Zuck!" Musk shouted.

Chen couldn't fathom why Musk, in a moment of celebration, fixated on the founder of Facebook. Perhaps the Twitter deal was Musk's attempt to lay siege to Zuckerberg's social media empire, or maybe he had some long-standing score he wanted to settle. Whatever the case, Chen didn't feel that he could ask.

"THE BIRD IS FREED"

The conversation between Chen and Musk lasted a few more minutes. As the celebrations trailed off, the Twitter executive could hardly focus on what was being said. He exited the conference room in a daze and started his trek back to the main office building. In the hall, he walked by a group of Musk's Morgan Stanley bankers, who celebrated the completion of the deal in a frenzy, whooping and laughing. Chen said nothing. He continued out a set of doors that separated Musk's restricted area from where regular Twitter employees were working on an open floor.

The threshold marked a stark difference between worlds. Behind him was the sheer joy of corporate conquest. In front of him, people were crying. They were executive assistants who had served the likes of Agrawal, Segal, and Gadde. As Chen was in the conference room watching Musk close the deal, four of Twitter's top executives had been fired.

>>> BEFORE CLOSING THE DEAL, Musk had already given directions for his first order of business once he officially took over. Musk determined that Agrawal, Gadde, Segal, and Sean Edgett, the company's general counsel, would be fired "for cause."

The legal distinction insinuated that the executives had done something nefarious to harm the company, and it would potentially allow Musk to avoid paying the $120 million in exit packages he owed to the four leaders once Twitter changed hands and its shares were delisted from the New York Stock Exchange. Those two events were triggers in the executives' contracts, allowing them to collect salaries and stock bonuses that they otherwise would have earned over many years of service.

Musk wasn't going to let the payments happen. By terminating them "for cause" he was snipping their golden parachutes. It was a slap in the faces of the people who had forced him to buy the company, and he was going to make sure they wouldn't be rewarded for their toils. The billionaire had his people draft letters outlining the terms of the executives' firings, and they were sent minutes before the deal closed at 4:00 p.m. He also sought to have them marched out of the building by security, a symbolic gesture akin to a medieval warlord putting heads on spikes after his conquest.

While Musk and his advisers may have thought they were being slick with their corporate assassinations, most of the executives were prepared. Gadde knew her thirty-minute conversation with Musk the day before—no matter how polite he was—would not change his mind. She stayed home on Thursday, leaving her San Francisco home only to attend a school function for one of her children. Agrawal and Segal had inklings as well—hence their departures from the office.

They had also prepared letters for Musk that outlined the terms allowing them to depart. Ever optimistic, Agrawal and Segal had written of their love for Twitter and their willingness to stay on. However, the triggers that allowed them to leave with their exit packages had been met, they noted. If Musk wanted them around, he'd need to ask them to stick around. By giving Musk notice of the triggers, the executives believed they had started a ticking clock that would allow them to claim their golden parachutes.

Only Edgett didn't realize he was in immediate danger. As Gadde's number two, the lawyer had spent more than a decade at the company, once testifying to Congress about Russian disinformation on the platform, but preferred to stay behind the scenes. While his last months at Twitter had been consumed by legal work for the deal and he served as the board's secretary during its many meetings, he didn't think he had done anything to land himself in Musk's crosshairs. Edgett arrived at the office on Thursday and treated it like any other day.

As the deal was closing that afternoon, he was meeting with Marianne Fogarty, Twitter's chief compliance officer, on the building's ninth floor. Though employees in costume were gathered a few hundred feet away and beginning their Halloween festivities, Edgett and Fogarty huddled in a glass-walled conference room in an area that only executives could access, perching on stools at a high table as they went over an internal compliance investigation. Twitter's general counsel reflexively checked his phone, then looked up at Fogarty in shock. He glanced quickly back down at his phone, trying to comprehend the email he just received.

"I think I've just been fired," he said in a hushed tone.

At that moment, Kathleen Pacini walked through the executive pen. Edgett popped his head out of the conference room and frantically waved her in, breaking the news of his dismissal.

"THE BIRD IS FREED"

They stared at each other, stunned. They had assumed that Musk would wait at least a day to fire anyone.

"Did you get fired?" Edgett texted Gadde, who immediately confirmed that she, too, had been hit. Agrawal and Segal were also gone. As Fogarty and Pacini watched Edgett type out frantic texts, they realized they had suddenly become some of the most senior executives remaining at the company.

Pacini shifted immediately into planning mode. Agrawal, a close friend, had warned her about this, and she began taking stock of her mental transition plan. With the firings of half the C-suite, hundreds of Twitter employees were now reporting directly to Musk, who she imagined would assume the role of chief executive.

Within five minutes of Edgett receiving the email, there was a knock on the conference room door. Twitter's security guards were standing outside and it was time to go. Earlier, the company's corporate security team had negotiated with Musk's people that should there be any firings, they would be the ones to escort out their former colleagues. They knew the building and the processes to recover any company items, they said, also knowing that it would afford the people they had worked with some dignity. Edgett gathered his things and said goodbye to Fogarty and Pacini, and then rode Twitter's gold-frosted elevators down for the last time.

As word of the firings spread, other members of Twitter's transition team began to trickle into the ninth-floor executive conference room. They began to hammer out the logistics—who would report to whom, and what the communications plan would be to convey what had just occurred to Twitter's workforce. Members of Segal's finance team, who were also on the ninth floor, were among the first to hear of the executive defenestrations, and the executive area quickly filled up with shocked and outraged faces. The crude dismissals were a wake-up call, demonstrating how Musk would handle the transition. *This is going to be brutal*, some thought to themselves.

By then, #TrickOrTweet was in full swing. Children dressed as Marvel superheroes, princesses, and monsters ran between circles of employees, collecting sweets and snacking on cotton candy. Several workers crowded in front of a portrait station, donning devil horns and silly hats for group pictures. The cafeteria was transformed with cotton spiderwebs as magicians roamed the floor, performing card tricks for anyone hoping to take their mind off the deal.

A broad-shouldered man wore a pirate costume and mask. Rumors began to circulate that it was Musk himself, infiltrating the party. Another person in a shark suit also drew whispers—could it be the new owner? The revelry had a slapstick quality that felt classically Twitter. It was distressing yet funny, like the internet had come to life.

Within minutes of the 4:00 p.m. transaction and subsequent firings, news outlets began to report that the deal was closed. Employees who stayed at the party frantically refreshed their Twitter feeds. "Parag and Ned are out!" they murmured to each other. "Vijaya too!" There was no internal email announcing the monumental sale or the changing of the guard, and employees who had felt they had been excluded by Agrawal's lack of communication were only plunged into further darkness.

A hundred yards away in the executive offices on the ninth floor, the remaining leaders trying to guide Twitter's transition were stumped. There was no plan for how to communicate with the employees, and without a green light from the new owner, they weren't even sure if they had the power to make decisions.

—

>>> MUSK AND THE goons stayed away from the Halloween festivities, celebrating with the Morgan Stanley bankers in the war room. They sipped Pappy Van Winkle bourbon procured by Michael Grimes, a small token after the deal's closing fees had made Morgan Stanley millions of dollars.

Although Musk was grinning over his glass, he wasn't satisfied with his conquest. He itched to start another fire drill. Firing the executives was a fine start, but he was eager to put his fingerprints on the product.

One of Musk's gripes was that Twitter's website required people to log in before they could peruse the timeline of fresh tweets. It seemed to be one of the problems that inhibited Twitter from attracting new users—no one really understood how the site worked before they started using it. Musk believed Twitter should entice people to try it, and so he demanded that its home page be transformed from a blank log-in screen into an "Explore" page that displayed a collection of trending topics and popular tweets. The move would surely increase traffic and interest, he thought.

Musk tasked Davis with finding someone to execute the task, and the Boring Company leader ran full speed to deliver the results, no questions asked. The decision wasn't informed by any research, user studies, or consultation with Twitter's engineers or product specialists, but rather Musk's gut. Had he asked people who worked on the product, he might have learned that the log-in page was a vital defense against spammers and bots that crawled Twitter's site to pilfer content, forcing users to log in with credentials to prove they had an account.

At 10:00 p.m., Davis called Sullivan. Although Twitter's product leader was one of Musk's most strident opponents in the executive suite, he had survived the first round of leadership firings. "This needs to be done tonight," Davis told Sullivan as he described the assignment.

Sullivan told Davis that Twitter had actually tested the idea before. There were tradeoffs, he explained to Musk's friend, and ultimately the company had decided against leaving an open timeline running for people who didn't have an account. Davis brushed off the warnings. They had nothing to do with how product development worked in Musk's world. Musk delivered an edict, and then it was done.

Sullivan shrugged. It was the kind of nonsense he'd expected from Musk all along. He ended the call with Davis and went to work finding employees who were still awake and could execute Musk's vision. He gave the task to Twitter employees in London, who were just beginning their day.

It wasn't the only change afoot. While Musk had given the impression to workers a day earlier at the Perch that layoffs weren't imminent, he was indeed planning for drastic cuts. He wanted to start layoffs immediately, saving money by shedding their salaries. He deputized some of his associates, including his cousins, James and Andrew, and asked them to develop lists of who to keep and who to jettison. There would be no mercy.

Some of the list making fell to Austin, who, like the Musk cousins, had spent a grand total of twelve hours at Twitter. The infrastructure engineer had been unaware of Musk's plan to immediately axe Twitter's top leaders, but he wasn't surprised. Musk had made no secret of his repugnance for

Twitter's executives, and most of the goons figured they would be gone sooner rather than later. The immediacy of the employee cuts, however, felt different. The rush could lead to some cruel and potentially damaging outcomes for the company.

Musk also wanted to start locking Twitter employees out of internal systems in preparation for the layoff. Several of his employees from Tesla and the Boring Company summoned Roth and asked him to cut off access to Agent Tools, Twitter's internal system that governed accounts. Employees with access to Agent Tools could reset passwords, suspend accounts, and update users' contact information—or, if they decided to go against Musk, sabotage the site by messing with high-profile accounts. Roth complied and started locking his coworkers out of the system.

That evening, Musk finally slowed down. After exiting the office, he tapped out a message on his Twitter app: "The bird is freed."

ACT III >>>

29 > Code Reviews

Sullivan woke up at 4:00 a.m. on Friday, October 28, and blearily opened his laptop to make sure Twitter's log-in screen had been changed. Sure enough, it displayed popular tweets and trending topics, just as Davis had demanded the night before. He called Davis to let him know that he had passed Musk's test.

Then he started typing up his resignation letter. Around 7:00 a.m., he sent his notice to Spiro and a few of Twitter's remaining HR executives. Then he packed up his laptop and headed into the office.

Although Sullivan knew he could never work for someone like Musk, he felt obligated to make an orderly exit. When he'd left Facebook, he had spent weeks writing up transition documents to describe his work for his replacement and handing off projects, so he planned to do the same for Twitter. Arriving at headquarters, he took stock of who else remained.

Nick Caldwell, his counterpart who led the company's infrastructure engineering efforts, was absent. Caldwell's wife had died suddenly in mid-October and the executive had taken leave as he coped with the aftermath. He planned to hold her funeral that weekend, and submitted a resignation letter that day. Dalana Brand, Twitter's chief people officer, was also preparing to quit. Sarah Personette, the advertising executive, knew that Musk's disrespect for advertisers made her job untenable, and she planned to leave, too. All three would resign, though Musk later claimed some of them had been fired "for cause" to deny them their exit packages.

Employees kept dropping by the executives' desks, asking when layoffs might come and whether they should quit. None of the bosses knew what to say—while they knew they were leaving, no one from Musk's side had acknowledged or announced the resignations.

Finally, one of Brand's junior HR employees stopped at Sullivan's desk. "Maybe you should stop coming in," she said with a tight smile. "I would hate to see you get walked out."

Meanwhile, Musk's cousins set up a command center in the ninth-floor cafeteria, which had mostly been cleared of Halloween decorations from the night before. Along with a handful of Tesla engineers, James and Andrew Musk turned their focus to culling Twitter's engineering ranks, which numbered around 2,500 people. They wondered aloud if they should ask their Twitter counterparts to help them. They decided against it—they knew the tweeps were inherently skeptical of them, and the feeling was mutual. But they did demand that Twitter's human resources employees hand over data from everyone's performance evaluations.

Musk's people believed they were smarter than Twitter's workforce. The Twitter employees, many of whom had yet to arrive at the office that morning, had contributed to a floundering company that failed its potential, they thought. Why should they be trusted to choose who to keep to build for a better future? They might try to protect their friends, or worse, select people antagonistic to Musk.

Time and time again, Musk reminded his team that Twitter had a workforce bloated with many people who did nothing. In meetings with his employees, Musk cited Tesla's autopilot team. It was fifteen times smaller than Twitter's engineering staff but still managed to hit his demanding deadlines and regularly ship updates to the software. Autopilot was a matter of life or death for Tesla's customers, he thought. Twitter was just a website with constantly scrolling text and media.

Musk directed his cousins to take a look at Twitter's code repositories, which formed the basis for its site, apps, and features, and determine which engineers had been contributing. Engineers would be judged by volume: those who had written the most code were people they might want to keep around.

"Print out 50 pages of code you've done in the last 30 days," read a Slack directive from one executive assistant to Twitter's engineers. Employees were told they should be ready to share their work in so-called code reviews with members of the transition team, or even Musk himself. They would be evaluated on their material for its effectiveness, clarity, and contributions to Twitter's overall operations.

The order sent a panic throughout Twitter's workforce. Engineers who had come into the offices in San Francisco and New York for Musk's first full day rushed to connect their laptops to printers. The devices began constantly spitting out sheets.

In Slack and in private messages, Twitter employees complained about the exercise. Even if someone could show they wrote a lot of code, volume wasn't necessarily an indicator of good work. Sometimes, the best code was short and elegant.

Even some of Musk's employees were skeptical. They had expertise in software for rockets and cars, but they had no idea how to build, maintain, or run a large social network. It was like asking a plumber to judge the work of an electrician—it simply didn't make sense.

Austin was uncomfortable with reviewing code. The job gave him an incredible amount of power over people's livelihoods at a company he didn't even work for. He tried to talk to as many Twitter infrastructure experts as possible—some who had worked at the company for a decade—in an attempt to understand the company's back-end technology. He knew he'd eventually have to communicate these complexities to Musk himself.

In fact, Musk soon asked Austin if he wanted to run Twitter's infrastructure. Austin hated the idea but was able to wiggle his way out of giving Musk a direct answer. He worked for Musk because he believed in building electric cars for the masses. He didn't want to work at a social network.

Online, tweeps mocked the code reviews. Twitter was a self-deprecating place, and employees often dealt with difficult times by posting memes on Slack or shitposting on Twitter. The workers feared for their jobs, and had just seen their four top leaders unceremoniously dumped on the curb, but were still able to laugh at colleagues running from printer to printer or searching for paper.

Leah Culver, an engineer who worked on Twitter's app for Apple devices, tweeted "Happy Friday all" accompanied by a photo of her holding up a stack of papers with printed-out code for the company's Spaces feature, which allowed people to host audio broadcasts.

```
//LockScreenManager.swift
//TwitterAudiospaces//
//Created by Leah Culver on 7/26/22.
//Copyright © 2022 Twitter, Inc.
```

Culver's selfie—and the code it exposed—spooked some of Twitter's remaining executives, particularly Damien Kieran, its chief privacy officer. That afternoon, after finding out about Musk's orders, he went ballistic. A new directive was issued: "Stop printing."

Under its consent decree with the Federal Trade Commission, Twitter had strict orders to guard the privacy of its users and its data. Each sheet of paper printed with the company's proprietary code was a potential privacy risk. Any engineer could walk out of Twitter's doors with a printed copy of company secrets or users' information. Inadvertently, Musk and his team had created thousands of potential violations of Twitter's agreement with the U.S. government, which could in turn levy millions of dollars in fines.

Kieran, an Irishman who was a former motorcycle mechanic, was responsible for determining if Twitter was compliant with global privacy regulations. He explained the problem to Austin and several other goons, who then communicated the concern up the chain. Engineers would not only have to stop printing out the code, but those who had already done so were now required to get rid of it, the chief privacy officer said.

Kieran set up paper shredders on the tenth floor and had Musk's guards stand by them. Before going home, engineers had to personally destroy any code they had printed by feeding their papers into the machines. Lines formed in front of the shredders as the muscled security detail watched over them like hawks.

—

>>> ROTH, the highest-ranking employee responsible for content moderation after Gadde's departure, assumed he was on the chopping block. But when Musk summoned Roth to meet him in one of the office kitchens that day, it was to put him to work. There was one thing he wanted Roth to do immediately: reinstate the Babylon Bee. The conservative satire site had

been banned from Twitter that March after it misgendered a government official.

Roth knew he shouldn't push back—Brazil's presidential election and the U.S. midterms were days away, and he wanted to hang on long enough to monitor the platform for misinformation during the crucial votes. But he couldn't help but test Musk's rationale.

"Is it your intention to change the policy on misgendering?" Roth asked.

Musk hemmed and hawed, unsure if he wanted to overhaul the policy. "What about a presidential pardon?" he asked Roth. "That's a thing in the Constitution."

Roth kept gently pushing. "What if someone tweets the same thing that you pardoned the Bee for?" he asked. If the satire publication got a special pass to tweet transphobic content, Musk would surely face outrage from other people who wanted to post the same things but kept getting in trouble. It wouldn't be fair.

Musk understood. There couldn't be different rules for the accounts he enjoyed, he admitted—that wouldn't gel with his plans to maximize free speech and let anyone say whatever they wanted on Twitter. The policy would have to be changed, Musk said.

Roth, who had helped develop the rule against misgendering in 2018 and spent large portions of his career studying the harassment of queer communities online, agreed to change the rules as Musk ordered. But he offered a word of caution.

"Your first policy move, then, would be changing a policy that corresponds with a highly politicized culture war in the United States," Roth said. "A lot of people will look at it and say, 'That's his first step—dismantling a policy that relates to the protection of marginalized groups.' You're already dealing with advertiser backlash. I think doing that would not really go the way you're hoping."

"Misgendering is totally not cool," Musk told Roth. But the billionaire wanted to distinguish between threats of harm and rude comments, which he thought should receive a lighter punishment.

Roth tried to refocus Musk on the bigger picture and offered other choices. Instead of immediately giving the Babylon Bee back its account, what if Musk thought about moderating content in general? Did he want to keep the labels that Twitter had used in the past to flag Trump's account when he shared misinformation about voting processes? How would Musk like to handle tweets that weren't immediately dangerous but still broke the rules?

Musk seemed to enjoy being consulted on bigger decisions. He leaned in and listened intently as Roth offered him options for altering speech on Twitter.

The labels, Roth explained, were more than a tool to let users know that a tweet contained misleading information. Once a tweet was labeled, Twitter prevented it from spreading, turning off retweets and likes so it couldn't go viral. "We're limiting reach, not speech," Roth said, introducing Musk to a catchphrase Twitter's content moderation staffers often used as a North Star.

The engineering solution to the content moderation problem appealed to Musk. Emboldened, Roth started pitching Musk on Project Saturn. Agrawal's plan for Twitter was to allow all kinds of content onto the platform, while imposing limits on how far it could spread, a vision that dovetailed with what Musk said he wanted. Roth knew that mentioning Agrawal's name would immediately sink the project, so he gave Musk an unattributed version of the idea.

"Why don't we build some of the product capabilities to make the rules less punitive, and reinstate accounts?" Roth suggested. It was a path forward that didn't dismantle the rules.

Musk agreed, and told Roth to get started building it, finally greenlighting the project that had stalled all summer under his looming acquisition. For now, the Babylon Bee would remain off Twitter.

The conversation clearly impressed the billionaire. That weekend, in response to conservative users who called for Roth to be fired, Musk replied, "We've all made some questionable tweets, me more than most, but I want to be clear that I support Yoel. My sense is that he has high integrity, and we are all entitled to our political beliefs."

It also made a deep impression on Roth, who had feared that Musk would hate him. Musk was clearly persuadable, Roth thought, and liked thinking through the nuances of technology. Maybe there was a path for him to stay at Twitter after all.

CODE REVIEWS > 277

>>> THE CUTS TO Twitter's engineering staff wouldn't be enough for the savings Musk wanted. He and Spiro summoned several of Twitter's finance, legal, and human resources executives to the second-floor conference room at 1 Tenth to discuss broader layoffs. Pacini went to sit down with Musk, along with Mary Hansbury, an employment lawyer, and Tracy Hawkins, an executive who managed Twitter's real estate and return-to-office plans.

Musk and Spiro told them that the dismissals needed to happen as soon as possible. But Musk didn't seem to have a clear plan. He wanted deep cuts and suggested about half of the company's 7,500 employees would have to go, but the billionaire wasn't decisive about how many jobs he wanted to eliminate or which departments he wanted to shrink. He said he wanted to keep exceptional people but had no obvious criteria for judging them.

Musk and his lawyer stressed that cuts should be made no later than Monday, October 31. On November 1, many employees were set to receive their vest, and Musk's side seemed to be looking for a way to avoid paying.

Pacini and Hansbury knew the haphazard, rushed cuts were not the right way to do a layoff. They were still sitting on the plans for Project Prism, Agrawal's mass layoff agenda from April, which laid out specific cost savings and balanced the staff reductions with severance payments. Managers had even drawn up lists of specific people to eliminate. Those plans could be resuscitated, the HR executives suggested. Mass cuts like the ones Musk wanted couldn't be rammed through in just a few hours. People would get hurt, and Twitter would probably end up breaking its contracts with employees or labor laws in the many countries where it had people.

"I'm used to paying penalties," Musk told the Twitter executives. "I'll still come out ahead."

"People sue him all the time," Spiro chimed in. His brash client wasn't going to be intimidated by the threat of legal action. All he cared about was speed.

That day, Spiro also gathered all of Gadde's former employees to explain how they would operate under Musk. There were in-house lawyers, public policy executives who kept the company in compliance with regulations around the world, and content moderation specialists who developed Twitter's rules.

He offered a quick reassurance that he was not at Twitter to disrupt anything that was already working well. But then Spiro quickly launched into his directive: cost cutting. He asked the executives where Twitter's legal costs could be reduced and who could be laid off.

The lawyer told the assembled executives that they reported to him now, and that he was making decisions as Musk's delegate. "Elon doesn't want to come in," he told them, subtly hinting that there would be no going over his head to the boss. To some of the executives, it seemed Spiro thought Gadde had simply been managing a team of lawyers akin to the law firm Spiro came from, and wasn't aware that she also oversaw lobbyists, policy experts, and content moderators.

Sinéad McSweeney, a vice president at the company with cropped blond hair and a lilting Irish accent, piped up. McSweeney oversaw Twitter's global public policy team from its Dublin offices and had worked at the company for a decade, keeping it in line with European and international law. Her role included constantly reminding her American colleagues of Twitter's international presence and the legal issues it encountered around the world. It seemed to her that Spiro had a similarly narrow view of Twitter.

What should be done with Twitter's public policy operations, she asked Musk's attorney. "I don't believe in having fifty people where five can do the job," Spiro responded.

He suggested that, because Musk was world famous, much of Twitter's public policy work would become redundant. "He can go in front of the U.K. Parliament tomorrow and can get to meet with any prime minister or king," Spiro scoffed. "It's a different world for us all now."

Everyone in legal should keep him informed of major decisions so he didn't wake up to ugly surprises, Spiro said. Their top priority should be cost cutting, and they should eliminate expensive contracts with outside lawyers immediately.

He and Musk didn't want to disrupt the machine, he added, and while he and the billionaire didn't always see eye to eye, they wanted things to work effectively. He closed the meeting by excusing Musk's absence again—the new owner was probably downstairs making more cuts to the workforce, Spiro said.

That night, Spiro met again with McSweeney and other members of Twitter's public policy teams to start hammering out layoffs. "The sky is not falling in," Spiro told them, adding that Twitter's workers should think of their world as expanding with new opportunities.

He then launched into a strange aside about Musk's politics. While his boss was a darling of the right, he told the assembled policy executives not to take Musk's public perception too seriously because he was, at heart, a centrist. Musk would be direct and transparent with them but ruthless about costs, Spiro warned. McSweeney should prepare to cut 25 percent of her staff, so Musk didn't come in and start slashing and burning himself.

The layoffs were to happen that weekend, Spiro added. "By Monday, this company won't have 8,000 employees," he said. "And by the Monday after that, there won't be 7,000 employees.

"Elon isn't an animal," Spiro added—if the laid-off workers needed to find jobs elsewhere, the billionaire would call his friends and find positions for them.

>>> TWITTER'S REMAINING EXECUTIVES SOON began to grasp how Musk operated. Twitter had been governed—and at times hampered—by its Socratic approach to making decisions, but Musk reigned like a king. He wanted to make all the decisions, no matter how small, and delegated only to the handful of loyalists who could translate his sometimes vague demands into action. He assigned projects to members of his entourage based on their interests. Spiro took on legal and policy matters, while Gracias was tasked with finance and sales. Davis was on the budget. Calacanis was told to sit in on product meetings, but Twitter engineers quickly began to view him as a joke when they realized Musk had little regard for him.

No one in the motley crew made a habit of saying no to the boss. Musk's family also appeared at Twitter. He seemed to rely on his son X as a personal security blanket, holding the boy as he weighed decisions and allowing his nanny to shepherd him out of the conference room when he cried. The billionaire's mother, Maye Musk, a seventy-five-year-old model, inexplicably attended some meetings, while his assistant, Balajadia, scampered in and out of conference rooms to deliver food, Diet Coke, or whatever the techno king desired. All of them kept Musk on task and feeling appreciated, reinforcing that he was the center of their universe.

Musk kept his interactions with Twitter staffers to a minimum, keeping most of them in the dark and leaving them to hear about his decisions through a game of telephone with his confidants that often warped his meaning. Although Leslie Berland had promised to host a company-wide meeting with Musk that Friday, it never happened.

He often derailed the conversations about Twitter's operations with the few Twitter executives he deigned to speak to, with long rants or random phone calls. In one meeting to discuss legal and privacy matters soon after the takeover, he abruptly paused the discussion as a call flashed across his phone. It was Shaquille O'Neal, Musk boasted. He stopped the meeting to take the call from the former basketball superstar, who had invested in Twitter as the billionaire scrounged around for cash to aid his deal. The Twitter executives in the room, who had worked grueling hours late into the night with Musk and his team, grew increasingly impatient as Musk chatted away, humoring the celebrity.

His new employees were rapidly learning that Musk hated excuses and explanations but loved to be admired. He got keen satisfaction from telling jokes—which meant that everyone around him had to be ready to laugh.

In one incident, a Twitter executive was brought in to meet Musk for the first time. He had previously worked at Google, and Musk, having learned of his background, launched into stories about his past relationships with the search company's founders and its current chief executive. He said he was angry at Sundar Pichai, Google's leader, for not putting antennas into Android phones that could connect with Starlink, SpaceX's proprietary satellite internet service.

Then Musk, riffing off his own story, offered up something else he had heard to impress the executive even more. He had a friend that worked on the company's key search products who had been told that Google intentionally kept its share of the search market under 70 percent to avoid antitrust regulation.

"Get it?" Musk said, with a smirk. "Sixty-nine percent?"

Musk looked around as his voice grew louder. "Sixty-nine percent!"

—

>>> THE SITUATION WAS not any calmer in Twitter's Manhattan office. Those who showed up at the high-rise in Chelsea—mainly sales managers and executives—were hooked to their group chats with their San Francisco counterparts, thirsty for any sliver of information. One vice president was told not to gather anyone in large groups so it wouldn't look like they were organizing against new management.

While Musk and most of his henchmen were in California, plans had been made for Twitter's new owner to fly to New York the following week to deal with the advertiser problem. The billionaire, however, had sent Gracias ahead to prepare and get an understanding of Twitter's sales operation and its relationships with the advertising agencies of Madison Avenue.

On Friday night, October 28, Gracias called two meetings with sales executives. "I'm Antonio and I've known Elon for many years," the financier said. "I've sat on his boards. I've worked closely with him and I'm helping him in this process."

Those on the calls weren't sure what to make of Musk's proxy. Gracias came off as arrogant. At one point, after a sales leader from Tokyo on the call introduced himself, Gracias volleyed back some words in Japanese and asked about a local restaurant. He had lived and worked in Japan for a bit, he said confidently.

Gracias gave empty assurances that advertising was important to the company. To his audience, it seemed that he, like Musk, didn't understand how Twitter made its money from those advertising relationships. Some 20 percent of brands accounted for more than 80 percent of the money the company brought in, said Jean-Philippe Maheu, the company's second-highest-ranking sales leader, making it the company's top priority to keep those few big spenders happy.

After Advertising 101, Gracias looked around the room and deadpanned: "So do you guys have any ideas for this? What can I tell you guys?"

After a brief, puzzled silence, the salespeople asked about layoffs. There was an air of self-preservation to it all, as the leaders, many of whom believed they would be dismissed in the cost cutting because they earned high salaries, tried to justify why those who worked for them should stay. After all, these were the people who kept brands spending millions of dollars a year on the social network. Gracias provided little guidance, simply telling them to keep "exceptional" performers.

Another executive also wondered about Musk's plans for content moderation. What should we tell advertisers about how we plan to deal with divisive or hateful content, she asked.

Gracias thought for a second and then parroted Musk's lines on "freedom of speech" and free expression. Elon's view is that Twitter's rules should closely follow what's allowed in the U.S. Constitution, he said. To illustrate his point, he noted that people should expect to see speech they didn't like, and cited the fact that even Nazis had the right to assemble and demonstrate in public spaces.

Some members of the call, who were Jewish, were taken aback. Sure, the Constitution afforded rights to citizens that couldn't be infringed upon by the government, but Twitter was a private company. It had no obligation to give Nazis a platform, and no brand would want their ads appearing next to Nazi content.

After the last meeting ended, Maheu approached the venture capitalist to push back on the timing of the firings. Terminating people before the November 1 vest would decimate morale and cause more fractures in the already tense relationship between Musk and the workers. "Don't do this," Maheu said. "This is the worst message you can send."

Gracias didn't seem particularly moved.

30 > Lords and Peasants

Twitter's workforce woke up Saturday morning in a state of panic. Managers compiled lists in tightly held Google Docs that they traded back and forth. For the workers they wanted to retain, leaders were required not only to describe what the person did but also include a description of why they were exceptional, so that Musk's lieutenants could evaluate them.

In meetings with Twitter's HR team, Musk remained noncommittal about the percentage of staff that needed to be eliminated. His cousins had recommended deep cuts to the engineering department, but his deputies gave targets to Twitter executives that seemed to shift by the hour. At one point, they told the sales team to eliminate a fifth of its staff, while the company's compliance department was told to eliminate 27 percent of workers. Trust and safety, which was dealing with an ongoing election in Brazil and preparing for the U.S. midterms, was told to cut 15 percent. Twitter's human resources team was instructed to cut itself in half, but the depth of the cuts were later increased until only a quarter remained.

Matters were complicated by the fact that Musk's friends regularly contradicted each other. Each of them would insinuate that their messages came directly from Musk himself, and yet none of them aligned. Spiro might give one number for the amount of cuts, but Davis had another. Even Musk would change his mind, influenced, in part, by the last adviser he spoke with.

Twitter's remaining leadership launched a rescue mission. They played shell games with their retention lists, transferring employees to other teams that had a spot or two open in an attempt to save people who were in the U.S. on visas, had dependents who relied on their healthcare, or had other personal reasons they needed to stay at the company. Pacini made it her mission to convince Musk to pay the November 1 vest.

First, Musk needed to be convinced that flouting the law and laying off half the company immediately would end up burning him. Employees in legal, HR, and finance whipped up a financial model that compared the amount of money Musk would save with immediate cuts to the amount he would have to spend to settle employee lawsuits and pay regulatory fines. The mock-up showed that, in the long run, Musk would actually save money if he did the layoffs by the book.

The other problem was actually getting the payouts to employees. Since it no longer had publicly traded shares, the company moved its equity program from Charles Schwab to Shareworks, a tool affiliated with Morgan Stanley that handled equity and cash bonus disbursement for private companies. But Pacini had to convince Musk to turn on Shareworks before employees could receive their payments.

—

>>> ON SATURDAY MORNING, Esther Crawford, the product manager who had boldly approached Musk at the Perch, was eating from a bowl of cereal at her kitchen table at her home near Berkeley, California, when she noticed a call from an unfamiliar number.

"Hey, what are you doing right now?" said the voice on the other side of the line. It was Sriram Krishnan, a former product leader at Twitter who had been let go in late 2019. Since his departure from the social networking company, Krishnan had reinvented himself as a tech investor, podcast personality, and general partner at Andreessen Horowitz, one of the firms investing in Musk's Twitter. He was helping with the transition, he explained.

"How fast can you get here?" Krishnan asked. "Elon wants you to come in right now."

While Crawford knew about Krishnan's checkered past at the company, she had little time to contemplate why he was advising Musk. She finished her breakfast, said her goodbyes to her family, and then drove across the Bay Bridge to the San Francisco office. There, she got her marching orders from Musk, who was flanked in his 1 Tenth conference room by Krishnan,

LORDS AND PEASANTS

Calacanis, and Sacks. She was going to lead a revamp of the company's subscription service, Twitter Blue.

Musk believed subscription revenue was the holy grail. His investor presentation had projected it would reach $10 billion by 2028, and he believed he could unlock Blue's true value with a new twist. The service would now give any subscriber one of Twitter's coveted verification check marks, Musk told Crawford, allowing anyone who paid to obtain a badge that had previously been meant for celebrities, politicians, and other notable figures and organizations.

Featuring a clear check mark outlined in blue, the verification badge had been a status symbol on Twitter. In the past, the company had given them out at its own discretion, distinguishing high-profile accounts from impostors. Justin Bieber and Barack Obama had check marks, as did Musk, and more than 420,000 other users.

Like many features at Twitter, however, the verification system wasn't perfect. Verified accounts could be illicitly bought and sold, and the company was opaque and haphazard about doling them out. Musk himself came to hate what he saw as Twitter's two-tiered class system of the verified and unverified. It especially chafed him that journalists, whom he often bickered with, could easily obtain the check marks. He couldn't understand why any reporter, particularly those he saw at newer or less reputable outlets, were granted the badges simply because of their profession.

"It's a system of lords and peasants," he said to those in the war room. He cited instances where anyone working for *Business Insider* or *BuzzFeed*, two online outlets, were waved through the program by Twitter's old management.

"Who the fuck is this person," he said, discussing a hypothetical verified *Business Insider* reporter with a low number of followers. "He just writes copypasta of something I tweeted and now he's verified. Fuck them, you know?"

To Musk, selling off the check marks was the ultimate democratization of the service. He also thought that charging for verification would help extinguish what he still believed to be the massive bot problem on Twitter. Those who paid for Twitter Blue would have to associate their account with payment information, a telltale sign in Musk's eyes that the account belonged to a real person.

Crawford thought Musk made some good points, and she, too, believed that the current verification system had its flaws. But she thought the idea that subscriptions would translate to billions of dollars in revenue was absurd. Internal research and Blue's own initial rollout had shown that very few people—the Twitter diehards—were willing to pay for extra features.

Still, she took the assignment with a smile. Crawford enjoyed seemingly impossible tasks, and this one at least gave her job security and proximity to power. Musk gave her a deadline of November 7 for the relaunch of Twitter Blue; she immediately began messaging and calling people to recruit them to her team. She had ten days to deliver or risk being fired.

>>> AT NOON, the human resources executives gathered to regroup. Musk had agreed to slow down the layoffs, accepting the cheaper option. He had also acquiesced to turning on Shareworks. But Musk had a demand, Pacini said.

The billionaire, racked with paranoia, had convinced himself that not all of Twitter's employees were real. His fears that Twitter couldn't discern between its human and bot users had mutated into the notion that the company also couldn't keep track of its employees. In meetings, he fretted about what he called "ghost employees" who might be collecting paychecks from the company without earning them. Before Twitter sent out any payments, it needed to conduct an audit to ensure all its employees were real. When Pacini introduced the idea on the noon call, several executives burst out laughing at the absurdity of the idea.

To deal with Musk's unusual demand, the HR team turned to Kaiden.

Musk's circle knew about Kaiden. He had met with Musk before the deal closed and was on the call when Segal rejected Gracias's audacious request for money from Twitter. With a team of finance folks, Kaiden got down to work, reaching out to managers around the world and asking them to confirm their employees were human. He and his team had two days until the vest date to confirm the identities of more than 7,000 full-time employees.

>>> MUSK'S GROWING PARANOIA wasn't reserved just for his outlook on Twitter's past management. On the evening of October 28, Musk was retiring to Sacks's five-story mansion on a street in San Francisco's Pacific Heights neighborhood known as Billionaire's Row. A few blocks down, a disturbed man who embraced far-right conspiracy theories about the 2020 U.S. election broke into the home of Nancy Pelosi, the speaker of the U.S. House of Representatives, just past 2:00 a.m. While she was not there, her husband, Paul, was home. The eighty-two-year-old called the police and confronted the intruder, who, after demanding to speak to "Nancy," attacked her husband with a hammer and fractured his skull.

The assault was major news. San Francisco police kept information under wraps as it investigated. But right-wing political personalities sought to exploit the situation involving the Democratic Speaker's family. In the days after, Donald Trump, Jr., posted memes on his Twitter and Instagram account playing up the baseless theory that Paul Pelosi had known and was engaged in a tryst with his attacker. Others chimed in, including the conspiracy-minded Georgia congresswoman Marjorie Taylor Greene and Roger Stone, the former Trump political consultant. The misinformation stewed online on Twitter and 4chan, the hateful message board popular with the far right, before being picked up by discredited conspiracy websites.

Musk followed it all. Amid the hard push to implement layoffs that weekend, Musk found time during his Sunday morning in San Francisco to respond to a tweet from former Democratic presidential candidate Hillary Clinton that condemned "the Republican Party and its mouthpieces" for spreading "hate and deranged conspiracy theories."

"There is a tiny possibility there might be more to this story than meets the eye," Musk replied to the former First Lady, with a link to the story from the *Santa Monica Observer* with the headline "The Awful Truth: Paul Pelosi Was Drunk Again, and in a Dispute with a Male Prostitute Early Friday Morning."

At first glance, the *Santa Monica Observer* could have been mistaken for a local news blog. But it was one of dozens of untrustworthy websites that had popped up in recent years to launder false information. In 2016, it had pushed the idea that Hilary Clinton had died and been replaced with a body double. As screenshots of Musk's reply proliferated, the company's executives sent frantic messages to one another. Could Musk really believe this trash?

Musk deleted the tweet within a few hours, but the damage was done. Advertisers, already nervous about the prospects of his ownership, had the clearest evidence yet that Twitter's new steward was not serious about policing misinformation. The company's sales execs worked the phones with clients in an attempt to get in front of the controversy, but to little avail.

31 > "Educate me"

Musk's private plane landed at 2:00 a.m. at New Jersey's Teterboro Airport on Halloween, the first Monday following his takeover. He had already been planning to visit the company's main East Coast office, but advertisers' concerns, exacerbated by his Pelosi tweet, made the trip all the more urgent.

At the Chelsea office, building staff and assistants had been told to prepare for their new leader's arrival, and they wanted to impress. But they had little to accommodate, much less entertain, Musk's toddler son, who had traveled with his father. There was no play area or place for him to nap, so they improvised, clearing out a storage closet that usually held chairs and tables and laying down blankets. The workers gathered various Twitter-branded throw pillows from around the office to build a makeshift bed, while someone went out and purchased a set of building blocks. The staffers watched as the child and his nanny settled into the small space, which remained guarded by two bulky security guards who followed Musk's progeny at all times.

Across from the storage closet, Musk began a series of meetings with the leaders from some of the world's top advertising agencies. With Personette's resignation confirmed, Musk was joined by Maheu. Berland had vouched for Maheu personally and told Musk about his connections with the largest advertising companies in the world.

Maheu immediately brokered meetings for Musk with WPP, Publicis Groupe, and Horizon Media. The goal was to reassure the ad agencies that Musk's Twitter wouldn't descend into a cesspool of misinformation and hate speech. The previous Friday, General Motors announced that it would pause advertising on Twitter, and other brands were rumored to be heading down the same path. The chaos was less than ideal for the multibillion-dollar corporations trying to sell people cars or convince them to watch a new movie. (General Motors, a competitor with Tesla, sought assurances that its advertising data wouldn't be shared between Musk's companies before it resumed spending.)

Maheu, a shrewd Frenchman who had developed deep connections to the ad industry over a decade at Twitter, knew he had to act fast to prevent the contagion from spreading. It would take only a matter of days before others would follow General Motors like lemmings, leading to steep reductions in revenue. He tried to make the gatherings orderly, but it was impossible.

In the meetings with advertisers, Musk surrounded himself with his motley entourage. There was X and Musk's mother, Maye, who was in New York for an event that evening. Calacanis joined, as did Michael Tucker, a music producer better known as BloodPop, who had written songs for Justin Bieber, Britney Spears, and Lady Gaga. Tucker's presence went unexplained and he sat in silence through some of the meetings. Advertisers were equally puzzled by their spontaneous introduction to the Musk circle.

Musk said the right things to advertisers and committed to a vision of continued content moderation. But Maheu also noticed a strange dynamic developing. Even though they were skittish, the ad agency leaders didn't challenge Musk but rather wanted to charm him. They asked him softball questions and cozied up to him. After all, this was the leader not only of Twitter but also Tesla and SpaceX, two multibillion-dollar companies that did little to no advertising. What if those companies decided they wanted to start running ads one day?

The flattery was on full display that afternoon when Bill Koenigsberg, the chief executive of Horizon Media—an ad firm that represented Hershey's and Burger King—parroted a question that had been asked repeatedly since Musk announced his takeover.

"My clients asked, 'Is he going to get Donald Trump back on the platform?'" he said.

Musk had grown tired of the question. So instead of giving his typical staid answer, he took out his iPhone, opened up the Twitter app, and drafted a tweet: "If I had a dollar for every time someone asked me if Trump is coming back on this platform, Twitter would be minting money!" He looked around the room.

"You're my content counsel," Musk said with a wide grin. "Should I tweet that or not?"

To Maheu, the billionaire was like a fifth grader asking his parents if he could light an M80 in the house. Calacanis, giggling, said yes. So did his mother. Koenigsberg, who clearly had no intention of rocking the boat, gave his approval. The music producer gave his thumbs-up.

"No," said Maheu, the lone voice of dissent, knowing that any mention of Trump would generate an unneeded press cycle. Musk eyed him briefly, then shrugged. He hit the tweet button.

>>> THAT NIGHT, Kaiden was still scrambling to confirm that all of Twitter's employees were real people before his Tuesday morning deadline. He worked late into the evening, pestering managers to reach out to each of their employees and confirm their existence and reiterating impatiently that no, he wasn't joking.

On Monday evening, the accountant blasted out an email to several managers about his ongoing "payroll audit."

"We need to know by 8 AM PT if you know the following employees and that they are really humans," Kaiden wrote. "If you don't know please dive a level deeper and ask your directs if they know." Attached to the message was a list of unaccounted-for Twitter workers. Kaiden knew he had to fulfill Musk's insane demand.

>>> IN NEW YORK, the sales team prepared to argue for their jobs. Tesla and SpaceX relied on word of mouth and consumer evangelism through Musk's rabid fanbase. With such strong support, Musk came to see advertising as a waste. "I hate advertising," he tweeted on a whim in October 2019.

But Twitter's lifeblood was advertising. Ad revenue made up 90 percent of the company's revenue, or some $5 billion annually.

After his meetings with Maheu, Musk was set to be briefed on the status of the company's ad business. He finally arrived at the meeting three hours after its scheduled start at 6:00 p.m., hurrying into a conference room with Twitter's top sales leaders—sans Maheu—and apologizing for his tardiness. The ad sales experts held their breath for a hostile Musk, but they found the billionaire to be far from it, perhaps humbled by the advertising backlash. They would have as long as they needed to explain the ads business to him.

"I have a party tonight, but it doesn't matter," Musk said, referencing a Halloween party he was expected to attend. "This is much more important.

"I'm really new to this," he continued. "Educate me."

What was supposed to be a thirty-minute gathering stretched beyond two hours as the sales leaders walked him through the basics of advertising. They broke down Twitter's biggest customers, its advantages, and its technical disadvantages when competing against Facebook and Google, the dominant players in the digital ad space. Musk was soft-spoken but asked probing questions. More important, he listened as the ads professionals presented. It was a side of Musk that many in the room, most who were meeting him for the first time, were unprepared for.

Still there were worries. Some of the Twitter execs wondered why he hadn't taken the time to do this type of diligence before buying the company. And they recalled that when Musk and his bankers went and pitched potential investors on backing his acquisition of Twitter, leaked presentation decks reported by the media said that Musk's company would bring in $12 billion in annual advertising revenue, more than double its current clip, by 2028. If he was just learning how Twitter kept the lights on, what were his figures based on?

When the meeting ended, Musk was whisked out of the room and into a car where his mother, Maye, was waiting. The next day, tabloids would circulate photos of Musk, dressed in a $7,500 leather armor suit stamped with a devil's insignia, and his mother, styled as Disney's Cruella de Vil, on the red carpet of a Halloween gala hosted by former supermodel Heidi Klum.

In group chats, some employees joked that the costumes seemed a little too on the nose. Others asked how many fifty-year-old men dress up for parties with their mother?

32 > A Blue Heart

By the morning of November 1, Kaiden felt like his work was done. He had verified that Twitter's employees were, in fact, real and Musk's fears of ghosts were delusions. Not one to use Slack for small talk, he was businesslike in his announcement: "Payroll is working on processing the vest now," he wrote in the workplace chat that morning. The message garnered dozens of thumbs-up and prayer hands emoji responses.

Musk was less thrilled. The payout would cost him $200 million as he enriched a group of workers he believed were lazy and undeserving. And while he had authorized Kaiden to conduct the audit, his team had not given the chief accounting officer the authority to make the announcement. The next day, the goons locked the accountant out of his laptop and he was escorted out of the San Francisco office. He walked out a hero.

On November 1, Spiro gave McSweeney new marching orders. Instead of cutting her team by 25 percent, she needed to find a way to dismiss 50 percent, he said. She had been in direct contact with Spiro throughout the weekend as they traded layoff lists, and was shocked at the sudden change.

Just days earlier, Spiro had been reassuring her and promising to help along the way, but she was being handed off to Sam Teller, Musk's former chief of staff who somehow had also been drafted into this mess. Teller asked for a crash course on Twitter's public policy operations, forcing McSweeney to repeat what she'd told Spiro on Friday. McSweeney ran through her spiel again.

McSweeney thought the magnitude of the layoffs made no sense, and she was worried about taking instructions from people like Spiro and Teller, who had no formal roles at Twitter. Their orders seemed likely to get her into trouble, particularly as she prepared to lay off employees in Europe, many of whom were protected by stricter labor laws than their colleagues in the United States and entitled to extended notice of a mass layoff. She wrote to the company's human resources and legal teams, notifying them of her concerns. But no one seemed willing to contravene Musk.

>>> MEANWHILE, Musk's trip to New York had done little to mollify advertisers. Interpublic Group, or IPG, a large advertising company whose agencies represent American Express, Coca-Cola, and Mattel, told its clients to temporarily pause spending on Twitter.

The news didn't sit well with Musk. He began to think there were other forces at play. He crafted a narrative—much like the ones he advanced in the early days of Tesla when he suggested that oil companies, short sellers, and news outlets colluded to bring about its failure—that activist groups were pressuring advertisers to stop spending. He became convinced these groups, including Media Matters for America, a left-wing media watchdog group, and the Anti-Defamation League, the Jewish advocacy group, were conspiring against him, funded by moneyed interests to prevent him from building the free speech platform for the masses.

These groups had voiced their fears to advertisers. "We are concerned that Mr. Musk's acquisition of Twitter may accelerate what ADL has seen repeatedly: the pushing out of marginalized communities from social media," Jonathan Greenblatt, the chief executive of the ADL, said in a statement on the day Musk completed his acquisition. But this wasn't necessarily a new phenomenon. Activist groups had held social media companies accountable for years with their studies of the platforms, and Musk's actions simply attracted their further scrutiny. Musk would later push to ban posts on his site that called for ad boycotts, arguing that they amounted to blackmail. The rule never went into place.

After the IPG decision, Maheu knew Twitter wouldn't be able to stop the dominos from falling. But it wouldn't be his problem for long. Gracias hadn't taken his advice on the layoffs kindly and Musk clearly disliked him after his pushback on the Trump tweet. So when the call came that he'd be escorted out of the New York building by Twitter's security staff, he was ready.

He left his laptop and carried nothing as he walked by the desks of the men and women he led across Twitter's sales teams. Maheu was the first person fired on the East Coast.

Berland would follow close behind. Sitting on a different floor from the sales staff, she was unaware what had happened to the sales exec. But that day, her communications with Musk and his goons had become sparse. She had vouched for Maheu to Musk, and his attempt to stop the billionaire from tweeting would not reflect well on her.

That afternoon, her phone rang with a call from Pacini. Berland stepped into a private booth. Pacini had been crying.

"I hate to do this, but I have to tell you," Pacini said, trailing off. "You're no longer employed at Twitter."

Berland waited a beat to collect her thoughts. It wasn't entirely unexpected, but this wasn't the way she had planned to go out. It was so cold.

"What's the context?" she asked.

"I don't have any," Pacini replied. "There will be a security guard by soon to escort you out."

The guard was already at the door of the booth. Berland left her computer and began her walk.

Unlike Maheu's walk, there were fewer people to see Berland go. The chief marketing officer waved goodbye to her former assistant, before taking an elevator down. After a short ride, she exited through a back entrance and was whisked into a car that drove into the drab Manhattan cold. As word of her abrupt firing spread through the office, Berland tweeted a simple goodbye: a single blue heart emoji.

—

>>> TWITTER PRODUCT MANAGERS, designers, and engineers knew they had days, if not hours, to prove their worth. Outside of Twitter Blue, Musk didn't appear to have a clear road map of products for employees to work on. Continuing the projects chosen by old management seemed risky—Musk had already made it clear he didn't respect Agrawal's vision.

Ideas were thrown out from everywhere, and projects were started and stopped as Musk's lieutenants jockeyed for position, all believing they had brilliant ideas to transform Twitter. With Musk tied up sorting out operations and layoffs, some of the goons portrayed themselves as decision-makers and issued product directives.

Calacanis asserted himself. He sat in on meetings with product and policy staffers, later tweeting out his impressions and findings as if he were speaking for the wider company. His posts hinted that he was taking suggestions from his Twitter followers and relaying them to the top. But Musk wouldn't tolerate anyone else running product. When he saw Calacanis's tweets, Musk dispatched someone to the war room to tell his friend to stop acting like he was calling the shots.

"To be clear, Elon is the product manager and CEO," Calacanis later tweeted, following the reprimand.

Desperate to save their jobs, workers started throwing mud at the walls. One team was spun up to work on paid direct messaging, a feature that would allow regular people to send private notes to so-called Very Important Tweeters. Mock-ups presented to Musk's entourage showed a user paying a few dollars to message the musician Post Malone, with Twitter taking a cut of the proceeds.

Another pet project for Musk was encrypting direct messages, which would prevent anyone except the sender and receiver from reading them. For years, the company had debated whether to encrypt the messages, sealing them off completely so that even Twitter itself couldn't access them or share them with law enforcement. Gadde wanted to offer more privacy to users but feared that encrypted messages would facilitate harassment and the exchange of illicit material. Engineers scrambled to kick-start the effort, code-named Night Parrot.

One night that week, a member of Musk's team called one of Twitter's cybersecurity engineers, who had been tapped to aid the transition. Could he come back to meet with the boss? The security expert sighed—he had just settled in after a long day at the office and taken a shower. He got dressed again, making his way back to work through San Francisco's empty streets.

Musk held court in his conference room, while two bodyguards with Texan accents loitered in the kitchen nearby. Around midnight, he finally beckoned the engineer in. Musk was cordial, asking about the worker's background and what he'd done at Twitter over the years, but the meeting felt like a job interview. Musk was deciding whether he could trust the engineer.

Then Musk's paranoia emerged. "Did Twitter read my DMs?" he asked. He appeared convinced that the former executives had been snooping on him during the lawsuit.

The engineer tried to answer as best he could. The company made it extraordinarily difficult for employees to access a user's direct messages, but some workers were allowed to do it when responding to reports of abuse or subpoenas. As far as he knew, no one had read Musk's messages—but it wasn't impossible.

Musk kept digging, certain the old regime had spied on him. Who could have had access? And how could he find out? The engineer delicately tried to steer the conversation back to encryption, saying Musk's worries were a good example of why Twitter should encrypt messages. The pit in his stomach widened, and he felt that whatever he said would be the difference between being promoted or fired. Finally, Musk moved on.

In other meetings, Musk toyed with the idea of adding paywalled videos—like the adult content site OnlyFans—allowing users to charge for premium content. Twitter would take a cut of those proceeds, though staffers immediately began expressing concerns about the potential for pornography and pirated films, which would degrade advertiser trust in the platform. Paywalled videos and the paid message features never launched, while encrypted direct messages were unveiled months later in a very limited capacity.

Employees scrambled to latch on to Musk's pet projects, messaging each other privately to try to join Night Parrot or find their way into Crawford's shadow project, which wasn't much of a secret as Musk and Calacanis tweeted about their idea of charging for verification badges. With Musk's stamp of approval, the Twitter Blue relaunch felt like the surest way to survive the coming layoffs.

Crawford had circulated a spreadsheet of the seventy employees around the globe who were working on paid verification. It became known as a de facto "safe list" and some members of her team began to add their friends and colleagues with the hope of throwing them a lifeline. The list started to flood with the names of people who were dependent on Twitter's healthcare benefits for themselves or their family members, or relied on their job to maintain their immigration status. Eventually, one manager put the kibosh on the effort and locked the document from further additions.

"The team needs to be light," the manager messaged the offenders.

Even Crawford, who had a direct line to Musk, wasn't immune from the rumors. After Maheu's and Berland's firings in New York, she became convinced that the executive terminations had also swept up her boss, Tony Haile.

As paranoia swirled about Haile's supposed firing, in which he was said to have been marched out by the guards, Crawford was gutted.

"Why would they do that?" she said to some of her Blue employees. "It's so unnecessary. Like, I'm a go-getter but I have limits. We're here to ship this thing and we're going to get this out. But if we quit, it will be on our own terms."

Haile, however, had not been terminated.

"Hey I'm still here," he wrote in a message in Slack to his team later that day. "There are rumors of me having been fired but as far as I know I still work here." (Haile eventually did quit.)

Unwilling to wait for the axe to fall before taking action, workers tried to connect with each other off Twitter-controlled platforms, which they knew were monitored by Musk's team. They added one another on LinkedIn, created external Slack groups, and traded phone numbers to create support networks in case they all lost access to their email accounts. They also began circulating "A Layoff Guide," which outlined workers' rights in the U.S. and gave tips on how to deal with requests from new management and workplace surveillance. Some even tweeted at Musk, asking him to lay them off in order to spare colleagues on work visas.

The human resources employees responsible for pulling off the massive cuts felt just as confused. It seemed to them that Sacks, Calacanis, Davis, and the other Musk lieutenants weighing in on the layoffs were trying to outdo each other, each hoping to emerge as the most ruthless. The lists of employees who would be dismissed kept expanding to keep up with their demands.

Birchall emerged as a lone voice of reason among Musk's advisers. The HR teams discovered they could go to him to argue for a team or an employee to stay, and he would listen to the justification. The rest of Musk's people seemed to delight in laying waste to Twitter.

Birchall, Spiro, and Teller tried to counsel their new Twitter counterparts about how to speak to Musk. Never pretend to know the answer when you don't, they said. They also recommended against ever having open-ended conversations with Musk. He should be presented with options, preferably no more than two or three, and be allowed to choose from them. The Twitter employees should then make their recommendation from among the options so Musk could assess their judgment.

Pacini took their recommendations to heart. She and the rest of the Twitter employees assigned to the layoff drew up options for Musk. Brian Bjelde, a human resources vice president Musk brought over from SpaceX, weighed in, explaining how the rocket company had handled severance in the past so that the options they eventually put in front of Musk would have the benefit of having been vouched for by someone who had spent twenty years working for the billionaire.

One was a bare-bones version that covered what Musk was legally obligated to pay, and the other was slightly more generous. The severance was uninteresting to Musk. He told Pacini to let Bjelde pick what he thought was best. Musk also remained noncommittal about the total number of employees to eliminate, leaving the decision-making to his inner circle.

As employees went searching for morsels of information about the layoff, they found that Musk's transition team hadn't been careful to keep their plans a secret. On Wednesday, workers discovered an open Slack channel called #tundra-ec-comp, in which human resources employees were talking openly about severance and cuts. The workers took screenshots and quickly circulated them in private chats.

"The severance calculations are updated based on the master list as of 12:30pm today," one HR rep wrote, noting that the list was "pretty much final."

The employees fixated on her last message. "The count on my sheet is 3738," she wrote.

The figure represented just about half of Twitter's 7,500 full-time employees, the most solid confirmation that many people were about to lose their jobs. The workers kept digging. That afternoon, one realized that the online calendars for some of the transition team were still public and found an invite on Sacks's schedule for Project Tundra, a discussion for the so-called "reduction in force." Also invited to the meeting were Gracias, Spiro,

Birchall, and Teller. After workers distributed screenshots of Sacks's calendar among themselves, employee calendars were suddenly made private.

Later that day, an engineer in the San Francisco office was just pacing the hallways when he came across one of the hired hands brought in from Tesla, who seemed engrossed in a phone conversation. The Tesla worker didn't seem to notice the Twitter employee approach as he hissed into his device.

"Cut off their access at 5 p.m. None of them are to be trusted."

33 > The Snap

By the morning of Thursday, November 3, employees had all but confirmed their firings by snooping on internal calendars and message boards. But they'd still heard nothing from Musk. Instead of telling his new employees that they would be laid off, Musk was busy sharing the news with advertisers.

Musk hopped on a call with Twitter's Influence Council, a group of more than one hundred advertisers and executives from corporations like General Motors, Mastercard, and Microsoft. The group held occasional meetings to give guidance to the company, and although they'd expected to gather at the beginning of 2023 for an all-expenses-paid getaway in Napa Valley's wine country, Musk had canceled the event.

"We are doing a reduction in force at Twitter and that's going to be sort of executed in the next few days," Musk said to open the call. "And it just doesn't seem right that if we did a big reduction in force, and then . . . now you're at a boondoggle in Napa."

Among those listening in were Linda Yaccarino, the head of global advertising at NBCUniversal, as well as Roth, one of the company's remaining safety and policy executives, and Robin Wheeler, a sales executive who was attempting to keep Musk on task.

Wheeler had been abruptly elevated after the departures of Maheu and Personette and was facing a bit of an audition in front of her new boss. As marketing officers who collectively controlled billions of dollars in ad spending listened, Wheeler stroked Musk's ego.

"We have the greatest, greatest product and engineering innovator of all time working at the helm of this company, and I'm so excited about what that means for our product road map," Wheeler said. Later, she claimed Musk's ownership had already caused Twitter to attract its highest-ever number of daily active users—like a homeowner watching her house burn down, marveling at how many neighbors were rubbernecking.

While the marketing leaders were deferential, they also peppered Musk with their concerns, asking about content moderation, product plans, and the billionaire's late-night tweeting habit. He tried to address them all, promising the company would make no major content decisions until "at least a week" after the coming U.S. midterm election. He also focused on his newly co-opted idea of "freedom of speech, not freedom of reach," which Roth had introduced to him during their meeting the week prior.

As he listened to Musk's pitch, Detavio Samuels felt skeptical. Samuels was the CEO of Revolt, a media company founded by Sean "Diddy" Combs that focused on Black audiences. He had read reports that usage of racial slurs had increased dramatically on Twitter since Musk's takeover, as trolls were emboldened by the new ownership to spew hate. Black communities made essential contributions to Twitter, he said, and Musk's approach to content moderation concerned him. He built upon Musk's metaphor to make his point.

"As a Black man, I don't want to walk down a neighborhood where people are whispering the N-word, whether I hear it or not. I don't want to walk down a neighborhood where I hear one person whisper the N-word. Maybe nobody else heard it, but I did," Samuels said. "As you continue down the path that you're going, I really recommend that you have conversations with this specific community and that we make sure that whatever solutions are created are solutions that make them feel safe and in a welcoming space."

Musk agreed Twitter should "make people feel comfortable." But then he suggested Samuels's entire premise might not matter.

"I don't know if you know this, but Puff is an investor in Twitter," he said, using a nickname for Combs. "You know, he's a good friend of mine. We text a lot."

Some of the Twitter executives had to resist the urge to bury their heads in their hands.

>>> NOVEMBER 3 WAS an abnormally warm November day in Manhattan, and the sun heated the roof deck at Twitter's New York office to nearly 70 degrees Fahrenheit. More than fifty employees headed upstairs to gather together one final time. They traded phone numbers and chatted about job opportunities, then arranged themselves on the patio furniture and posed for a picture. There were a few forced smiles.

Everyone was convinced it was their final day at Twitter. Earlier that day, employees had noticed that their days of rest, the monthly no-meeting, recharge days, had been removed from their calendars. The new owner was sending a message.

Musk's transition team had also ordered the shuttering of the internal company directory, known as Birdhouse, which listed people's titles and chain of command. Without it, workers wouldn't be able to see who was left at the company, or, conversely, determine who had been fired.

By 5:00 p.m. that afternoon in San Francisco, some employees had thought they had survived another day. Still, workers around the globe, some who stayed online long past the end of the working days, remained vigilant.

"Has the red wedding started? 👀" one London-based marketing employee joked in a public Slack channel, alluding to an episode in the popular television show *Game of Thrones* where several main characters are surprisingly and brutally murdered.

"Cant see org chart on birdhouse?" another worker wrote at 5:04. "Did they change that?"

just now, right?

directory is gone too.

It breaks regularly

Thirteen minutes later, they got the email they were all expecting: "Update regarding our Workforce"

> In an effort to place Twitter on a healthy path, we will go through the difficult process of reducing our global workforce on Friday. We recognize that this will impact a number of individuals who have made valuable contributions to Twitter, but this action is unfortunately necessary to ensure the company's success moving forward.

> To help ensure the safety of each employee as well as Twitter systems and customer data, our offices will be temporarily closed and all badge access will be suspended. If you are in an office or on your way to an office, please return home.
>
> We acknowledge this is an incredibly challenging experience to go through, whether or not you are impacted. Thank you for continuing to adhere to Twitter policies that prohibit you from discussing confidential information on social media, with the press or elsewhere.

The email, signed by a faceless "Twitter" and not Musk himself, explained that those who were safe would receive confirmation via their work emails. Those who were not would be notified of their terminations to their personal email addresses by 9:00 a.m. the following morning.

But while Musk had been unconcerned with the scope of the cuts, he had insisted on getting final authority over the message that was sent to employees. Pacini had written three versions of the email for him. The first was quintessentially Twitter 1.0—it apologized for the cuts and wished employees well. The second was a middle ground that laid out the news with few platitudes. And the final version was the most cold and curt. That was the one Musk had picked.

When the email hit their inboxes, some workers were congregated in the Lodge, a wood-paneled common area inside Twitter's headquarters that was decked out with two miniature log cabins where employees sometimes relaxed or held conference calls. The Lodge served beer alongside coffee, and video screens lining the walls often broadcasted sports games. Shortly after, security guards came in and cleared everyone out of the round tables and cozy cabins. The building was closing, they said.

Twitter employees headed downstairs to await their fate. On the ground floor of Twitter's headquarters was a gourmet food hall called the Market, which sold groceries and grab-and-go lunches. Crucially, it also had a small bar in the back. Workers flooded in, overwhelming the lone bartender who had no idea about the bloodbath that loomed from above. As she frantically poured beer, the overcrowded bar felt like the last day of camp, with everyone commiserating about the summer while they waited for their parents to come pick them up. Some of the workers from Musk's other companies trickled in, too, gathering on one side of the room.

Everyone kept their phones gripped in their hands, frantically scrolling

Slack as they drank. Although Musk's message had promised that terminated employees would receive email notifications, many of those who were cut were not even afforded that small dignity. The layoffs were unprecedented in Twitter history, and its engineers had simply not had enough time to properly build an off-boarding system that would send out the required emails and cut off access to all internal systems simultaneously. The system ran haywire, and employees found that they had abruptly been locked out of their work computers or email accounts.

Some were banished from Slack, while others maintained access even though they had been locked out of email. The employees who remained vented their sadness and frustration by posting blue heart emojis in Slack, while others shared memes. One person posted an image of a scene from *Avengers: Infinity War* in which Thanos, the main antagonist, gains powers that allow him to eliminate half of all life in the universe with the snap of his fingers. This was Twitter's own mass extinction event. Employees began referring to it as their "snap."

As employees chimed in about their own layoff experiences, or piped up to say they still appeared to be safe, tweeps began flooding Slack with the salute emoji. The stern-faced symbol became a favorite among Twitter employees as the perfect encapsulation of the solidarity and sense of duty they felt as they ground ahead at work that year under the distractions of Musk.

"Can't log into emails. Mac wont turn on," one London-based tweep tweeted, sharing photos of his laptop screen and his email log-in page. "But so grateful this is happening at 3am. Really appreciate the thoughtfulness on the timing front guys."

The layoffs were like a brushfire that spread across an open field. They arrived in New York and along the East Coast, and then workers in London and Dublin received emails saying that they were in roles that have been "identified as potentially impacted or at risk of redundancy." Some in California and Seattle received termination emails around 11:00 p.m. In Tokyo, employees worked through most of their shifts on Friday before receiving notices toward the end of the day.

Inside the Market, employees kept glancing down at their phones. As some of them received their termination notices, they shouted to the room that they were out.

"I just got the message," one yelled.

"It was nice working with everyone," said another.

As the night wore on and the layoff emails kept landing, Twitter's workforce trickled across Market Street to the Beer Hall, a dark taproom that had hosted countless goodbye parties for departing tweeps over the years. They continued drinking there until the late hours of the night. Some workers didn't get their layoff notices until they stumbled home, while others woke up to the messages the following morning.

Even those who were working on Musk's pet projects weren't safe. After 9:00 p.m. on Thursday evening, Crawford summoned her team to join their nightly "standup" meeting, in which they shared updates about the Blue Verified project.

"I recognize this is a crazy moment where Tweeps are losing access and the layoff is in progress," she wrote in a Slack group for the team. She included a broken heart emoji and a link to a video conference line.

After the meeting started at 9:30 that night and Crawford began going over the work that had been done, two employees suddenly dropped off the call. The workers found themselves unable to get back on because they were locked out of their emails and computers. Realizing what had happened, they texted their Blue teammates to tell them that they were out. As word of the layoffs rippled through the meeting, people's faces sank, but Crawford pressed on.

34 > The Aftermath

Twitter's remaining employees opened Slack on the morning of Friday, November 4, like survivors emerging from their bomb shelters after a nuclear explosion.

Managers tried to take stock of the wreckage. Some of them didn't know which of their employees remained and asked them in Slack to raise their hands via emoji reaction to show they were still around. Ella Irwin, a leader on trust and safety, announced to her team that she was still employed, before linking to a document and asking hundreds of people to fill it out as a demonstration that they were still there.

The depth of the cuts became clearer as employees were able to piece together who was gone and who remained. About half of the company's full-time workforce was eliminated. The company's infrastructure organization, known internally as Redbird, lost about 80 percent of its engineering staff. Dozens of product managers were laid off across the company, while teams that focused on human rights, experiences for disabled users, and marketing were all reduced to skeleton crews. The compliance department, which monitored the company's adherence to global regulations, was expecting to lose a bit more than a fourth of its employees, but ended up losing half its staff.

The teams that dealt with civic integrity and content moderation were also slashed, even though Twitter was juggling an important election period. The result of Brazil's presidential runoff election was still being contested online, while the U.S. was preparing for its own midterm elections in less than a week. Musk and his friends had decided to drastically reduce teams that labeled political misinformation. They fired some of the only people who knew how to operate tools that ferreted out misleading tweets about the elections.

Some of the people who remained had survivor's guilt, wondering why they had been spared in the haphazard cuts while friends and colleagues got the axe. It didn't make much sense. Some teams, like communications, were cut from hundreds down to a handful of people. Others were untouched. Some people who were known to be poor performers didn't get the email, while engineers seen as essential to operations had their access unceremoniously cut off. Some expectant mothers and workers on parental leave lost their jobs. So did some on visas.

Former employees organized in encrypted group chats and private Slack rooms to commiserate and contemplate legal action against Musk over their abrupt dismissals.

By Saturday morning, Dorsey, who had said little about the takeover since he anointed Musk in his April tweet thread as the "singular solution I trust," felt compelled to weigh in. In the months since he signaled his approval of the deal, he had tweeted only a few times, mostly about Bitcoin. But even Dorsey wasn't blind to the salute emojis, goodbye tweets, and devastation that filled his timeline. "I realize many are angry with me," he tweeted. "I own the responsibility for why everyone is in this situation: I grew the company size too quickly. I apologize for that."

> I am grateful for, and love, everyone who has ever worked on Twitter. I don't expect that to be mutual in this moment . . . or ever . . . and I understand. 💙

It was Dorsey's first public admission that the growth he had pursued as chief executive had put the company on an untenable path, a conclusion long accepted by Twitter's previous executives and board. Left unsaid, however, was how he—and Twitter's board—empowered a man with $44 billion to callously slash and burn the people for whom he supposedly felt "love."

Dorsey's message came after about twenty employees in Twitter's office in Accra, Ghana, began receiving notices that they were all being laid off and would receive a month of salary as severance. Most had been hired only in the past year, after Twitter opened its outpost there in April 2021.

As news broke of Twitter's treatment of its employees in Accra, Lara Cohen, a vice president of partnerships who was also laid off, attempted to bring the situation to the attention of Dorsey and tagged him in a tweet. Dorsey replied in a direct message to Cohen asking her what exactly she wanted him to do.

"Maybe you can use your influence to do better by these people," she responded. "They deserve better than this. We made a commitment to them."

Dorsey seemed unmoved. "I'll certainly look to hire them," he responded.

35 > Verified or Not

For Esther Crawford, Twitter Blue represented an opportunity. In private, she told those working on the project with her that charging users to become verified was an idea with limited possibility for success. At the same time, she saw it as a chance to impress Musk and gain influence. Musk's fans and lieutenants had hammered the company as bloated and lazy.

"This is our chance to show the best of what Twitter has to offer," she told her team. It was also a chance to build karma with Musk. If they could accomplish this one task now, she suggested, they would be able to negotiate for what they wanted—raises, new positions, new products—after gaining his trust.

While she saw the upside of getting Musk's approval, the absurdity of being asked to launch a product built to his specifications in less than two weeks amid widespread layoffs was not lost on Crawford. In one group gathering, she presented her team with customized mugs that wouldn't have been out of place at a tech-themed Bed Bath & Beyond. "Chance made us coworkers, crazy psycho shit made us Tweeps," read the mugs' inscription.

Depending on who you asked, Crawford was either a Musk loyalist or a rank opportunist. She certainly wasn't the only one in either case. Musk's fanboys were everywhere, and some emerged in the wake of his acquisition to celebrate and critique their coworkers who lost their jobs. After the Snap, an air of self-preservation hung heavily over Twitter. One engineer, who was desperate to cling on to her healthcare, privately criticized Musk and his ideas to her colleagues, while commending the billionaire in public Slack channels, claiming that "Elon is the Steve Jobs we need." Others believed that the Musk shake-up could be their opportunity for a promotion or job change that had been hindered by Twitter's previous bureaucracy.

Most of the employees Crawford brought to the Blue Verified team came to view the project as pointless at best and, at worst, something that could drastically undermine trust on the platform. Paying for verification badges would lead to impersonation and undermine the whole principle of the check marks, they thought. Crawford shared their worries. But it was what Musk wanted. No amount of reasoning or lessons from Twitter's past would convince him otherwise.

—

>>> BEFORE THE CHECK MARKS, impersonation on Twitter—as with the original @ElonMusk account—was pervasive. While the company's founders knew it was a problem, cracking down on accounts pretending to be famous people was always a secondary concern as they struggled to keep the site online. By June 2009, however, the situation became untenable.

Tony La Russa, the World Series–winning manager for the St. Louis Cardinals, became fed up with a parody account that made jokes about unsavory team incidents—including the death of one of his players—and sued Twitter for hosting the content. The fake tweets were "derogatory and demeaning" and damaged his image, La Russa's lawyers argued. In response, Twitter took down the imposter and rolled out a solution: Verified Accounts.

The company began slapping the verification badge on the accounts of athletes, politicians, musicians, and public officials. With the introduction of the blue seal, Twitter became the first social network to verify its users, sparking one of its most lasting impacts on internet culture. Other services began verifying prominent users—including Google+ in 2011, Facebook in 2012, and Instagram in 2014—creating coveted social media status symbols and spreading a class system across the modern internet.

While many people saw the check mark as a designation of fame, it became an important part of Twitter's public utility. It marked the real accounts for brands like McDonald's and Coca-Cola, making the platform much more attractive to advertisers as well as to government and emergency service accounts, which provided enormous utility for those looking for train times or searching for information in the wake of a natural disaster.

>>> MUSK'S BLUE VERIFIED program was going to destroy the whole system. If Musk's paid subscription service allowed anyone to be verified, then no one would be, in the traditional sense. The utility of the check mark would evaporate. As one Blue Verified worker later put it: "It was such an obvious trainwreck, that the main job of everyone on the team was to make sure it was the safest trainwreck possible."

Crawford and her team tried to develop safeguards that would protect the usefulness of verification. On October 31, they presented two options to Musk and his team. In the first, there would be two badges: people who were already verified would keep their filled-in badges designating importance, while those who had purchased them through Blue Verified would get transparent badges. To illustrate the differences, they created mock-ups of two tweets. One, from the legacy verified @JoeBiden account, advocated for people to vote. The other, from the Blue Verified @JoeBiden account, tweeted it was "starting nuclear war with Russia." The only thing to distinguish between the two accounts was the transparent badge on the fake one.

The second option from the Blue team showed both accounts with the same type of verified badge. But in addition to that badge, official accounts, like @JoeBiden, received another label to clarify their importance. In Biden's case, his account was delineated as a "United States government official."

The attempts to make clearer distinctions between legacy and paid verified users, however, was not embraced by some of Musk's friends. The first option "feels like a second-class citizen," Sacks wrote in an email that night, adding that it would "disincentivize purchase." He expressed more support for the second option, but cautioned against using the labels beyond government officials.

Sacks's email kicked off a reply-all discussion. Crawford noted that settling on the aesthetic was a "high priority decision" because it would dictate what the team would move to build. Soon the email thread ballooned with responses from engineers, Sriram Krishnan, and Musk's own assistant. With the boss watching, everyone felt like they needed to weigh in.

"There are too many cooks in the kitchen here," he wrote. "Over time, we can introduce tiers with more features, but the goal right now is to maximize the number of verified users as fast as possible to improve the user experience for everyone (verified or not) on Twitter."

—

>>> MUSK'S HYPERFIXATION ON Blue extended beyond the design. He also engaged in seemingly endless deliberations about how much it should cost. Meanwhile, advertisers were bolting for the exits and Twitter was bleeding revenue. Sacks insisted they should raise the price from its current tag of $4.99 a month to $20 a month. Anything less felt cheap to him, and he wanted to present Blue as a luxury good. In emails, he compared Blue to a designer purse.

"Chanel could make a fortune selling a $99 bag, but it would be a one-time move," he wrote. "A 'promotional offer' may not be the position we want. A luxury brand can always move down-market, but it's very hard to move up-market once the brand is shot."

Calacanis, however, vehemently disagreed. "It should be $99 a year," he insisted. During one meeting, he launched into a spiel about how Twitter users were more likely to open their wallets for a $100 per year subscription if it seemed slightly cheaper at the $99 price, as though he had just watched a YouTube video explaining the basics of consumer psychology.

Musk seemed more swayed by Calacanis's price than Sacks's, except for the fact that he hated the number 9. Tesla never used the number on its website, and he saw any attempts to play mind games on customers as tacky.

"That's dumb. We don't do it and it makes no sense," he chastised the podcaster.

"Fine, it can be $100, but it should be $99," Calacanis relented.

Musk also turned to his biographer for advice. "Walter, what do you think?" Musk asked during a meeting about pricing.

"This should be accessible to everyone," Isaacson said, no longer just the fly on the wall. "You need a really low price point, because this is something that everyone is going to sign up for."

To those in the room who designed or built Blue Verified, the discussions were baffling. There was Musk, a supposed genius leader who had built billion-dollar companies, soliciting advice from a small inner circle of advisers who had little experience working on social networks. Sure, they used Twitter, but these rich men were not representative of the hundreds of millions of people around the world who logged in to the platform every day. They were all making gut decisions based on their experiences in their own Twitter filter bubbles—and Musk lapped it up.

Musk had largely come to peace with his price of $100 a year for Blue. But during one meeting to discuss pricing, his assistant, Jehn Balajadia, felt compelled to speak up.

"There's a lot of people who can't even buy gas right now," she said, referencing skyrocketing inflation. It was hard to see how any of those people would pony up $100 on the spot for a social media status symbol.

"But think of everyone with an iPhone," Musk responded. "If you can afford an iPhone, you can definitely afford this."

He paused to think. "You know, like, what do people pay for Starbucks? Like $8?" Before anyone could raise serious objections, he whipped out his phone to set his word in stone.

"Twitter's current lords & peasants system for who has or doesn't have a blue checkmark is bullshit," he tweeted on November 1. "Power to the people! Blue for $8/month."

—

>>> CRAWFORD GREW CLOSER TO MUSK, texting with him regularly, and members of his inner circle started opening up to her. In the chaotic early days of November, Balajadia sat Crawford down to coach her on managing Musk. She was one of the few women Musk trusted, and while she technically worked as an operations coordinator for the Boring Company, she was treated as a glorified assistant. She monitored his calendar, followed him at some public events, and set up his laptop for him when Musk needed to do work that couldn't be done on his iPhone. Balajadia also made some public appearances on behalf of the Boring Company and, in 2018, presented a "proof of concept" plan to the Culver City Council to build a 6.5-mile tunnel from SpaceX headquarters in Hawthorne, California, to West Los Angeles. The tunnel was never built, but Balajadia continued to devote her life to Musk.

"I can tell you're going to be around for a while, so let me tell you something," Balajadia said. "Elon is special in this world. It is our job to protect him and make sure what he wants to have happen, happens. We need to protect the mission."

To Crawford, the message was eerily similar to those she grew up hearing in a Christian cult in Oklahoma. But instead of a prophet, Balajadia had Musk. She had spent large chunks of time away from her young child so she could travel with him. But the mission merited those sacrifices.

Crawford came up with her own tactics for dealing with Musk. Knowing that she needed his approval, she avoided upsetting him at all costs. Maybe she could live up to her Old Testament name, protecting her people from a king who planned to exterminate them. She quickly learned that she could challenge him, but typically in one-on-one settings where he was jovial and willing to learn from the person in front of him. Individually, Musk could be charming, willingly engaging in discussion and listening to the expertise of his counterpart. Put him in a larger group setting with people outside of his inner circle or those he didn't trust, however, and Musk's ego ran wild. He could never be seen as inferior or uninformed. The people who survived in his orbit learned this quickly.

In one meeting, days after the deal closed, Musk was sitting in his second-floor conference room scrolling Twitter as some of the transition team chatted about product ideas. After reading a tweet about the FBI and Hillary Clinton—a constant fixation for the right-wing internet—he announced to the room that he had drafted a tweet about Clinton, hoping to elicit some laughs. Crawford stood up. "You can't do that right now!" she said dramatically. They had product matters to discuss and she didn't want it to be a distraction. Maybe do it later, she suggested, before bursting out laughing. The histrionic, joking approach worked. "Are you my tweet adviser now?" he asked, raising an eyebrow at Crawford. He never sent the tweet.

Publicly, however, Crawford was seen as a cheerleader. Through the chaos and callousness of the early days of the takeover, she remained outwardly optimistic, giving others the impression that she had become Musk's devoted fan. On the night of November 1, after working long hours with the Blue team, she and her colleagues decided to tweet a joke. She had brought an eye mask and silver REI sleeping bag to the office for nap breaks, and after a product designer named Evan Jones had snapped a photo of her actually sleeping, she asked him to retake the picture before he posted it. He climbed on an office couch to get a better overhead angle of her in a sleeping bag, then tweeted the photo. Crawford then retweeted it with her own message: "When your team is pushing round the clock to make deadlines sometimes you #SleepWhereYouWork."

The image rocketed around the internet. To Musk's detractors, it symbolized the cringeworthy hustle culture that dominated the tech world and normalized the idea of working at all hours to appease a corporate overlord. To Musk's supporters, the photo showed the impact a once-in-a-lifetime innovator was having on a company that needed a revolution. In reality, it was a joking, if not shrewdly calculated bit of self-promotion by Crawford. She and Jones faced a maelstrom of criticism across Twitter, and the product designer wanted to delete his post. But Crawford demanded he keep the photo up. This was her time to shine.

>>> THE HUSTLE WASN'T completely fake. The Blue team stayed up late fielding requests from Musk and his transition team, and created a twenty-four-hour rotation where projects could be passed to other employees around the globe to ensure that work never stopped.

Crawford tried to share her concerns about burnout with Musk. Some people on the Twitter Blue team had taken to monitoring their elevated heart rates on their Apple Watches and sharing them with their colleagues to make light of their situations. "I don't want to push the team to die over this," Crawford told Musk.

"Well, push them to just before they die," Musk responded, laughing. While those who heard the exchange understood it to be a joke, they also wondered if there was an element of reality to it. They had heard the stories over at Tesla and SpaceX.

Given the quick turnaround, the Blue employees knew they'd likely launch a minimally viable product and could only pray that there wouldn't be any major bugs or issues. At times, they worried they were going too fast and could run afoul of regulations. In one meeting, Crawford raised a potential pitfall. The company's lawyers had warned that there were consumer protection laws outside the U.S. that would require the company to refund subscribers who were later suspended by the company for breaking its rules. We might be liable to give some money back, Crawford explained.

"Don't care," Musk said. "Fuck it. Do I seem concerned about legal battles? Move forward. Get this thing done."

The only thing that seemed to hold Musk somewhat accountable was a fear of public shame on Twitter. In one meeting he blurted out that House Representative Alexandria Ocasio-Cortez, a Twitter power user and darling of the progressive left, would roast him if the Blue relaunch went poorly. Calacanis chimed in to agree, stating that he thought Ocasio-Cortez would run for president one day.

At Tesla and SpaceX, engineers often joked that the only laws they respected were those of physics and Musk's own demands. At Twitter, the Blue team got a crash course. In the same meeting, Musk wanted to review the descriptions of Twitter Blue that would be used online and in the Apple App store. "It should be: 'Rocket to the top of replies, mentions, search, and topics,'" he said, reading through some of that copy. "Remove the comma before 'and.' I find it troubling."

Crawford, a fan of the Oxford comma, tried to defend it. "I find the opposite with commas," she said with a smile. Some in the room shifted nervously in their seats.

"Too bad, I'm the law," Musk replied.

—

>>> AS THE NEW TWITTER BLUE features came into focus, so did the fears about how it would be exploited for impersonation. What would happen if an account pretending to be a local fire department declared an emergency? Or a fake account posing as a politician or election authority spread misinformation about an upcoming vote?

Stopping election interference was of paramount importance to Twitter employees. Brazil, one of Twitter's largest markets outside the U.S., held its presidential runoff election the weekend after Musk walked into Twitter. The populist incumbent, Jair Bolsonaro, was unseated by Luiz Inácio Lula da Silva but said he would challenge the results. As Musk dismantled Twitter's trust and safety team, hashtags in Portuguese that questioned the veracity of the vote trended on the platform. Employees who remained on civic integrity initiatives gamed out situations in which Bolsonaro, who Musk claimed to sometimes speak with, could spread disinformation using paid verified accounts.

Twitter users and employees also raised concerns about the upcoming U.S. midterm elections, which were set to occur on November 8. The Federal Bureau of Investigation shared those fears. About a week before Election Day, an agent from the bureau's Foreign Influence Task Force reached out to the company to better understand Twitter's plans for Twitter Blue.

Musk was briefed on the outreach during a November 3 meeting and was told that the FBI was interested in whether there would be changes to currently verified official government, state or federal, election accounts. The agency also asked whether fraudulent accounts posing as government officials would be able to become verified.

"They are generally concerned with any changes that might impact election misinformation operations during the election," Crawford said.

Musk's response didn't exactly breed confidence. "We can launch Blue, but wait until after the election to change ranking," he said. Musk had planned to prioritize tweets from Blue subscribers in the algorithmic Twitter timeline, giving his paying users an advantage when it came to getting the most attention on the platform.

Musk also said he planned to give an "official badge" to all government entities "of significance" to indicate that they were authentic accounts. But Musk didn't define what he meant by "of significance."

Employees knew it was impossible to filter through the thousands of verified accounts that could potentially deal in official U.S. elections and determine if they should receive "official" billing just five days before the vote.

—

>>> ON THE EVENING OF NOVEMBER 4, as the company was being gutted by layoffs, Crawford and her team submitted the final changes for the Twitter Blue launch to Apple's App Store. Initially the subscription service would be available for purchase only through Apple devices, partly because the rush meant there had been less focus on the Android app, and partly because

Musk believed he could piggyback off Apple's ID and payment systems. Everyone had to have a unique user ID to use and pay through Apple devices, and by relying on that, Twitter would have to do less verification of users on their own, he argued.

The billionaire called Crawford to personally congratulate her on the effort. It was a validating moment. The product manager had proved to Musk that people at Twitter could get shit done. But his praise didn't erase the pit in her stomach as some of her colleagues who worked on the project were being terminated.

She was also still fretting about Musk's plan to launch paid verification on November 7. Despite the warnings from the FBI, Musk didn't seem to register the problems with his agenda.

That Friday, an employee took to a company-wide Slack channel and tore into the timing of the launch.

"Why are we making such a risky change one day before elections, which has the potential of causing elections interference by malicious groups that could abuse the new rules to their advantage in the spread of misinformation?" they asked.

Crawford made one last attempt that weekend in a private chat with her boss. "Do you want to be blamed for the outcome of this election?" she asked.

"Well, when is it?" Musk replied.

"It's in two days," Crawford said, stunned that he hadn't clocked the date that she and her team had been warning him about since the start of the project.

Musk paused, processing. "Oh, I didn't realize," he said after a moment. "Okay, yeah, it's fine. We can wait. Why don't we wait?"

The launch was moved to November 9.

36 > Elections

By November 4, the Friday after the Snap, Musk was in a foul mood. He was no stranger to scrutiny, but the chaos of the Twitter deal had elevated him to a new level of exposure. Musk was the platform's main character, whose every decision was scrutinized and pilloried by the site's users. Twitter was worth far less than Tesla and SpaceX, but it was a center of culture, current events, and news, and there were few subjects people on Twitter liked discussing more than Twitter itself.

Agitated, Musk handled the criticism the only way he knew how: tweeting through it. In a stream of consciousness that flowed through his account starting Friday, he tweeted 105 times over the course of the weekend. Annoyed by the onslaught of critique, he changed his profile description from "Chief Twit" to "Twitter Complaint Hotline Operator."

At first, he turned his attention to Twitter's advertisers. After a week spent trying to reassure them, Musk blamed "activist groups pressuring advertisers" for Twitter's "massive drop in revenue" and chalked it up to his enemies attempting to destroy free speech. Then he turned to bullying, promising on Friday evening to unleash a "thermonuclear name & shame" of advertisers who dared to stop spending money on Twitter.

Through the weekend, he continued with masturbation jokes about Mastodon, the Twitter alternative that users were threatening to migrate to in the wake of his takeover. He aimed barbs at the comedian Kathy Griffin, who decided to turn her Twitter account into a parody of Musk's in an attempt to test the limits of "free speech" and was subsequently banned by the billionaire. And he responded to a tweet that featured a quote from a white nationalist. Musk was always unfiltered, but his weekend felt especially manic.

On Monday morning, employees woke up to a tweet from Musk that shocked them.

"To independent-minded voters," Musk wrote on Sunday, November 6, two days before the U.S. midterm elections. "Shared power curbs the worst excesses of both parties, therefore I recommend voting for a Republican Congress, given that the Presidency is Democratic."

The tweet was the latest indicator of Musk's rightward shift. In June, he had predicted a "massive red wave" in the midterm elections, and in August he appeared as the special guest of Republican House Minority Leader Kevin McCarthy at a GOP retreat in Wyoming, where he glad-handed politicians who, a few years earlier, could have easily been targets of his ire for their connections to Big Oil and climate change denial. He had found a new political home, and was hoping to push his followers in the same direction.

Twitter's past leaders had done everything they could to make it seem like the platform favored no political candidate or party. Dorsey had privately donated to past U.S. presidential candidates like Tulsi Gabbard and Bernie Sanders, but he had never used his account to tell citizens to vote for a particular party or candidate. Like many companies, Twitter had operated a political action committee and donated equally to Democratic and Republican candidates—but the company shuttered it in 2020 after it banned political advertising, arguing influence should be earned and not bought.

—

>>> ON THE MORNING of Monday, November 7, the in-house lawyers who had survived Musk's layoffs logged on to an 8:30 a.m. video call for a briefing from Spiro. As they called in, filling in little boxes on a computer screen, Spiro popped up on the call wearing a tight T-shirt with ARMY emblazoned across the front.

Looking at the sullen faces, he knew the forty people who had logged in from Twitter's legal and public policy teams were wary of him. They wanted assurances. So he tried to speak to them on a level that he thought they could all understand. He told them that the layoffs were over, and a new chapter was beginning at Twitter. Cost cutting would continue, but the company would reduce its spending by canceling contracts, not chopping more heads.

"Listen, I know it's tough, but if everyone can just hang in there," Spiro said. "What you all need to remember is that Elon is a guy who's created electric cars and launched rockets." His belief in his boss's super powers seemed absolute. A positive outcome was on the horizon, he promised.

"We're all going to make so much money out of this," he continued, as some of the faces on the video chat contorted. The lawyers asked about who would lead them. Musk's lawyer had spent time at headquarters in the days following the takeover, asserting his authority over legal and policy decisions, and some onlookers wondered if Spiro was angling to lead the legal department himself.

"You're all leading yourselves," Spiro said. "Each team has a leader and I don't believe in boundaries or walls. There's no rule saying we need a general counsel or a chief legal officer. You're not living in that world." Besides, it wouldn't be useful to appoint someone who could end up gone within three weeks if they pissed off the boss, he continued.

Twitter's lawyers tried to keep their emotions in check but messaged one another in private as Spiro spoke. Some said that Musk's plans to not appoint higher-ups made sense—the billionaire had made his disdain for management clear in his layoffs, and seemed to want to make the decisions himself. Another made her opinions on Musk's hotshot counsel very clear: "This guy's a fucking douche."

The lawyers badgered Spiro about what their roles would be and how Musk planned to run the company. Musk would address them directly in due time, Spiro said. "Be careful what you wish for. If I had thrown you in front of Elon, it wouldn't have gone well for half of you," he added.

The lawyers should drive their own work from now on without waiting for someone to give them directions, Musk's attorney continued. "Don't be afraid to ask questions," he advised. "No one is going to get fired, although you may want to use your judgment as to whether to ask a question in front of fifty people." Spiro knew the best way to convince Musk of something was to speak to him in private.

His bravado seemed to grow as he navigated the Q&A. Some asked about the suddenness of the layoffs, and whether labor laws had been violated in the process. Others asked about how Musk planned to comply with the raft of new global regulations for social media companies that were expected to come into effect soon.

"Here's the thing, I've known Elon for a long time," Spiro said, looking into the camera. "Elon is willing to take extraordinary amounts of legal risk." Some of the people on the video call put their heads in their hands, in disbelief that a seasoned lawyer would say something like that out loud. "He's launching rockets into space," Spiro said. "If the worst we get is a query from the FTC, that's not the big leagues for us.

"People who work at companies are not supposed to talk to media. Those who are doing it and think I'm dumb and don't know it, I do know it and will watch it develop and put my foot down if I want," Spiro said, referencing Musk's willingness to crack down on anyone who leaked to the press. The call ended soon after.

>>> THE LONGER THE Blue team worked under Musk's direction, the more they realized his decision-making was driven solely by gut instinct. Musk's unparalleled success in building two world-changing companies had given him—and his allies—the belief that he was the alpha when it came to product decisions. No one was better or more qualified, and he made that readily known.

Despite running his other companies, Musk seemed like he was constantly available, if not in person, then over text or email, and he wanted to call every shot, no matter how small. After the initial round of layoffs that removed many, many executives and managers, Musk had more than a hundred people reporting to him.

The intensity of it all—the eighteen-hour working days, the middle-of-the-night emails and texts, the whiplash changes of heart—began to weigh on them. Even Crawford was starting to feel the heat. She walked a fine line between pleasing Musk and limiting his potential damage to the website, and often remarked to some of her reports that it was becoming too much. After one meeting in which Musk demanded they once again remove any official designations from individual verified accounts, she told a teammate:

"I don't want the Dunning-Kruger effect to get in the way of good decision-making here."

She and other employees began to understand that Musk's expertise in other areas didn't necessarily translate into running or understanding Twitter. At its core, SpaceX was a physics problem. Tesla was a manufacturing challenge. But Twitter was a social and psychological problem. Beyond the engineering challenges of keeping one of the most trafficked sites online, the company boiled down to empathizing with other human beings and understanding what kept them coming back to the platform; sharing their unfiltered thoughts; or clicking, engaging, and reading whatever came across their feed. It became clear that Musk struggled to understand how anyone else used the platform.

"It's human nature to over-index on your own lived experience," one Twitter Blue team member said, after leaving the company. "But Elon's power and money means nothing stops him from making bad decisions based on his own limited experience."

>>> WHILE TWITTER'S REMAINING employees attempted to prepare the platform for the midterm elections, Musk focused on his more immediate concern: cost cutting.

"Regarding Twitter's reduction in force, unfortunately there is no choice when the company is losing over $4M/day," he tweeted a few days earlier. Left unsaid in his tweet was the $13 billion in debt he saddled on the company, a ticking time bomb that he believed might detonate in a worsening global economy. He had tasked Steve Davis with looking for other ways to cut Twitter's expenditures, and the yes-man immediately set up meetings with members of Twitter's financial planning teams.

In one meeting, a sickly looking Davis, who by that point was already sleeping in the office to accommodate Musk's frenetic pace, asked one finance employee where he would start trimming. The employee was surprised.

"You mean, like teams?" the worker asked, having seen half the company depart a few days ago in the Snap.

Davis laughed. "We already did that!" he replied. It wasn't just people they needed to go after. "Now we have to cut another $500 million from the budget."

It was a jaw-dropping number. In 2021, the company's general and administrative business costs—which accounted for global office rents, employee perks, food, and supplies—totaled about $580 million. Musk and Davis were now looking to excise the equivalent of that budget from the company. Data center costs, employee meals, software expenditures—nothing was off-limits.

>>> EARLY TUESDAY EVENING, Musk entered his second-floor conference room at 1 Tenth as some of the early midterm results began to trickle in. Early returns suggested that the "red wave" the Tesla chief had predicted was turning into a trickle, though he seemed more focused on the meeting in front of him with Twitter's finance leaders, who had been expected to give him a diagnosis of the company's spending, sales forecasts, and potential savings. More than thirty people gathered around a custom-built oak conference table to introduce themselves to the new boss and explain what they did. And as one employee spoke up to explain that he was the head of revenue forecasting for the company, Musk paused him.

"I wanted to talk to you, what's your forecast for the general economy in 2023?" the billionaire asked. "Because I spoke with Ari Emanuel today and he said the economy is going to be down massively."

The name drop caused a few in the room to raise their eyebrows. Musk was citing advice from a Hollywood mogul. The head of revenue forecasting tried his best to explain that Twitter had seen drops in advertising revenue due to the war in Ukraine, the pandemic, and racial justice protests around the police murder of George Floyd, and noted that large global events had a way of impacting how advertisers allocated money. Musk nodded along before asking how Twitter handled the Great Recession of 2008. After someone chimed in that the company wasn't making money back then in its early days, Musk went on an extended discussion on how the recession almost led to the end of Tesla.

"I'm traumatized from it," he told the room, reiterating his belief that the global economy was about to go into free fall.

Then he underlined his commitment to propping up his new purchase. "Also this isn't public yet, but I sold more Tesla stock that will go toward supporting Twitter," he said. In filings with the SEC earlier that day, Musk disclosed that he sold 19.5 million shares of Tesla for nearly $4 billion.

For most of the two hours, Musk listened as employees walked him through Twitter's costs. He was somewhat reasonable to the gathered employees, many of whom were interacting with him for the first time. This wasn't the blustering bomb-thrower from Twitter but someone who seemed to have more control over his executive functions. They also knew that he would not behave like Twitter management of old. Musk wanted to cut, and cut deeply, and wondered aloud why employees had been coddled up until now.

As one director outlined Twitter's lease commitments to office spaces around the world, Musk frowned and squirmed in his seat. The company had offices from San Francisco to London to Singapore, and some of these leases had been signed to lock-in rates for the long term. Musk couldn't believe it.

"I don't care what the contracts say," he said.

"But some of the leases have been signed for the next thirty years," said one director. "We can't negotiate. They've already been agreed to."

"Everything can be renegotiated," Musk replied. "We renegotiate or it goes to zero."

With Davis listening in, Musk's words became the God-sent directive. Employees responsible for relationships not only with office management companies, but also software vendors, cloud computing vendors, and even employee benefits providers, went back to their partners to ask for steep reductions in price. It didn't matter that the old management had signed a contract. There was a new owner now and they couldn't do anything about it.

It was a mob-like move. To many of these third-party vendors, Twitter, a major company with incredible brand recognition, was a marquee client. That was particularly true for the office landlords in major cities like San Francisco and New York, where office rents had fallen through the floor during the pandemic. As the work-from-home trend took hold, there was less demand for office space and Musk and his goons knew it. If a landlord wouldn't reduce the rent, they would just stop paying the bill. Musk knew the company was too big to be evicted from its headquarters, and in smaller offices like Boulder and Seattle, it was a means of breaking these contracts.

As the meeting dragged into the evening, some employees worried that their window for voting that night might be over as polls closed. During the gathering, Musk, on one of his tangents, told the group that he was voting Republican to bring the country more to the center. While he had tweeted to this effect earlier, the employees were still amazed to hear their chief executive advocating for a political outcome so blatantly at work.

Voting records would later show that Musk didn't vote at all.

37 > Zombie Attack

Crawford arrived at Twitter headquarters early on the morning of November 9. With the revamped Twitter Blue set for its unveiling, her moment had come. She was the first to arrive from her team and set up in Caracara, a tenth-floor conference room, as the sun rose over downtown.

But Musk, without even setting foot in the office, had already thrown a wrench in her plans. He'd been agonizing over whether or not there would be other badges to delineate a person's or organization's identity that would be used in addition to the blue check—a measure Crawford wanted to ensure that verification didn't break completely. Of particular influence in the billionaire's thinking was Isaacson, who had pushed Musk to not have anything beyond the check marks that came with paying for Blue.

"I've been talking to Walter about this," Musk had said in an earlier meeting. "And Walter is very adamant that the power of Twitter is that it's democratic where world leaders and regular people can stand shoulder to shoulder. He said we don't want to create a class system and yeah, I trust Walter. He's a very smart guy."

Despite this, Crawford and her team had been able to convince Musk that certain very, very high-profile accounts should be given "official" badges. She began rolling them out on Election Day. The gray, transparent badges sat under an account's username and, as Crawford would describe in a tweet, were a way to "distinguish between @TwitterBlue subscribers with blue checkmarks and accounts that are verified as official." Essentially, the "official" label would serve the function of the old verified check mark, ensuring that users could tell when an account truly belonged to a celebrity, brand, or politician.

By Wednesday morning, however, Musk had come to hate it. The label was being ridiculed on the site, and Musk absorbed the blows. One user, an anonymous fan account for the rapper Nicki Minaj called @NipTuckReload, which had 5,000 followers, tweeted that the labels were "ugly." The billionaire took a screenshot of the post and emailed it to Crawford and her team, who immediately scrambled to create an alternative before Musk did something drastic.

But they were too late. By 8:38 a.m., Musk tweeted that he had "killed" the label and declared they would be no more.

"Please note that Twitter will do lots of dumb things in coming months," he posted six minutes later. "We will keep what works & change what doesn't."

—

>>> THAT MORNING, Damien Kieran called McSweeney in Dublin. The privacy lawyer had been working around the clock with Lea Kissner, Twitter's top security executive, and Marianne Fogarty, its head of compliance, to complete a regular audit that was due to the Federal Trade Commission. The FTC kept a watchful eye on the company's privacy practices and expected quarterly updates about how its privacy work was progressing.

As the lawyer in the group, Kieran was responsible for signing off on those audits and declaring that their contents were accurate. In October, a former security executive at Uber had been found guilty of lying to the FTC during a privacy inquest at the ride-hailing firm, and although he had yet to be sentenced, he faced the possibility of years in prison. The trial had been closely watched by other Silicon Valley executives, and Kieran was well aware that he could find himself in the same position if Twitter made any mistakes in its audit.

But Musk's mass layoffs had made those mistakes all but inevitable. Twitter had to document which employees were responsible for specific privacy programs, and many of them had vanished during the purge. The team finalizing the FTC report was understaffed and was struggling to finish. Kieran, Kissner, and Fogarty had begun discussing whether they should put their names on the document at all and had sought advice from a lawyer.

The suggestion he received wasn't optimistic: Kieran and the other executives should resign before they signed off on an audit they weren't confident in.

On the call with McSweeney, Kieran was cagey. He had received legal advice regarding Twitter's compliance with the FTC, he told her. Was she interested in hearing it? He wouldn't tell McSweeney exactly what was going on but told her it was a "precarious situation."

As the executive responsible for making similar declarations about data privacy to regulators in the European Union, McSweeney wanted to know what was going on. She agreed to convene a meeting with other workers in Dublin, so they could hear what the lawyer had told Kieran and decide what they should do.

—

>>> MUSK JOINED the Blue launch team in Caracara that afternoon, sailing through the glass door to the conference room and plopping down at the center of the table, next to two doughnut boxes that had sat open since early morning. The engineers and designers who had gathered in the room quieted down and straightened up.

"There are a whole bunch of people dying to show what a huge idiot I am," Musk said. He decided to keep the launch quiet, worried that an announcement would lead to an overload of Twitter's systems.

Still, he expected people and organizations, who had longed for the verified badge or who wanted to support his cause, to sign up in droves. "It's better to have a little bit of confusion reign for a soft launch so we don't break the system," he told employees. And he was worried about bad actors, predicting a wave of impostors, scammers, and pranksters that would be like a "zombie attack" in *World War Z*, the 2013 action horror flick starring Brad Pitt. By midmorning, news outlets began to take notice of the new Blue features and published stories about the relaunched service.

For more than six hours, Musk sat in the room and kept a watchful eye. This was his game-changing Twitter service, one that he believed would wean the company off its almost complete dependency on advertising, and he was there to shepherd it through, like the liftoff of a SpaceX rocket or the first Tesla vehicle delivery.

Crawford and others in the room, including Jonah Grant, a longtime engineer who had been pulled into the project, were beside themselves. Musk had eliminated their official badge that morning, arguing to the room that "aggressively stopping impersonation" would make it unnecessary. But he didn't explain what kind of aggressive measures should be taken. Days earlier he had eliminated swaths of employees responsible for trust and safety, and the product manager asked how exactly Twitter's remaining content moderators would come up with solutions to stop fake accounts. She laid out a potential scenario where an account subscribed to Blue pretended to be a local government official who didn't have a large following. How would a content moderator determine which was the legitimate account?

"We should consider what is the magnitude of the thing that happens before we take action," Musk replied. "I want to pay close attention and react with agility, but not react with a fear of a thing that may or may not occur."

It was an absurd answer. Those who had worked at Twitter knew it was nearly impossible to patrol the platform for every harm in real time, even with automated technologies and thousands of employees and contract content moderators, many of whom had been laid off by Musk's team. This was a service used by millions around the globe. Someone impersonating the New York City Fire Department may be flagged and reported early, but what if it was a police service in a village in India? Or a Ukrainian politician during the war? Or a U.S. embassy in the Middle East?

Crawford and Grant continued to lightly needle Musk on potential pitfalls, but rather than suggesting definitively that she foresaw a problem, she presented ideas as hypotheticals, careful to never directly challenge him. "I think the thing that will be interesting to see, too, will be state-sponsored actors, do they jump on this bandwagon?" she said, a few half-eaten doughnuts sitting behind her laptop. "Because those are some of the more sophisticated—"

Musk cut her off. Of course they'll jump on this, one hundred percent. "What we're doing here with verified being attached to a credit card and a phone is the worst possible thing to do to bad state actors," he said, waving his hands like an orchestra conductor. "Not because they don't have the money. But because they don't have the phones and cards. That's what they don't have. You can't just produce a million phones or cards a day."

No one in the room said a word, but many wondered what he was talking about.

"Governments are terrible at execution," Musk continued. "The government is the DMV. That's who we are fighting here."

Musk said it with such confidence that no one dared challenge him. But for those who had worked in tech as Russia meddled in the 2016 U.S. election, his answer was nonsense. The Twitter Blue program would not stop state-sponsored propagandists from making nonverified accounts, which required no credit card or phone number. And who was to say that a motivated state actor couldn't procure a U.S.-based phone number and credit card to buy one of Musk's verified badges? In 2016, Russia's Internet Research Agency used credit cards to buy $200,000 of advertising on Facebook to sow social unrest in the U.S.

Musk underscored his belief later in the meeting. He didn't hold the Kremlin's hacking capabilities in high regard because they hadn't disrupted Starlink, the SpaceX-developed satellite internet service that was used by the Ukrainian army in its war against Russia. "These could be famous last words, but thus far the Russian attempts to hack Starlink have been not very good," he said, casually revealing war secrets to a bunch of civilians.

When Musk left the room a few minutes later to talk with his assistant, those who had remained quiet during his bizarre explanation sprang into life. Here they were, half a day into rolling out Twitter Blue, yet they were still addressing fundamental concerns around real-world harm. There were fears about what would happen in foreign countries, when government entities or officials lost their verification. "All these other people around the world rely on a label of some kind, a badge of some kind, to tell them this is the real account to listen to," said Crawford. "And they will be hijacked."

Christian Dowell, a lawyer who oversaw the launch, chimed in, suggesting that verified accounts could direct SWAT teams to people's homes. "Somebody could die actually," he said.

ZOMBIE ATTACK

>>> WHEN MUSK REENTERED THE ROOM, the chatter died. People clacked away on their laptops monitoring the launch, while Musk stared at his phone. No one had expected Musk to be with them for so long, yet he continued to sit at the table, seemingly with no other obligations or places to be. He picked through the snacks, at one point eating half of a doughnut in a single bite, and at times blurting out whatever crossed his mind.

"If I'm looking at my Twitter feed right now, the worst shit I'm seeing is from legacy verified," he said, looking for a response from the room.

Grant sensed an opportunity. "What do you mean by worst?" he asked.

"Like . . . the most annoying replies. Like replies I don't want to see," Musk replied, stumbling for a rationale. "Not from the new ones, but from the legacy douchebags."

Musk had been hoping to kneecap the reporters, celebrities, and news outlets that he found useless and annoying. "I fucking hate legacy verified," Musk blurted out. Grant just stared at him.

Yoshimasa Niwa, a twelve-year Twitter veteran and a master of its Apple app, tried to get Musk to understand the harm he could cause by selling check marks. Niwa was from Japan, and he had seen a random Twitter account use a new artificial intelligence program to create a fake photo of a flooded area in his home country during a recent storm.

"They said, 'Oh my god, escape from here!'" Niwa explained to Musk, highlighting it as a life-or-death situation. "And I remember believing that was actually true." Imagine if they now had a verified badge, he said.

"Safe to say we'd suspend that account," Musk replied. "And we'll keep their eight bucks. It may not seem like much but people really don't like losing their eight dollars. So we'll see what happens here."

Niwa stared at him, nodding, though visibly frustrated. By the time Twitter would move to suspend such an account, it would have already caused harm.

"We have to be adventurous, so we're going to be adventurous," Musk continued.

A few minutes later, he fashioned his hands into a pair of finger guns. "We're going to be shooting from the hip in real time," he said.

A software engineer calling in from Toronto posed a question to Musk about the official badges. "What would you consider a serious incident that would require us to put back such a label or some other differentiation between accounts?" he asked.

Musk intertwined his fingers and paused for a few seconds. He stammered. "If there's like death or serious injury or something like that, um, you know, uh," Musk said, fidgeting. "Something beyond annoyance or mild confusion—that would be enough. Let's just see what happens after we resist the initial wave of attempted fraud. There's going to be this initial *World War Z* attack where forces are going to test us in this way. We don't want to react to that. We first want to stop the zombie wave. Or can we stop the zombie wave? I think we will be successful and then see where it settles out. I think we need to hold steady for a while."

It was an incomprehensible answer, but the employees had given up on reasoning with Musk. There was no way to inform content moderators that they should act only when they saw tweets "beyond annoyance or mild confusion" or wait for someone to be grievously harmed or killed.

>>> MUSK'S RUSH TO get the product out led him to decide that there would be no review process for accounts signing up for the new Twitter Blue. If a user provided credit card information, that was enough to receive a check mark and all the other added features.

And that was exactly what happened. One user pretended to be the gaming company Nintendo, bought a check mark, and then shared an image of Mario, the lovable plumber from its Super Mario games, giving the middle finger. Another Blue Verified account mimicked the account for American Girl, the famous toy brand, and tweeted that one of its dolls, Felicity, "owned slaves." There were hoax Tesla announcements, a fake O. J. Simpson admitting to murder, and a post from an impostor Senator Ted Cruz, which declared his love for eating babies. The fake verified accounts caused so much confusion that the stepmom of Brooklyn Nets point guard Kyrie Irving, who was also his agent, called Twitter's Partnerships Team to get the company to take down a tweet from an account pretending to be an ESPN reporter. The post claimed the basketball superstar had been released from his team.

ZOMBIE ATTACK

In the Blue war room, Musk found some of the fake verified accounts to be hilarious. He threw back his head and howled when he saw the Mario tweet, which stayed up for more than two hours before the account was suspended for impersonation.

Grant showed Musk a tweet that had started to go viral. "It took me less than 25 minutes to set up a fake anonymous Apple ID using a VPN and disposable email, attach a masked debit card to it (with the address being Twitter's HQ), and get a verified account for a prominent figure," wrote @JackMLawrence, a PhD student. "Just think what a nation-state or bad actor could do . . ."

Musk shrugged, apparently seeing no issue. "Twenty-five minutes is a long time," he said. "Thanks for the eight bucks dude."

The Blue team employees were perplexed. Musk had been harping on a massive impersonation attack that aimed to embarrass him. How could their leader think it was a laughing matter? The fake posts, no matter how funny, were causing reputational damage not only to the people and organizations being impersonated, but to Twitter itself. And yet, Musk enjoyed the madness.

Toward the end of the day, Calacanis joined Musk and the Blue team to monitor the final hours of the launch. The two men compared the new revamped verification system to the opening up of a country club or the cutting down of VIP ropes at a nightlife hot spot, leading them to name-drop all the exclusive places they had been together. To a group of half-interested engineers, Calacanis name-dropped the Burning Man camp of former Google chief executive Eric Schmidt while Musk casually said they had recently been to actress Bella Thorne's home.

The billionaire left Caracara on Wednesday with a sense of measured success. He had manned the launch like a proud general directing his troops to a beachhead, garnering more than 78,000 signups to the relaunched Blue by that evening. While there were some blips, the casualties were accepted as the cost of the mission and laughed off by Twitter's owner. He walked out of the conference room just before 6:00 p.m. in a light mood, chatting with Crawford about why she thought there were so few employees in the office.

"It's just, for me, bizarre to see an empty office in the middle of the day," he said to his product manager.

—

>>> BY THAT NIGHT, however, something changed. The cheer of the day had evaporated. Musk had gotten wind that some of the remaining Twitter execs had discussed the possibility of leaving en masse to protest his leadership. There was also the matter of the audit that was due to the Federal Trade Commission by morning. Twitter's report was in shambles—the company could no longer account for which employees were looking after its privacy practices, and the entire setup would need to be overhauled.

In Slack, one employee had warned that workers would be required to "self-certify" that their projects complied with FTC rules, essentially taking on the legal liability that had belonged to their executives. "Elon has shown that he cares only about recouping the losses he's incurring as a result of failing to get out of his binding obligation to buy Twitter," the employee wrote.

Musk had also started to fixate on the costs of running Twitter and worried about what he believed would be a widespread economic downturn in the next twelve months. So he decided to do something he hadn't yet done since taking the reins at Twitter: email all his employees. "There is no way to sugar coat this message," he wrote, warning them of the "difficult times ahead."

"The road ahead is arduous and will require intense work to succeed," Musk continued. He warned of a "dire economic picture." Employees, who had been allowed to work from home for years, would be required to come into the office at least forty hours a week. It was a radical change. Dorsey had promised they would never have to come back to an office, and some workers had taken the opportunity to move around the world.

Musk was known for late-night email blasts at Tesla and SpaceX—and had released a similar panic among employees at his automobile manufacturer earlier that June—but for Twitter employees, it was especially jarring. This was the first time they were hearing from their new owner.

Musk sent another message later that night. "Over the next few days, the

ZOMBIE ATTACK

absolute top priority is finding and suspending any verified bots/trolls/spam," he wrote. It wasn't a laughing matter anymore.

—

>>> BY THURSDAY MORNING, Kissner, Kieran, and Fogarty had all resigned. It was a sign that the company's top remaining leadership had little faith in Musk, and it was of particular concern given the trio's responsibilities. Their resignations indicated that Twitter was in deep trouble with the FTC and that the executives had chosen to depart rather than place themselves in the crossfire.

Their resignations came before Kieran had a chance to meet with McSweeney and convey his worries, but she was horrified when she learned of the executives' departures. Soon, their responsibilities would become hers.

Spiro stepped in to do damage control. The FTC had already come calling about the executive resignations and demanded that the company explain how it would keep up with its privacy obligations. In an email to Musk on Thursday, Spiro said that employees were wrong to believe they could face criminal liability if Twitter fell out of compliance with its FTC consent decree. "That is simply not how this works. It is the company's obligation. It is the company's burden. It is the company's liability," he wrote, copying the entire workforce on the message.

By the time he arrived at the office on Thursday, Musk was in a brooding mood. His email did not get the resounding response he was hoping for, and he stomped around the office, wondering aloud why there were few butts in seats. The impersonation attempts had also continued through the night and the global press made a point to report on every one. Musk was acutely aware of the problem, and though he thought they were funny, the growing pressure over the fake accounts began to weigh on him. In a telltale sign of his annoyance, he coped by posting.

"I love when people complain about Twitter . . . on Twitter," he tweeted with two crying laughing emojis.

A little after midday, Musk sent another jolt through the workforce. With only twenty minutes' notice, he called for a meeting with employees, sending people scrambling to see if they could dial in and watch Musk's first remarks to Twitter as its owner. In the hour-long discussion, the billionaire painted a bleak picture for the company's future, and himself as the savior seeking to rescue it.

"We just definitely need to bring in more cash than we spend," he said. "If we don't do that and there's a massive negative cash flow, then bankruptcy is not out of the question."

He told employees to prepare for "a year or two of serious recession," mentioning "recession" no less than twelve times, and said he believed the company continued to be "overstaffed."

"You may have read that I sold a bunch of Tesla stock. The reason I did that was to save Twitter. Not because I lack faith in Tesla. I think Tesla stock is going to be worth an immense amount in the future. I sold the stock for Twitter. To keep Twitter alive."

Employees came away from the meeting in a frazzled state. They had clenched their bodies at the idea of being "overstaffed," and saw it as a sure sign that there'd be further layoffs. Musk's self-aggrandizement grated. He had loaded the company with debt. This was his doing.

—

>>> AS MUSK LECTURED his employees, the Blue impersonation onslaught continued. That morning, a Blue Verified account pretending to be Eli Lilly, the multinational drugmaker, tweeted, "We are excited to announce insulin is free now." The post, which garnered more than 3,000 retweets and was up for at least six hours, led the actual company to put out an announcement that the message was wrong and that it was taking steps to deal with fakes. Eli Lilly's stock dropped more than 5 percent that day.

Business leaders from major advertisers started to ring Twitter's sales team, telling them that they would pull their advertising unless the company did something about the fakes. In one call, executives from Nike threatened they would never advertise on the platform again if something wasn't done about Blue.

Then came the election meddling. While Musk had delayed the launch of Blue until after Election Day, some races, including the hotly contested race for Arizona's governorship, were still being tallied. The race featured Democrat Katie Hobbs, Arizona's secretary of state, against Kari Lake, a Trump-backed, conspiratorially minded Republican who had vocally supported Musk in his Twitter takeover and offered to roll out the red carpet if Musk moved the company to the state. With Lake a half percent behind in the vote count, a verified account purporting to be the Republican candidate popped up and posted: "It is with heavy heart that I must concede to my opponent, @katiehobbes." It was a fake, but it still racked up thousands of engagements—proof that Musk's follies could hurt anyone, even his political allies.

For all his previous bluster about naming and shaming advertisers who pulled off Twitter, the corporate threats stirred Musk's fear. Losing hundreds of millions of dollars in revenue in a snap would significantly impair his business, potentially making it tougher to raise more equity funding in the company, raise or sell off existing debt, or even just stay solvent. He sulked before turning to an engineer.

"Turn it off," he said, just after 5:00 p.m. in San Francisco. "Turn it off!"

Those who had already left for the day or signed off were summoned back to the San Francisco office or to their remote workstations. Having pushed hard to launch the damn thing twenty-four hours earlier, they were being told to reverse in an instant and reel it all back in. More chaos. Those who saw Musk could sense his tension and instability, clearly bothered by the calls he had had with the other business leaders. One person remarked: "That was the first time it seemed like he had been manhandled by someone else."

Some of the Twitter engineers on Blue joked that they knew it would come to this. It was like someone at a restaurant ordering a well-done steak, despite the loud protestations of the waiter, and then sending it back once the kitchen had prepared and delivered the meal to specification.

The engineers scrambled to figure out how to turn off Blue, and once again turn on the "Official" labels for brands and sponsors. That way, these advertisers would have a way of dealing with any impersonation attempt by any account that had already signed up for Blue.

Over the coming weeks, Musk's mood swings and periods of depression would become routine. The ad threats sent Musk on a tailspin and exposed something particularly human about him—he could be terrified into submission.

>>> OVER THE NEXT FEW DAYS, some people on the finance team also started to receive some strange requests for ad campaign approvals. They were last-minute asks, and while it wasn't atypical for a Twitter advertiser to force through a contract to start a big ad campaign the next day, this client was unusual: SpaceX.

"Team, I'm setting up a last-minute campaign for SpaceX (for Elon)," wrote one sales employee, in a message to the finance team, and linking to a series of contracts. The documents showed that SpaceX's ads would take over Twitter's Explore page with a promotional package that typically cost clients hundreds of thousands of dollars a day. "This is a very hot potato coming from Elon," the sales staffer wrote in another message.

Members of Twitter's finance team raised their eyebrows. They knew Twitter was losing advertising dollars hand over fist as Musk hopped from one controversy to the next. Suddenly, Twitter's new owner was pushing advertising dollars from one of his companies to another in what appeared to be an attempt to prop up his new toy.

>>> THE DEPARTURES OF Kissner, Kieran, and Fogarty set off a sense of hysteria among Twitter's employees and management, who wondered if they, too, should pull the rip cord. Pacini and Roth handed in their resignations. Robin Wheeler, a vice president who was one of the highest remaining sales people, also quit. She had been blindsided by the departures, particularly of Kieran, who had been in meetings with her that morning and gave no indication that he planned to leave.

Roth, comfortable that he had ferried the platform through the election, was ready to give up. While he'd been able to cajole Musk away from his plan to reinstate the Babylon Bee, he felt that Musk needed constant minding to keep him on a good path. As soon as one of Musk's confidants got his ear, the billionaire was off on a new, crazy venture, eagerly undoing whatever measured plan Roth had coaxed him to accept.

The messes with the FTC audit and Blue also weighed heavily on the safety expert. Musk hadn't been willing to listen to anyone's advice about the obvious pitfalls with Blue, and it seemed that it was only a matter of time before Roth would be asked to break the law on Musk's behalf.

At 12:29 p.m., Roth sent his resignation email to human resources. Then he got in his car and drove back to his home in the East Bay. Over the next few hours, he received text messages from Musk, Spiro, Krishnan, and a number of others in Musk's inner circle begging him to reconsider.

Musk called Roth directly, and the two debated Roth's decision for half an hour. But Roth stayed firm. He explained his reasons for leaving, but tried to be kind—he was worried about what Musk might do to him if he lost his temper. "I love Twitter as much as you do," Roth said. Musk agreed to let him go, later replacing him with Ella Irwin, a trust and safety exec who had initially resigned during the early days of the takeover.

Unlike the other executives, however, Wheeler was swayable. She took a call with Birchall, who asked her to put some faith in Musk, and offered her the chance to lead sales at the company. Within hours, she did a 180, and was soon organizing a call for Musk's goons so they could convince other waffling sales execs to stay.

On one call, to which Musk did not dial in, one Twitter sales executive had become fed up with the empty promises. Between the layoff and resignations, some brand safety and integrity teams had been decimated or eliminated. "Do you guys have a plan? Are you guys going to replace these people?" she asked.

Spiro was baffled. *Of course he had a plan.* SpaceX and Tesla solved much harder problems, so why couldn't Musk's team find a way to automate content moderation, he said to the group. Gracias chimed in to ask folks to cut Musk some slack. "Can somebody be a little sympathetic for Elon right now?" he said, playing the victim card for the world's richest man. "He's doing everything to fix this company but people aren't trusting him."

Musk would try to convince this particular sales exec in a direct call later, as would Birchall, separately from his boss. Musk's fixer said that he realized that Musk's team had come in like wrecking balls and tried to explain that Musk's lack of understanding and charm came from the fact that he was "on the spectrum." (The previous May, Musk hosted an episode of *Saturday Night Live* where he spoke about having Asperger's syndrome.)

"He doesn't understand how his actions can be read," Birchall said. Perhaps the exec could write up some feedback and email it to Musk directly? "His love language is direct feedback," Birchall said.

Birchall also let the executive in on a secret: the goons had been toying with the idea of a loyalty pledge or an ultimatum for workers to stay at the company. The sales leader was stunned. "This was one of the worst things you can do," she said, noting there were still a good number of people who were undecided on whether they should stay at the company and who were waiting for reasons to believe. But if you give them an ultimatum before you've shown them anything, they're just going to jump, she predicted.

The exec came away from the interactions feeling like she was being pitched on cult membership more than a job. Before officially quitting, she wrote a note with "direct feedback" for Musk and sent it the following night. She never got a response.

Spiro called McSweeney that night. Over the video conference, he told her that she would take on some of the work that Kieran and Kissner had abandoned, as well as Roth's role. "I trust your judgment," he told McSweeney. "I want you to take on a bigger role."

While Musk's lawyer spun it as a promotion, McSweeney knew better. Roth had overseen a large team and had crucial responsibilities, including making tough content moderation calls during election cycles. She had too much on her plate already, and had regularly been working twelve hours at a time since Musk took over. There was no way she could do Roth's job, too. And she had no interest in becoming the person to take the legal fall if things went sour with the FTC, as Kieran had been. Still, she told Spiro that she would help as much as she could.

38 > Fired for Shitposting

The sudden departures of the executives reinforced what Musk believed when he took over Twitter: none of the old guard could be trusted.

He sicced his lawyers on Twitter's former leadership, hoping to uncover justifications to deny them their golden parachutes or get them into trouble. In the days following the deal, Spiro and his law firm, Quinn Emanuel, sent letters to some of the remaining executives, asking them to preserve documents and prepare for interviews with lawyers for the purposes of an internal investigation.

The notes were ominous, and while they didn't accuse those receiving them of any wrongdoing, they seemed designed to flip the recipients into informants. "We believe you may have relevant information given your role in working with the subjects of the investigation," one of the letters read. The inquiry could include Twitter's past financial reporting, the company's efforts to calculate and eliminate fake accounts, security issues raised by Zatko, the allocation of corporate resources, and conduct related to the recent merger. "It was a fishing expedition," one recipient said.

Those who saw the letters or participated in the interviews with Quinn Emanuel staffers disagreed about what Musk wanted. Some believed the billionaire was seeking revenge on anyone involved in Twitter's litigation against him and was using this investigation to suss out who he should blame. Others thought he was looking for ways to relitigate the deal and potentially recoup some of the sale price by proving that previous management had defrauded him. Most of the executives, though, saw the investigation as Musk's attempt to find ways to justify his "for cause" terminations of Agrawal, Segal, Gadde, and others and deny them millions of dollars in payouts.

His goons dug for dirt everywhere. They were especially interested in the company's spending before Musk, leading Davis and others to train their focus on Segal. The chief financial officer had approved large budgets for lavish corporate perks, including the suite for Warriors games at San Francisco's Chase Center. When Musk found out, he went ballistic, lampooning Segal for spending so much on the box. His lackeys tasked Segal's former assistant with a full accounting of her boss's budgets.

Musk's lawyers weren't particularly thorough, however. Nick Caldwell, the former engineering leader who resigned while tending to his wife's funeral arrangement, later received an email noting that he had actually been fired for, among other reasons, approving the budgets for Twitter's lavish OneTeam celebration in January 2020. There was only one problem. He had joined the company six months after that event.

>>> THE FAILURE OF Blue and the mass resignations made Musk doubt his purchase. The following day, November 11, he had his fateful encounter with the senior data scientist, who called him out for his Paul Pelosi tweet. He flew home that weekend to Austin to brood.

Musk's growing paranoia was palpable over that weekend. Led by Christopher Stanley, an information security engineer who had come from SpaceX, workers examined a spreadsheet that had been created soon after the deal and listed the names of high-skilled or long-tenured employees to determine which workers to retain during layoffs.

Stanley, however, repurposed the list and used it to ascertain which employees had the know-how to take down Twitter's site or apps. From the beginning, Musk had been fearful that a rogue employee or two could sabotage the site, embarrassing him.

Several Twitter employees pulled Stanley aside in an attempt to reason with him about the strange internal investigation that focused on top company talent. If Musk was so worried about saboteurs, giving employees an easy way to leave the company might help, they suggested. Unhappy workers who had escaped the layoffs could either leave without severance or stay and suffer—giving them a way out could reduce Musk's risk. Or Musk could try winning his new workers over instead of terrifying them, they suggested. Stanley seemed sympathetic, but didn't agree to pass their advice on to his boss.

While some tweeps were miserable, others were happy to crack the whip on Musk's behalf. One, a senior director of software engineering named Luke Simon, had been fairly vocal about his unhappiness with Musk before the takeover. Once the deal was signed, however, he found a home—and an opportunity—in the restructuring, flashing a thumbs-up in a selfie with Musk on the day the billionaire arrived at the office. He made layoff lists and worked overtime to impress Musk and his associates. In a Slack group with other managers, he wrote on the morning of Saturday, November 12, that he was concerned about "silent" quitters who had not been eliminated in the Snap but were still collecting paychecks.

Managers like Simon were in the process of bringing back workers they had just laid off a week ago, as they realized there were essential jobs that needed to be done and no one around to do them. Simon had four people on his list to bring back so they could help with brand safety initiatives for flighty advertisers, but he wasn't happy about it.

"The engineers I am bringing back are weak, lazy, unmotivated, and they may even be against an Elon Twitter," Simon wrote. "They were cut for a reason.

"So we need to think of these people as just needing to be around until the knowledge transition is completed," Simon continued.

Less than a minute after Simon's post, a furious employee took a screenshot of the Slack message and shared it to #social-watercooler.

"@lsimon care to share your thoughts with colleagues?" wrote an engineer who shared the screenshot, adding Simon to the Slack channel. The message started a pile-on against Simon, who never spoke up and eventually locked his Twitter account as the image of his post made its way outside Twitter's walls.

Musk wasn't worried about the backlash against Simon. But he was concerned about the leak of the messages. It only hardened his belief that the tweeps were out to get him. On Sunday night, he instituted a code freeze, convinced that there were shadowy figures within the company that would intentionally break the site. The freeze, which required that Musk himself sign off on even the most basic changes to the site or its apps, prevented anything from being done at the company.

—

>>> THE DAY AFTER his confrontation with the data scientist, Musk tweeted that Twitter had been running slow in several countries. He suggested the problem was caused by the company's previous management, blaming them for building the Twitter app in a sloppy fashion that caused it to lag.

"Btw, I'd like to apologize for Twitter being super slow in many countries. App is doing >1000 poorly batched RPCs just to render a home timeline!" Musk wrote on Saturday. The tweet was wonky enough to soar over the average user's head, but to Twitter's engineers, it was blatantly false. In #social-watercooler, they called out Musk's mistake, showed he had conflated various technical terms, and poked fun at his lack of expertise with laughing emojis.

As engineers continued to mock Musk on Slack, one of them decided to publicly rebuke him. "I have spent ~6yrs working on Twitter for Android and can say this is wrong," tweeted Eric Frohnhoefer, quoting Musk's original message, on Sunday afternoon.

Frohnhoefer, a staff software engineer who lived in San Diego, had believed in Twitter's core tenet to "communicate fearlessly to build trust," and felt his response was in the spirit of that ideology. After tweeting, Frohnhoefer left home to go to Starbucks, thinking nothing more about it.

But during his coffee run, a colleague messaged him to say he had received a response. Actually, there were several.

At first, it seemed like Musk was open to the feedback—he tweeted back to Frohnhoefer asking the engineer to correct him. But Musk had taken the public criticism as an insult, and he launched into a tirade.

"Twitter is super slow on Android. What have you done to fix that?" he wrote. The engineer responded, continuing the tit-for-tat for another few rounds into Monday.

That morning, Musk ended it. "He's fired," he tweeted, before deleting the message. Later that day, Frohnhoefer shared that he had been locked out of his computer and terminated. Musk would later tell employees that he would have accepted it if Frohnhoefer had pointed out his errors in private, but tweeting publicly to embarrass him had gone too far.

"Criticize privately, but praise publicly," he said to some of his staff, clearly without any self-awareness that his tweets about Twitter's speed were indictments of the people who worked there. In one meeting after, an engineering executive asked employees to stop tweeting at Twitter's new owner.

Musk's firing spree didn't stop with Frohnhoefer. Others were terminated that day, including Sasha Solomon, a staff software engineer based in Portland. On Twitter, she was outspoken, posting about her daily life, job openings, and her French bulldog, Bosworth. Solomon lived her life on the social network, and when Musk started his acquisition overtures, she tweeted about that, too. In April, when Musk initially registered his interest, she tweeted "he better not be buying my favorite coffee shop too." After the Snap, she began to lay more heavily into Twitter's new owner.

As Musk battled online with Frohnhoefer, she chimed in and responded to one of his tweets. "You did not just layoff almost all of infra and then make some sassy remark about how we do batching . . . like did you bother to even learn how graphql works," Solomon tweeted, referencing the programming language she had specialized in. Like Frohnhoefer's posts, her tweet circulated widely. And like Frohnhoefer, she was locked out of her work computer and fired. "Lol just got fired for shitposting," she wrote in one final flourish. "I said it before and I'll say it again. Kiss my ass elon."

That post garnered more than 3,500 retweets and 28,000 likes, and eventually floated across Musk's feed, sending him on the warpath. When his cousin James attempted to approach him about another engineering matter that afternoon, Musk gave him a more pressing task: "Find anyone talking to me like this and fire them." The younger Musk could barely get a word in before leaving with his new task.

A tall man with floppy brown hair, James carried himself with a sense of boyish urgency. He kept quiet in meetings, dodging the spotlight that shone brightly on his cousin, and worked hard to clear a path for Musk's wishes to be done. If Musk gave him a task, James would bound off to do it, excitedly asking Twitter employees for help to make sure his older cousin got what he wanted.

In order to find these dissenters, James and his associates turned Slack into a surveillance tool. They got their Twitter counterparts to grant them administrative powers on Slack and searched for keywords like *Elon* and *Musk* to see what the employees had written about their new boss. Spiro also asked one executive if Twitter's security team had the ability to track what people said on Blind, the anonymous workplace message board, and perhaps find ways to unmask any anonymous dissenters or rabble-rousers.

In some instances, Musk's lieutenants appeared to have an unusual amount of insight into what employees had told each other during hushed conversations or on Twitter accounts that were set to private, meaning that no one except their followers should have been able to see their posts. HR executives imported from SpaceX stepped in to handle some firings, fueling the paranoia among employees that Musk's people were snooping on them.

These episodes made employees fearful about how far Musk's surveillance could reach. Some workers worried his loyalists would access their direct messages on Twitter to look for dirt. Typically this privilege was reserved for rare cases where Twitter executives aided criminal investigations, and it was avoided if other methods could be used to obtain the necessary information. Under Musk, however, some of the goons gained access to "God Mode," enabling them to see the public and private Twitter activity and data of any user.

Employees watched aghast as Frohnhoefer and Solomon were terminated. Some abstained from posting anything on Slack, suspicious that their online activity would now be more closely monitored. Still, there were others who were emboldened by the capricious cuts and the treatment of their colleagues. In typical tweep fashion, a handful dealt with their sadness by shitposting.

"The beatings will continue until morale improves," wrote one engineer who worked on Twitter's search feature.

"So is this like a Candyman situation?" another employee joked, referencing the 1992 slasher film. "Mention Elon three times and we get deactivated?" He was unaware that Musk's cousin and his lackeys were doing exactly that and tracking mentions of the billionaire's name.

By the end of Tuesday, about thirty employees had been fired for dissent. Some had tweeted openly about their distaste for Musk over the months. Others had talked about Twitter's new owner to their colleagues on Slack. A handful had no idea why they were let go. One, a twelve-year company veteran who had been one of Twitter's top engineers, had simply written on the work messaging platform that it was better to be laid off than to quit because a layoff would come with severance. "Don't resign, be fired," wrote Yao Yue. "Seriously."

All those who had been classified as dissenters got the same email, notifying them of "violating company policy" without specifying what that policy was.

Among those hit was Kiko Smith, an architect who oversaw the designs of Twitter's data centers. After more than two decades in corporate America, she had seen it all. She had worked for Enron Energy Services, an arm of the disgraced power giant that went belly-up in 2001. Smith was a tough woman and was known for championing female engineers after she herself had succeeded in male-dominated spaces her whole career. When the acquisition happened, she was optimistic that it could spur the changes she knew were needed to improve the company's infrastructure, and she kept her head down, with no reason to engage in the Slack banter that captured many of her younger colleagues. Even so, she was fired with no explanation.

A few days later, James Musk called up Smith. Several colleagues had gone to bat for her, calling her skills crucial to the company. He admitted he had no idea why she was terminated. "We are so sorry," he said. "We don't know why you were on that list and we would greatly appreciate it if you came back."

Smith couldn't help but laugh at the absurdity. Still, she didn't hesitate. Her husband had cancer and they needed the Twitter-provided health insurance. Chaos or calm, it wasn't really a choice for her. She accepted.

"I would like to apologize for firing these geniuses," Elon Musk tweeted a few days after the dismissals. "Their immense talent will no doubt be of great use elsewhere."

>>> CONTEMPT FOR MUSK'S "with me or against me" approach crystallized. In the New York office, someone put an image of Frohnhoefer's Twitter argument with Musk on a television near one of the micro kitchens. Those who saw it were unsure if it was a celebration of the engineer's boldness or a warning sign, like a decapitated head on a pike outside a medieval village.

In San Francisco, the events of the early week had taken Musk's distrust of Twitter's employees to new heights. In meetings that week with some of the borrowed hands from Tesla and SpaceX, he said he wanted 80 percent of the remaining people out of the company by the end of the week. Musk and some of his coterie kept comparing the company to Telegram, a messaging service with 800 million active users that had only thirty core employees. They wondered why Twitter couldn't be as streamlined as that.

Those around Musk like Sheen Austin, the Tesla infrastructure engineer who was now effectively running Twitter's infrastructure, and Musk's cousins Andrew and James, thought the 80 percent expectation was nuts. They had already worked to cut and cut and cut. While they, too, thought Twitter was plagued by nonbelievers hanging around taking a paycheck, they also knew their lives would become exponentially harder if Musk got his way. And facing a holiday period, which was crucial for advertisers, and the looming World Cup, which was always a Twitter traffic bonanza, they predicted it could get very painful.

So they presented a new idea to Musk, one that had been floated by Twitter's workers in the past. Perhaps, Musk could provide an incentive to leave. Many companies offered buyouts when trying to reduce head count, with guarantees of health insurance and paychecks for a few months, giving those on the fence about working a reason to raise their hands to leave. That way, Musk's advisers said, Twitter could shed the deadweight and ensure that people who were at Twitter actually wanted to be there. They proposed offering a severance payment equivalent to three months of salary.

Twitter's owner had initially opposed offering buyouts, feeling that they were rewarding the people who hated him, but he eventually came around. This was the only way of ensuring that whoever remained was loyal. As some of the hired hands left the office that Tuesday night, they felt they had convinced him to do the right thing for the company.

FIRED FOR SHITPOSTING

That night, Musk boarded a private plane bound for Delaware, where he was set to testify the next day in a Tesla shareholder lawsuit. Before the plane touched down, he sent an email at 3:00 a.m. Eastern. Titled "A Fork in the Road," the email was sent to all employees and included a link to a Google form where they would complete a survey.

> Going forward, to build a breakthrough Twitter 2.0 and succeed in an increasingly competitive world, we will need to be extremely hardcore. This will mean working long hours at high intensity. Only exceptional performance will constitute a passing grade.
>
> Twitter will also be much more engineering-driven. Design and product management will still be very important and report to me, but those writing great code will constitute the majority of our team and have the greatest sway. At its heart, Twitter is a software and servers company, so I think this makes sense.
>
> If you are sure that you want to be part of the new Twitter, please click yes on the link below:
>
> formes.gle
>
> Anyone who has not done so by 5pm ET tomorrow (Thursday) will receive three months of severance.
>
> Whatever decision you make, thank you for your efforts to make Twitter successful.
>
> Elon

In his message, Musk had ignored an important piece of his loyalists' recommendations. Instead of asking those who wanted to leave to raise their hands, he did the opposite, making people actively commit to wanting to work for him. Musk wanted a pledge of loyalty.

>>> WORKERS AWOKE THE next morning in shock. If Musk was looking for some kind of awakening or sign of intent from Twitter's employees, the move was poorly executed. Some workers, who were on leave or vacation, didn't see the email before Musk's twenty-four-hour deadline passed. Others never clicked the link, believing it may have been a scam. "In response to Elon's email last night, we are working to get out an FAQ for you today," read one Wednesday morning Slack message from one of the remaining human resources employees. "In the meantime we want to confirm that the link provided is not a phishing attempt."

The resulting FAQ document was somehow less self-aware than Musk's original email. It said workers would be expected to "maximize working from an office" and be prepared to commit to working early mornings, late nights, and weekends. If Musk preferred the stick over the carrot, this was his way of swinging a Louisville Slugger covered in rusty nails.

By Thursday morning, Musk was not getting the desired response. By 10:00 a.m. in San Francisco, some undecided employees received invitations to meetings hosted by some combination of Musk, his advisers, and executives from Tesla or SpaceX. The meetings were a last-minute sales pitch intended to convince employees to stay.

In one gathering with the corporate finance team led by Birchall and Bret Johnsen, SpaceX's longtime chief financial officer, employees were told they would have a massive impact on society. "We are finally going to stand up for democracy and free speech in the United States," said Johnsen. "We are going to save democracy."

Some of the people on the video call had to quickly switch off their cameras, blocking Johnsen from seeing them as they burst out laughing. It was lost on them why Johnsen used the word *we*. He wasn't a Twitter employee and it wasn't even clear if he used the service. When someone asked if he was planning on being their new CFO, he said he already had a job.

As the 5:00 p.m. deadline approached for employees to declare themselves "hardcore," Musk personally logged in to a meeting with infrastructure engineers to convince them to stick around. During the call, he emphasized his past successes.

"I know how to win," he said, alluding to the financial successes of Tesla and SpaceX. Under his guidance, the engineers who built those companies had become very wealthy.

"If you want to win, you should join me," Musk said.

Musk continued to speak through the 5:00 p.m. deadline. By that point, some of the engineers had seen and heard enough. They hung up, not even bothering to stay for the end of his pitch. Twenty-five minutes later, Musk registered the gravity of his ultimatum the only way he knew how: comedy.

"How do you make a small fortune in social media?" he tweeted. "Start out with a large one."

As the newly fired employees began to announce their departures, Twitter came to life with waves of salute emojis and goodbye posts. #RIPTwitter began trending around the globe as users chimed in with their favorite memories, top tweets, and eulogies for the site. If Twitter was indeed on its deathbed, posters wanted to scratch that itch one last time in a whirlwind of nostalgia and gallows humor.

It took several days for HR to figure out who had quit. The team responsible for the task drew up a master spreadsheet that included all employees and went through it name by name, comparing it to Elon's new list of "hardcore" people. They had to reach out to dozens of employees personally to ask them if they actually meant to resign. Even weeks later, there were still workers in the spreadsheet who the company was desperate to keep because of their specific skill sets but whose status was uncertain.

Although Musk was frenzied about catching saboteurs, his "fork in the road" plan opened a floodgate for departing employees to meddle with the company. Because it wasn't obvious who had resigned, Twitter's security teams scrambled to figure out whose access to internal systems ought to be cut. Unlike the mass layoff, in which workers were suddenly locked out of their laptops, employees who decided to leave ended up keeping access to internal systems for weeks until Twitter finally determined that they were gone and revoked their credentials.

When the dust settled, more than 1,100 people, or about 31 percent of the workforce, left at the fork in the road, opting into the buyout or not responding to the email at all. It wasn't the 80 percent that Musk had pushed for, but he seemed satisfied. While he initially had doubts about giving people a way to leave Twitter, he was pleased that some workers had pledged themselves to him. In a rare moment of grace, he allowed the company's human resources team to send out separation agreements to the departing employees and accepted the resignations of a few more workers who had waffled over the decision, allowing them to collect the promised severance.

For those who stayed, the loyalty pledge had different meanings. There were those who truly believed in Musk and his vision or saw a chance to advance their careers as management cleared out. Others were stuck—they needed healthcare coverage, a work visa, or hadn't saved enough to go without a regular paycheck while they hunted for a new job. Opting into a hardcore Twitter was their most viable option.

After Thursday's hardcore deadline, the remaining employees were notified via email that Twitter's offices would be closed on Friday. Then Musk contradicted himself. Shortly after 1:00 a.m. on Friday, he demanded that "anyone who actually writes software, please report to the 10th floor at 2pm today." Some employees, thinking Musk was testing their commitment, booked flights from across the country to San Francisco.

Some of Musk's lieutenants spent that day calling people they hoped to coax into returning. James Musk, Stanley, and others divided up lists of people they wanted to bring back, including some who had been ham-handedly fired in the dissent purge. In one call, Stanley asked a dissenter if they believed in their boss.

"I want to know: Are you loyal?" he asked them.

>>> McSWEENEY WAS ONE of the employees who rolled her eyes and ignored Musk's message. She was confident that Irish law protected her from taking a silly loyalty oath, and she would continue working under her employment contract. But she soon received an email from HR accepting her resignation and was locked out of her work computer. Her lawyers quickly wrote to Twitter to say she had not resigned, kicking off legal negotiations over whether McSweeney still had a job. The process would drag on for months as McSweeney grappled for reinstatement.

With Roth and much of the executive leadership gone, Musk began shaping the Twitter platform in his own image that Friday. In October, he had promised he would form a "content moderation council" to oversee major content decisions and account reinstatements when he bought the company. But after he threatened his "thermonuclear name & shame," Musk saw no need to please anyone else anymore, as U.S. ad revenue was trending at 80 percent below internal expectations.

That morning, he reinstated the accounts of the Babylon Bee and Jordan Peterson, a Canadian conservative media commentator who had been previously banned for violating Twitter's rules around misgendering.

"New Twitter policy is freedom of speech, but not freedom of reach," Musk tweeted, suggesting that negative and hateful tweets would still be allowed on the site but not amplified by Twitter's algorithm. It wasn't exactly a new concept to social media, and was one that Twitter had largely been practicing. But Musk claimed it as his own.

Musk's purchase of Twitter and decision-making emboldened others to come back to the platform as well. Among them was Andrew Anglin, the founder of the neo-Nazi website the Daily Stormer, who was suspended in 2013 after violating the site's hateful conduct rules. Andrew Tate, the misogynistic former kickboxer and internet personality who was later charged with rape and human trafficking, also came back with gusto, as did Kanye West, who had been suspended in October for tweeting that he would go "death con 3 on JEWISH PEOPLE." (Musk would go on to suspend West for posting a photo of a swastika, though he would welcome back the rapper again later.)

The unbannings of Friday morning emboldened Musk, who then began to reconsider the site's most famous banned user. Musk had faced questions about the @realDonaldTrump account from the moment he said he would buy Twitter.

If Trump was going to be a major player in the next presidential race, Musk told those around him, then he should be on Twitter. Not allowing him on the platform would amount to election interference and deny the American people Trump's messages and campaign statements. And imagine the kind of drama and engagement that would come with Trump's reintroduction, he told some of his lieutenants.

Without consulting anyone on the evening of Friday, November 18, he ran a poll asking if he should "Reinstate former President Trump." Within twenty-four hours, more than 15 million accounts had voted, with 51.8 percent in favor of bringing back the former president. "The people have spoken," Musk tweeted.

Unsuspending Trump wasn't just a matter of clicking a button, however. As one of the most followed accounts on the service—with some 88 million followers at the time of his January 2021 ban—@realDonaldTrump was an anomaly. In reinstating him, Twitter's systems had to rebuild all those connections, re-creating the hundreds of millions of follower relationships, tweets, and engagements that existed prior to the suspension. The site became noticeably slower as the systems tried to re-architect the connections and deal with the influx of traffic of people coming to check if @realDonaldTrump was really back online. At one point, Twitter almost crashed.

There was another problem: Musk had no idea if Trump wanted to return. Following his suspensions by Twitter and Facebook in January 2020, Trump joined Truth Social, the social network he partly owned and staffed with his former political allies, saying he would not go back to Twitter.

Musk tasked Irwin, the content moderation head who had replaced Roth, with starting the conversation with Trump's handlers. A corporate veteran whose two decades of experience involved managing abuse and risk at banks, an online dating company, and Amazon, Irwin had no idea where to begin.

Musk eventually scrounged up a number for Stephen Miller, the former president's far-right adviser on immigration policies, who he thought could provide a bridge between Twitter and Team Trump. Over the next several weeks, Trump's staffers indicated their boss may be interested in coming back, and Irwin and her team explored what that would entail. In anticipation, Twitter fixed a log-in issue for the former president, switching out an old White House email address for a personal one for Trump.

For all of Musk's statements about giving equal treatment to all users, the kid-glove treatment doled out to Trump was unique. Twitter's owner knew of the attention that would come with the reemergence of the former president on the platform. Still, Trump and his team remained cagey about working with Musk, and despite regaining access, the former president would wait several months before making a comeback tweet. Trump had learned to live without Twitter.

39 > Zero-Based Budgeting

Musk could see the walls closing in. Advertisers, who in the past accounted for more than 90 percent of Twitter's revenue, were just not coming back, even in the crucial holiday months. They had been worried about content moderation at the outset of Musk's takeover, and his actions only compounded their fears. As it became evident that Twitter wouldn't hit its advertising targets, sales associates who typically closed million-dollar deals with Fortune 500 companies, including Disney and Netflix, were told to focus downstream and see if smaller businesses like the New Jersey Dental Association would want to run ads.

By the end of the fourth quarter, Twitter's revenue for the last three months of 2022 was a little more than $1 billion, down 35 percent from the same period the previous year. Even worse, the business showed few signs of recovering, making it tough for his banks, which had underwritten the loans for Musk's Twitter deal, to repackage and sell off that debt. They were left holding the bag.

That $13 billion in loan obligations became a millstone around the company's neck. While Twitter's owner warned about the possibility of bankruptcy, the situation was entirely of his own making. The company did not have that kind of debt on its balance sheets when he bought it, and Twitter's interest payments alone would balloon to more than $1.5 billion a year. The company was incurring more than $3 million a day in interest, with its first quarterly interest payment fast approaching at the end of December.

Musk and Birchall would explore the possibility of using his Tesla shares to back new margin loans to pay off Twitter's most onerous debt, and would again think about raising fresh capital from outside investors. But with Tesla's share price dropping as investors feared Musk was spending too much time on Twitter, and the social network's financials in the dumps, neither option was great. The only way out in Musk's mind was through deep cuts to Twitter's expenditures. And he had just the man to do it.

>>> ONE MAN WHOSE loyalty to Musk was never questioned was Steve Davis. With shifty eyes and prominent features, Davis had a nervous energy about him that always seemed as if it needed to be diverted, if not into work, then some other activity. After moving to Washington, DC, to open SpaceX's office there in 2011, he opened a series of frozen yogurt shops, and then a bar that was one of the first in the nation's capital to accept Bitcoin.

"I like to do something new about every six to nine months and a lot of them fail," Davis said in a television interview. Around that time he also earned a doctorate in economics at George Mason University.

When the boss came calling about needing assistance during the early days of the Twitter transition, Davis didn't hesitate. It didn't matter that he and his partner, Nicole Hollander, had just had their first child. They would all come along for the ride. He moved his family—including the newborn—into a conference room where they all slept to cut down on hotel expenses. Hollander, a real estate asset manager, took on responsibilities managing Twitter's global office portfolio, sometimes sitting in meetings while nursing her baby or cleaning its onesies in one of the office kitchen sinks. At one point some office staffers were asked to do the family's laundry, washing Davis's underwear because he had run out while staying in the office.

Davis was a case study in Muskian loyalty, and he dedicated himself to addressing Musk's displeasures, among them Twitter's rent bills. With fifty offices in more than thirty cities around the world, the company spent about $130 million annually for workspace, a figure that drove Musk—and by extension, Davis—insane. In meetings and emails, Davis would compare Twitter to SpaceX, which had five times the number of employees but paid $26 million in rent a year.

To those who dealt with the company's real estate, it wasn't apples-to-apples. SpaceX had offices in cheaper locations, including its launch areas in rural Texas, and in industrial areas for manufacturing. Twitter's offices, however, were in higher-priced metropolitan areas or university hubs, with the intent of attracting talent who lived in those places.

Still, Davis fixated on the costs and came up with a novel solution to get rid of them. "We just won't pay landlords," he told Twitter's executives. "We just won't pay rent."

—

>>> AMONG THE REMAINING TWEEPS, Davis was seen as the most volatile member of the transition team. While Spiro and Birchall did Musk's bidding, they could at least be talked to and reasoned with. Davis was not the same; if a directive came from Musk's mouth, it was seen as the word of God, with little room for interpretation. When Twitter employees joked about Musk leading a cult of followers, they saw Davis as the one who'd mix the Kool-Aid and taste test it for maximum potency. Even Musk's inner circle viewed his blind obedience with trepidation.

In mid-November, Davis roamed the office with a wide-eyed look, as if he hadn't slept for days. He was still focused on the company's rent costs and walked up to two engineers who had been chatting on the open floor, butting into their conversation. "Do you live in San Francisco?" he asked. One said no. The other replied affirmatively, offering up his monthly rental payment.

"Would you live at the office for half of that?" Davis asked in response. The engineers looked at each other, unsure of what to say, before they shook their heads. Twitter pays millions of dollars a month to rent office space, Davis said. "And I want to turn some of that into apartments."

The two employees thought he was joking but decided to play along with what they believed was a thought experiment. One lived in New York and came to San Francisco often for work. "If they let me fly business class, I would consider staying in a hotel room in the office," he said.

"Do you know where I'm staying?" Davis asked. He told them he was living in the office with his partner and their newborn.

Hoping that other workers might be as dedicated as he and his family, Davis eventually developed his concept for the "Twitter Hotel." Later that month, Irwin, the new trust and safety leader who lived in Seattle, ran into

Davis during one of her visits to San Francisco. Like a child excited for show-and-tell, he took her for a full tour of the Twitter Hotel.

From the tenth floor, they got into an elevator, riding it down to the eighth floor, where Davis and his partner had already been instructing staffers to disassemble and move desks. They also locked down the floor and prevented normal employees from being able to access it with their badges.

On the tour, Davis revealed a conference room with a blowup mattress and video game console on the floor, while other meeting spaces had been outfitted with blackout curtains. He opened another door that offered a more complete room with an actual bed, though there was no attached bathroom.

Davis would tell those who doubted him that Musk stayed here. They, too, could have that privilege, sharing a communal bathroom with the billionaire as they brought glory to Twitter.

To Davis, this was a positive. At the Twitter Hotel, hardcore folks could forge camaraderie with one another and loyalty to the company. They might even get to run into their dear leader, if they were so lucky.

Irwin declined the offer to stay on the eighth floor, to which Davis later added exercise equipment and built out a mini gym for Hollander. Other visitors weren't as fortunate. It became company policy for out-of-town visitors to stay there and some worried they would never be reimbursed by the company if they stayed in an actual hotel. At best, the Twitter Hotel would save the company a few million dollars in employee accommodations a year. This wasn't as much a cost savings ritual as it was a test. To work at Twitter, you had to be hardcore.

>>> DAVIS AND MUSK continued to cut. To employees, it seemed they had moved beyond saving money and were enthralled by seeing how much pain they could inflict on workers. The same distrust of Twitter's previous management persisted as Davis reviewed contracts with property management companies, internet providers, or food service vendors. He operated under the belief that the people who had come before him were morons who had gotten ripped off by everyone.

First to go was the food. Before Musk, Twitter provided full meals to its

ZERO-BASED BUDGETING

employees. In the era of Silicon Valley largesse, the practice was commonplace among tech firms. Workers would show up for breakfast, bringing them to the office earlier, and stay at the office for lunch, preventing them from using valuable time to leave to procure food. Some companies even offered dinner, getting employees to work later into the evening as they waited for it to be served.

Musk came from another era. His disdain for free meals was compounded by the fact that fewer people were coming to the office than he liked. On November 13, he tweeted, lamenting that "almost no one" was coming in. "Estimated cost per lunch served in past 12 months is >$400," he wrote.

It was an eye-popping number underscoring the supposed incompetence of previous management, but it was unclear where Musk got it or how it had been calculated. "This is a lie," Tracy Hawkins, Twitter's recently departed vice president of real estate and work transformation, tweeted in response. "I ran this program up until a week ago when I resigned because I didn't want to work for @elonmusk. For breakfast & lunch we spent $20-$25 a day per person. This enabled employees to work thru lunchtime & mtgs. Attendance was anything from 20-50% in the offices."

Whatever the numbers on the meals, Musk was convinced they had to go, though the cuts didn't stop there. Davis and Musk ended relationships with a cleaning company that provided around three dozen janitors and other staff to Twitter's San Francisco, Los Angeles, and New York offices.

In San Francisco, Musk had consolidated all the workers to sit on two floors, creating a sweatbox that approximated a frat house during Greek rush week. The halls reeked of under-deodorized armpits and the lingering odor of pizza boxes and Chinese takeout containers, which piled up, mixing with junk that fans of Musk sent as gifts to the office. Mike Lindell, the pillow businessman and noted election denier who had become an adviser to Trump, was one such fan. He sent copies of his book *What Are the Odds? From Crack Addict to CEO*, as well as his signature MyPillows, which some employees used for naps.

As trash cans went unemptied, carpets unvacuumed, and bathrooms uncleaned, workers gritted their teeth. The restrooms became a particular problem, as Musk's piling of people onto fewer floors caused the toilets to constantly be in use. In New York, the stench of the bathrooms overwhelmed some parts of the office, while some employees complained about cockroaches flitting in and out of drains. In San Francisco, some employees got used to avoiding the bathrooms by running to nearby coffee shops or restaurants, while others had to bring their own toilet paper from home. In one bathroom, someone rigged a metal coat hanger to a stall handlebar to use as a makeshift spool for a toilet roll, a luxurious gift for their colleagues.

—

>>> MUSK MADE HIS imprint on the offices in other ways as well. For Musk and his goons, there was a reward to the Twitter takeover that went beyond any type of potential financial return. They were the conquering Romans, who battered down the fortifications of what they saw as a liberal bastion in the most liberal American city and won. Twitter and its employees were their spoils, and they were going to enjoy them.

On November 22, Musk tweeted a video of shelves full of Twitter-branded apparel he had come across while wandering the San Francisco office. "There's an entire closet of #woke t-shirts," he said, panning to a gray tee with white lettering that read #StayWoke.

To Musk, "the woke mind virus" had infected places like Twitter. "The conquest of Wokerosi is complete," tweeted Sacks, resharing Musk's video.

For current and former employees at Twitter, particularly its Black workers, Musk's message was disturbing. The shirts had been created after the Ferguson protests by Blackbirds, a Black employee resource group, and had been popularized by Dorsey, who proudly wore one at public events to bring attention to the disproportionate amount of police violence faced by African Americans.

Musk cut off funding for employee resource groups like Blackbirds, while also making other symbolic changes. He ordered the teardown of a Black Lives Matter mural, which had been put in the San Francisco office following the George Floyd protests.

Musk's team put up their own decorations. Near the tenth-floor conference room that he often inhabited while in the office, they put up a Galerie de Meme, or meme gallery, framing printouts of some of the billionaire's favorite juvenile internet jokes. Other interior design changes seemed like attempts to stoke Musk's ego. Before the acquisition, Twitter hung images of famous or popular tweets in its ninth-floor cafeteria. One was eventually replaced with an image of Musk's April 2022 tweet in which he joked he'd buy the Coca-Cola company and put cocaine back in its main beverage, a post that garnered nearly 5 million likes. Near the entrance of the commons area, someone also hung a collage of great free speech moments in human history. There were photos of the U.S. Constitution, John Milton's *Areopagitica*, Berkeley's 1960s Free Speech Movement, and Musk carrying his sink into Twitter.

Spiro was also thinking about how he could leave his mark. The lawyer had told his colleagues he had no plans to run Twitter, but he was quietly considering it. He dreamed up a plan to export Musk's free speech vision globally. He'd pitch Starlink to repressive countries, giving them internet access through Musk's company. Then he'd add a new provision: if they wanted Musk's internet, they'd have to allow Twitter, too. Spiro thought he could pull off a similar maneuver with Musk's plan to add payments to Twitter by providing easy online transactions to get countries that restricted online speech hooked on the platform. He filed paperwork on behalf of the company to pave the way for it to process payments.

But some of his fellow lawyers at Quinn Emanuel were concerned about Spiro. The race to prepare for the aborted Twitter trial had taken a toll, and Spiro hadn't paused for breath since. He had a heavy caseload and was juggling too much with his added responsibilities at Twitter. His coworkers weren't sure when he slept and thought Musk's confidant was acting recklessly, working at an unsustainable speed, and heading for a crash.

>>> TWO DAYS BEFORE Thanksgiving, Musk wanted to know what the strategy was for relaunching Blue. He gathered the Blue team in Caracara, while those who had already flown home for the holiday dialed in for what became a two-hour meeting of stream-of-consciousness rambling. For some unexplained reason, Musk sat in the conference room in a tuxedo.

"I've been thinking about this every fucking day," he said. "Every fucking day. My product judgment is really fucking good."

While the discussion was supposed to focus on relaunching Blue and preventing the wave of impersonations that had broadsided the system the first time, it devolved into whatever crossed the billionaire's mind. He talked about the SR-71 Blackbird, a favorite stealth plane of his that avoided Soviet Union–launched missiles during the Cold War, noting that its agility was exactly how the team should behave. "We're about speed, speed, speed," he said. (Musk's son X Æ A-12 was named, in part, after the Lockheed A-12, the precursor jet to the SR-71.)

It was important to move fast, he continued, because the bottom was about to fall out of the economy. Advertising was getting crushed and Blue was his way out of ad dependence. "I just got off the phone with Disney," he said. "The reason they had an emergency CEO change is they're going to have severe financial issues next year. Disney. Even fucking Disney."

"It's gonna be bankruptcy city in 2023," Musk continued.

Just as fears about a global economic downturn had influenced him to try and pull out of the Twitter deal, the talk of "bankruptcy city" informed Musk's operating decisions at the company. By the end of November and into December, he would repeatedly tell his associates he would not be paying rent anywhere. At one point, Musk told a finance worker that Twitter would send checks to landlords "over my dead body."

During the next two months, Twitter skipped out on the $6.8 million owed for December and January at its San Francisco headquarters, and avoided paying for its offices in New York, Seattle, and other cities, leading to a slew of lawsuits. In London, the Crown Estate, which oversees property belonging to the king and owned the building that Twitter occupied, filed suit.

Despite the lawsuits, Twitter stayed in some of the buildings Musk had deemed necessary for operations. It maintained its spaces in San Francisco and New York, consolidating remaining workers onto fewer floors. In some places, like Boulder and Seattle, the company simply accepted eviction and let its employees deal with the fallout. In some cases, workers were given less than forty-eight hours' notice to retrieve their belongings before they were locked out. In Singapore, workers were walked out by building management.

Davis and Hollander oversaw some of those changes, and as Twitter downsized, they saw an opportunity to make a little more cash for Musk.

ZERO-BASED BUDGETING

Enlisting the help of an online auction company run by a Twitter employee's husband, they began listing items for bid, including a $30,000 Twitter bird statue and a $25,000 espresso machine that was once used by Perch baristas, as well as office chairs, industrial fridges, and projectors. The world's richest man was trying to make a few extra bucks on a corporate garage sale.

>>> ON MONDAY, NOVEMBER 28, Musk awoke to news that Apple's App Store had rejected the latest update to the Twitter app. Without attempting to discern Apple's reasoning, he launched into a full-scale assault using his favorite weapon. "Apple has mostly stopped advertising on Twitter. Do they hate free speech in America?" he wrote at 9:45 in the morning in San Francisco. An hour later he tweeted, "Apple has also threatened to withhold Twitter from its App Store, but won't tell us why."

Up until that point, Apple, which typically spent more than $100 million a year on advertising on the platform, had been a friendly partner to Musk's Twitter. Earlier that month, as the Blue team rushed to relaunch the subscription service and develop a new app, Musk had made some calls to his connections at the Cupertino, California-based iPhone maker, leading Apple to expedite their typically rigorous app review process. It was the first time Twitter's engineers could remember receiving such treatment. Some members of the Blue team heard rumblings that Apple had been excited about also making money off the subscription product, as Apple took a 30 percent cut of any purchases made on iPhone apps.

But cracks were developing in the relationship. The company was concerned about the direction of content moderation under Musk and had followed other brands in significantly reducing their advertising spend on Twitter. It was also worried about reports that Twitter was exploring an OnlyFans service for paywalled videos for adult creators. Apple's App Store banned apps with "overtly sexual or pornographic material," and while Twitter had always toed the line—allowing for nudity and consensual pornography—the company had always attempted to place that adult content behind warning labels or outside of algorithmic recommendations, and certainly never attempted to make money off it.

Apple's concerns led to increased scrutiny, and soon its app reviewers began finding more porn in basic spot checks of the app. While Apple warned Twitter of its findings, those concerns went unheeded in the chaos of Musk's early reign. Eventually, the App Store rejected one of Twitter's updates on the basis of its findings, and demanded improvements before it could be released.

Musk viewed the App Store rejection and the apparent pullback in ad spending as a declaration of war by the electronics manufacturer. "Did you know Apple puts a secret 30% tax on everything you buy through their App Store?" he tweeted that morning, injecting himself into a debate that had already been raging in Silicon Valley between app developers and Apple.

In a meeting at Twitter that day, Musk became conspiratorial. He said Phil Schiller and Tim Cook, Apple's head of the App Store and CEO, must be in bed with the shorts, baselessly suggesting that the executives had aligned themselves with the short sellers who often bet against Musk's companies. To those in the room, it wasn't clear if he was joking or not. After all, Twitter was a private company. Musk also suggested that if Apple took Twitter out of the App Store, he could tell his Twitter followers to stage a protest at Apple's headquarters in Cupertino.

But Apple wasn't exactly looking for a fight either. The company and its CEO, Tim Cook, didn't like the idea of being drawn into a public spat with the erratic entrepreneur, who had once attempted to pitch Apple on buying Tesla. While Cook viewed Twitter's new owner with trepidation, Apple eventually made overtures to invite Musk to their campus that week.

Musk showed up at the $5 billion Apple Park that Wednesday, November 30, walking into the famous building designed by Norman Foster to resemble a spaceship, before being ushered into a conference room with Cook for a one-on-one meeting. Cook explained that Apple, like other advertisers, was worried about Twitter allowing more hate and misinformation, which Musk denied. After discussion, the Apple chief said the company would continue to advertise on the platform and that there were no plans to remove Twitter from the App Store as long as it complied with its rules. Then he made a special peace offering: Apple would only take a 15 percent cut, instead of 30 percent, from Twitter Blue subscriptions. Musk was satisfied, walking out of the meeting with Cook to a man-made pond in the middle of the campus.

But the victory was short-lived. That day, the FTC sent an exhaustive list of demands to Twitter, insisting that the company account for its privacy program after all the responsible executives resigned. The agency asked twenty-seven pointed questions, angling to figure out who would be left holding the bag, and zeroing in on Musk. The FTC demanded to know Musk's responsibilities at Twitter and asked for a copy of every email, Slack message, and memo that referenced the billionaire. The regulators seemed suspicious that Musk was running the privacy program into the ground and—despite Spiro's assurances that no employee could be held legally accountable—intent on pinning the blame on Twitter's owner. The FTC gave Twitter two weeks to respond.

>>> ON FRIDAY, DECEMBER 2, Musk prepared for what he believed to be a momentous occasion for his Twitter. For days, he had ranted about the executives who had preceded him at Twitter and like many of his conservative fans, Musk subscribed to the theory that Twitter had purposefully censored conservatives and promoted Democrats.

To prove his point, Musk had worked out a deal with Matt Taibbi, a former *Rolling Stone* journalist. He would give Taibbi unfettered access to Twitter's internal files and communications, allowing him to root around and document the liberal bias that had supposedly infested the social media company. Musk's only stipulation was that Taibbi had to publish his findings on Twitter itself before running any stories on Substack, an independent newsletter platform where he had a large following. Musk guaranteed his desired angle by selecting Taibbi—who had become a reactionary partisan. He would never side with Twitter's former management.

Musk's demands to provide Taibbi with access stunned the company's security employees. Taibbi wasn't an employee, and allowing him to riffle through users' information could violate Twitter's privacy obligations under its consent decree with the FTC and Europe's General Data Protection Regulation. But they didn't dare push back.

The first installment of the Twitter Files, about the company's 2020 decision to briefly block the *New York Post* from sharing its story about files excised from Hunter Biden's laptop, came with a strong endorsement.

"This will be awesome," Musk tweeted in anticipation, along with a popcorn emoji.

"It is a Frankensteinian tale of a human-built mechanism grown out the control of its designer," Taibbi wrote in an extensive Twitter thread that night, which included several screenshots of internal emails and Slack messages among Gadde, Roth, and other Twitter executives. Taibbi seemed to be drafting the story as he tweeted, taking long pauses between each tweet and occasionally deleting messages from the thread, but his thread, which was onerous to read, still spread rapidly, garnering more than 145,000 retweets.

Despite his efforts to reveal a plot to put Biden in the White House, the messages Taibbi posted seemed to prove the opposite point: they showed staffers desperately trying to understand the decision to block the *New York Post* story, and pushing back on the decision.

"This is an emerging situation where the facts remain unclear," Roth cautioned in one of the messages.

"I support the conclusion that we need more facts to assess whether the materials were hacked," Jim Baker, a deputy general counsel who reported to Edgett and Gadde, wrote in another.

On Saturday, December 3, Twitter employees realized that Taibbi's thread was just the beginning of the Twitter Files. Workers on the security team had been ordered to grant internal access to another right-wing journalist, Bari Weiss. The addition of another outsider to Twitter's systems only raised more legal concerns.

As one of the senior-most lawyers left at the company, Baker inserted himself into the mess. To make sure the journalists didn't inadvertently access user data that Twitter was legally obligated to protect, he insisted on vetting materials before they were handed over to the reporters. Over the weekend, as Taibbi and Weiss made increasingly vast and vague demands for any documents related to the stories they wanted to pursue, Baker became a bottleneck.

Frustrated, Weiss asked who was responsible for releasing files to her. When she found out it was Baker, "my jaw hit the floor," Weiss said. She quickly alerted Musk.

Previously, Musk had bashed Baker, a former general counsel for the FBI, alongside Gadde as an architect of Twitter's censorship regime. The billionaire was skeptical of Baker's deep government ties and figured he was another shadowy figure of the Deep State. After several purges of employees who weren't loyal to his cause, Musk assumed that Baker was long gone.

But his own lawyer, Spiro, had kept Baker off the layoff lists. The former FBI lawyer had quietly ridden out Musk's first month of ownership, keeping a low profile in the Washington, DC, office. As Musk impatiently awaited the next Twitter Files installment, he discovered that Baker was still inside the company, holding up the release of the files.

Livid, Musk demanded that Baker fly to San Francisco for a confrontation. Meanwhile, Birchall ordered Twitter's human resources team to start drafting a separation agreement for Baker so that Musk could fire him. Birchall claimed that workers from SpaceX's legal and government relations teams were conducting an internal investigation into Baker's conduct, and the result of their investigation would determine whether or not the lawyer would be offered severance.

When he arrived in San Francisco, Baker sat down for a meeting with Musk himself. "You're fired," the billionaire told him. But Baker was adamant that he had done nothing wrong, and walked Musk through his reasoning for examining the files before handing them off to Musk's pet reporters. He was protecting Musk, helping him avoid regulatory crackdowns.

Baker believed that Musk was listening and thought he might have saved his job. But in the end, he was escorted out of the building. Soon after, Spiro was gone from Twitter, too. Some employees thought the disagreement over Baker was the final straw. But Spiro wasn't confident that Musk would give him full leeway to enact his free speech plans at Twitter, and was leery about Musk's plans to continue layoffs after Spiro had told employees the cuts were over. Although he'd positioned himself to lead the company's legal and policy operations, Spiro told people close to him that he'd always planned his time at Twitter to be temporary. He went back to litigating for Musk and began to prepare for an upcoming trial in a lawsuit brought by Tesla shareholders.

>>> ON SATURDAY, DECEMBER 10, Musk erupted with frustration. Twitter was still hemorrhaging money and he decided to take the budget into his own hands. He summoned workers in San Francisco to the Caracara conference room at headquarters and insisted that finance executives and team leads from around the world dial in on a conference line.

Then he opened up a spreadsheet that documented Twitter's total expenditures and began to read it, line by line. As he ticked through different parts of the business, Musk demanded that the employees responsible for the spending explain their budgets to him. If Musk agreed the expense was necessary, the spending could continue. But if he didn't, the team's expenditures would go to zero immediately.

As Musk insisted on hearing directly from various business leads, frantic text messages pinged through Twitter's rank and file. More and more people piled into the video conference, ripped away from their Saturday grocery shopping and weekend chores.

Musk was relentless as he drilled into the details. He discovered that Twitter was paying for 15,000 licenses for a piece of software, even though at its peak, it had employed less than 8,000 people. Now, with fewer than 2,000 employees, the license fees were even more ridiculous. He ordered that someone renegotiate the contract with the software provider.

Twitter was also paying almost $20 million in domain registration fees, which kept it in command of twitter.com and other corporate websites. "Can someone explain why this is so high?" Musk asked.

Rebecca Falk, a lawyer who oversaw compliance and risk mitigation, piped up to answer. The payments had to do with security, she explained—Twitter was spending a lot to keep its domains from being stolen or compromised.

"Tesla isn't spending this much," Musk retorted.

Falk kept repeating the same sound bite: "It's important," she told him. "It's for security."

Musk grew angry as he kept insisting that the lawyer delve deeper into the nitty-gritty. But Falk, a lawyer who wasn't deeply steeped in the nerdy world of domain registrars, didn't seem able to give Musk the level of detail he wanted. Other executives who couldn't answer his questions had demurred, promising to look into their costs and get back to him. But Falk kept insisting that Twitter should pay its domain fees.

"The more that comes out of your mouth, the dumber it sounds," Musk finally told her.

Finally someone else on the call intervened, nudging Musk to move on. He continued to rip through the spreadsheet. Some of the gathered executives ached on Falk's behalf. Others who oversaw much smaller budgets thought Musk was finally exposing the excesses that had been tolerated for too long under Dorsey and Agrawal. *How are you in a position to spend that much and you don't know what it's for?* they wondered.

The call became a war zone as executives fought for their fiefdoms, trying to preserve what they had and suggesting cuts from other people's teams. As Musk continued, more silly spending came to light. The company was spending hundreds of millions of dollars on its contract for Amazon Web Services, the cloud-computing program run by the online retailer—but it also had a similar contract with Google to use its Google Cloud service. There were offices with two hundred desks but only sixty employees.

At one point on the call, as Musk probed the company's spending on web infrastructure, an executive said he would need to speak with Falk about a solution. "She's no longer with the company," Musk retorted, implying he had fired the lawyer. (In reality, Falk remained at Twitter for another year, departing in November 2023.)

The discussion, which lasted for six hours, was a wake-up call for Musk. He had thought that Davis had already eliminated every possible expense, but he had determined for himself that Twitter was still spending tens of millions of dollars that it didn't need to. In his mind, there was no accountability. Musk believed he would need to become even more hands-on in his management of Twitter.

40 > "I'm Rich, Bitch!"

On November 18, shortly after his resignation, Roth published an essay in the *New York Times* opinion pages, laying out a strategy for cracking down on Musk. The billionaire was unaccountable and pushing for a vision of free speech that was biased and untenable, Roth argued. His changes were "sudden and alarming," the former executive wrote, and Musk planned to rule according to his whims.

After Musk had embraced Roth and publicly told his followers to rely on someone they viewed as a censor for reliable information about what was going on at Twitter, Roth wanted to distance himself. With his op-ed, Roth did just that. But he also blindsided and infuriated Musk, who couldn't tolerate public critique—especially in the pages of a mainstream publication he loathed. Musk unfollowed him immediately after the op-ed was published.

As Irwin stepped into Roth's role, she made it her mission to eradicate child sexual exploitation material from Twitter. Throughout late November and early December, she and Musk tweeted publicly about the effort, which Musk called "priority #1."

On December 9, Twitter claimed its efforts were paying off. During the first month of Musk's ownership, it suspended nearly 300,000 accounts for violating its rules against child sexual exploitation, a 57 percent increase, the company claimed in a tweet. Musk claimed Twitter's former leadership had knowingly allowed the problem to fester.

"It is a crime that they refused to take action on child exploitation for years!" Musk raged. His insult prompted a rare retort from Dorsey, who responded, "This is false."

The next day, as Musk kept one eye on his Twitter feed, Roth's critics had begun to pick through his old tweets and online history. One of them unearthed his thesis from graduate school.

In his thesis, Roth had written about the fact that underage users often logged onto Grindr, a gay dating app known for hookups, and posited that the app should find ways to keep them safe rather than trying to ban them or ignoring them altogether. "Rather than merely trying to absolve themselves of legal responsibility or, worse, trying to drive out teenagers entirely, service providers should instead focus on crafting safety strategies that can accommodate a wide variety of use cases for platforms like Grindr—including, possibly, their role in safely connecting queer young adults," Roth wrote, arguing for more safety measures to protect young people online.

But Musk, who had been waiting for an opportunity to punch back at Roth, seized on the thesis as evidence that the former executive wanted to see teenagers form sexual relationships with adults. "Looks like Yoel is arguing in favor of children being able to access adult Internet services in his PhD thesis," Musk tweeted.

In no time, Twitter users were flooding Roth's account, accusing him of pedophilia. Then came the death threats. As online sleuths dug up his personal information, Roth's phone started ringing, and it wouldn't stop. Then the address of his Bay Area home he shared with his husband appeared online. Soon, the *Daily Mail* picked up on the story and published a story about the harassment, complete with details about Roth's house.

The onslaught was like nothing Roth had experienced before, even when the attacks came directly from Trump's White House. Back then, Twitter had posted a security guard outside his home. But now, Roth was unemployed and had no corporate resources to fall back on. He considered buying a gun to defend himself if someone broke into his home, but he felt too uncomfortable with the idea of owning a weapon.

Among the threats that pinged constantly on his phone, Roth saw a message from Del Harvey, Twitter's first trust and safety employee who had worked alongside him to craft the company's rationale for banning Trump. Roth and Gadde had barely spoken since her firing, and the relationship between the two was frayed. But Harvey had asked Gadde to help and she complied, arranging a place for Roth and his husband to stay while they figured out what to do.

Roth felt stung that Gadde had allowed Harvey to be her emissary instead of reaching out directly, but he didn't hesitate to take the offer.

Several weeks later, as the harassment died down, the couple returned to pack up their things. Neither felt safe staying there any longer, and they decided to list the house for sale. They sold it at a loss, disclosing to potential buyers that the property was marked by Musk's online harassment.

>>> WITH SPIRO'S DEPARTURE from Twitter, Musk needed new deputies. He imported a new raft of employees from SpaceX to fill the void, including Chris Cardaci, the company's vice president of legal, and Tim Hughes, its senior vice president for global business and government affairs. Cardaci took over Twitter's legal team, while Hughes stepped in to run its compliance department. In a meeting with Twitter's remaining lawyers, Davis told them that Hughes was the best fit for the role because he worked on Starlink while at SpaceX, which had some dealings with the FTC.

"We have to be flawless when it comes to compliance," Davis said, and in a rare lapse from his die-hard cost cutting, agreed that Twitter could hire people to work on the company's FTC issues. The regulator was indeed circling Twitter. Its questions appeared to be a clear pretext for pinning any criminal liability directly on Musk, should the agency determine Twitter was not in compliance with its consent decree. While Spiro had brashly dismissed the FTC, saying Musk was unafraid, the issue required legal attention. Twitter had blown past the agency's deadline by several weeks, and had little to show for it.

"The company does not have unlimited resources, and it has been working diligently to generate a response," Twitter wrote over and over again in a confidential letter to the FTC in December, responding to each of the agency's inquiries with the same excuse. One thing was unavoidable: Twitter had no choice but to admit that Musk was the only person in charge.

"Mr. Musk has served as the sole director and as Chief Executive Officer, President, Treasurer, and Secretary," the company wrote. "Mr. Musk has general supervisory authority over the business of the company, including influencing the design, establishment, and implementation of a comprehensive privacy and information security program."

"I'M RICH, BITCH!"

>>> ONE OF THE FIRST THINGS that Musk tweeted upon purchasing Twitter in late October was a nod to the Babylon Bee. "Comedy is now legal on Twitter," he wrote less than twenty-four hours after signing on the dotted line.

Musk had always wanted to be seen as funny, and on December 11, after spending the day berating workers in the Twitter office in an agitated state, he headed to San Francisco's Chase Center, home of the Golden State Warriors, to blow off some steam. Instead of basketball, though, Musk went to see his friend, comedian Dave Chappelle, who was playing a sold-out show.

"Ladies and gentlemen, make some noise for the richest man in the world," Chappelle announced as Musk walked onstage, mic in hand.

It was the end of the comedian's show and he had brought on his buddy as a set-ending showstopper. Musk raised both arms to the sky, then waved after an initial set of applause. Then came a cascade of boos. Musk paced the stage, looking at his host for some sort of cue amid the whistles and hollers.

"Cheers and boos, I see," said Chappelle, trying to defuse the situation in the 18,000-person-capacity stadium. "Elon . . ."

"Hey Dave."

"Controversy, buddy," said Chappelle.

"Weren't expecting this, were ya?" asked the billionaire.

"It sounds like some of those people you fired are in the audience," the comedian said, rescuing a few laughs. The boos picked up once again as Chappelle lit a joint and made fun of the cheap seats. Musk stammered, with the mic held limply near his lips. The pair eventually walked off stage, but reappeared for an encore with Chris Rock and Chappelle's entourage to make one last attempt at rehabbing Musk's image. Chappelle handed him the mic and encouraged him to yell a classic tagline of his. "I'm rich, bitch!" Musk screamed. A flurry of boos.

Musk barely slept that night. By 4:00 a.m. he was tweeting about defeating "the woke mind virus" and COVID, but the Chappelle incident was still on his mind. While he would often come across insults on Twitter, he could usually insulate himself from the negativity. Chapelle's show, however, was real life, and it gave Musk an unfiltered taste of the growing antagonism that had been building against him over his purchase of Twitter, his politics, and his general online antics. He blamed the incident on San Francisco's "unhinged leftists" in a response to a Twitter user. "Technically, it was 90% cheers and 10% boos (except during quiet periods), but, still, that's a lot of boos, which is a first for me in real life (frequent on Twitter)."

After hordes of people ridiculed that post and his fragile ego, Musk deleted it. Comedy was legal in real life, too.

>>> BEYOND THE REPUTATIONAL HIT, Musk's tweets about Roth would have financial impact. His increasingly erratic posts were like large neon signs to Tesla's public market investors that he was not spending time at his electric car maker, and shareholders began to worry. In their minds, Tesla and Musk were synonymous, and without his attention the company could easily fall by the wayside. When markets opened on Monday, Tesla's stock fell more than 6 percent during trading.

The stock closed down another 4 percent the following day, briefly erasing Musk's title as the world's richest human.

Internally, Birchall pleaded with employees to have some sympathy for Musk. "Elon is not draconian," he told them during a meeting. While his behavior might give the opposite impression, his actions were rooted in believing Twitter was the best social platform, Birchall said. Roth had questioned Musk's commitment to safety, but he'd made Tesla the safest car on the road, he continued. "He wakes up every day trying to make humanity better," he insisted.

The noise filled Musk's Twitter feed and mind. Those who met with him during that time worried if he had gone off his antidepressants, which he talked about openly, or perhaps started taking something else. In his brief moments of free time, employees saw him in Caracara—which he had now fully commandeered—seated at the table with a Diet Coke, scrolling and scrolling. Eventually, Musk's team installed frosted glass to shelter the boss from the worried glances.

>>> ONE OF THE few things that seemed to have a positive impact on Musk during that time was his son X Æ A-12 (pronounced "X Ash A 12"). While other children his age would have been in daycare or preparing to enter

"I'M RICH, BITCH!"

preschool, the Twitter office effectively served as his toddler's childcare. Musk and X's mother, Boucher, declared him the "chosen one" to those around them and painted the boy as a messiah-like character who would learn from his dad. If Musk was the architect of a more techno-perfect future, they thought that X could potentially take over the Tesla, SpaceX, and Twitter empire that had been laid out before him. "X is growing up in the business world so he can know more than his dad," Boucher told one Twitter executive.

Musk doted on X, and told his lieutenants that the child, who frequently stayed up late into the night waiting for his father, was up to absorb the lessons of how he ran his businesses. But to some of Twitter's staff, the child's presence felt like a wild experiment in parenting. Beyond Musk, his nanny, and his bodyguards, X appeared to have no one else to interact with, certainly no one his own age. Musk would bring him into meetings and the toddler would talk and occasionally curse, a trait learned from his father, leaving workers confused as to why they had to speak over a child while explaining something to their boss.

Musk had always been a conflicted father, and despite fathering at least eleven children, he struggled to raise them. In the earlier days, his eldest children, a pair of twins and a set of triplets, would often come to his various offices—much like X—and he was able to keep them close as he worked. But now that they were older, some had become tired of their father's lack of attention and constant work. They grew distant—none more than Jenna, who was seen by her father as a neo-Marxist, brainwashed by her progressive schooling. In conversations, he told those close to him that none of his teenagers wanted to see him these days, making those who heard the comments wonder if the time he spent with X was to ensure yet another relationship with a child wouldn't deteriorate.

—

>>> THE WEEK OF DECEMBER 11, Musk sent X Æ A-12 down to Los Angeles via his private jet to be with his mother. When the child got there, a car driving him was followed by a stalker who thought the billionaire was inside. When the vehicle pulled over at a gas station, the boy's security guards confronted the disturbed individual and took footage, which Musk later

posted on his Twitter account. He and his family were being threatened, he said, and it was because of a Twitter account that tracked the whereabouts of his private plane.

When Musk bought Twitter, he directly cited the @ElonJet account as something he would not ban. "My commitment to free speech extends even to not banning the account following my plane, even though that is a direct personal safety risk," he wrote after his first two weeks in charge. Behind the scenes, however, the account became a regular topic of discussion, with Irwin, the trust and safety head, advocating for its removal because it broadcast location information. While Musk initially demurred, the incident with his son caused him to change his mind overnight.

Twitter moved to ban @ElonJet on December 14, with Musk hastily crafting a policy to state that any account sharing another's "live location information" would be suspended. It was unclear what this meant. Since its inception, people had posted information about themselves and others in real time at sporting events, protests, or simply moving through everyday life. If a user happened to tweet a photo of Tom Cruise walking through the doors of a McDonald's and tweeted the location, would they be banned under the new policy?

The company not only banned @ElonJet, but other accounts that shared links to other sites with public flight data. That included the Twitter account for Mastodon, a competing social network that shared that people could find the @ElonJet account on their site, as well as more than half a dozen journalists at *The New York Times*, *The Washington Post*, and CNN, who reported on what was happening with the suspensions.*

"They posted my exact real-time location, basically assassination coordinates, in (obvious) direct violation of Twitter terms of service," Musk later tweeted, the wheels of conspiracy turning in his head.

Musk's associates, including Calacanis and Sacks, would attempt to dissuade him from going further. Musk had gone off the rails, connecting a bunch of seemingly unrelated events into a larger plot of subterfuge. He grew extremely dark, telling those around him that perhaps he made a mistake in buying Twitter. The financial burden and the focus it took away from

---

* Among those banned was Ryan, one of the authors of this book.

his other companies, including Tesla and SpaceX, was becoming too much. "He came in with a savior complex, as if he was coming to save humanity," said one former Twitter executive. "Suddenly you turn around and people actually see you as the devil and not a hero. That's a difficult thing to deal with."

But the night after the bans, Musk couldn't stay offline. When Katie Notopoulos, a reporter at *BuzzFeed News,* started a Twitter Space to talk about the suspensions in a live audio discussion, a few of the suspended journalists, including Drew Harwell of *The Washington Post* and Matt Binder of *Mashable,* were able to join. They exploited a glitch in Spaces, which was built on a different database infrastructure than the rest of the social network, and began attempting to make sense of Musk's thin-skinned hypocrisy. Musk eventually joined the space and huffily tried to offer a defense. "Showing real-time information about someone's location is inappropriate," he began, before ripping into his longtime enemies.

"There is not going to be any distinction in the future between journalists—so-called journalists—and, and regular people," Musk said. "Everyone's going to be treated the same. You're not special because you're a journalist. You're, you're just a—you're just a Tw—citizen."

As the reporters tried to press Musk on what exactly he meant, Musk raised his voice and then abruptly left the call. Minutes later, he ordered Twitter's engineers to close all Spaces, and later tweeted that the company was "fixing a Legacy bug." There were limits to free speech after all.

41 > Self-Doubt

The 2022 FIFA Men's World Cup was an opportunity for Qatar to put itself on the map. On Sunday, December 18, the country was the center of the world, as soccer stars, including Argentina's Lionel Messi and France's Kylian Mbappé waited in the innards of a golden, egg-shaped stadium for the most important game of their lives to kick off.

Musk was also there, in a luxury box with other powerful people who had traveled to the Persian Gulf for the game. But he wasn't there for the entertainment. Musk hated sports. In his youth he spent more time reading, tinkering with computers, or playing video games. His purchase of Twitter shifted those priorities. Sports was a major source of Twitter usership from NBA trade rumors to Olympics coverage to Super Bowl chatter, and large live sporting events were a boon to the company's user and revenue figures as its sales team could sell special ad packages or show advertising against highlights or content licensed from the major global leagues.

Among sporting events on Twitter, though, the Men's FIFA World Cup was the biggest. Consisting of a month of dozens of soccer matches, the event, which takes place every four years, had consistently set traffic records for the company. In 2010, the first World Cup with a major Twitter presence, the company warned users to expect major outages, which happened when the Netherlands national team upset Brazil in the quarterfinals in South Africa. The 2014 event in Brazil led to some 672 million tweets, the highest number of posts related to a single event at the time. Over the years, as Twitter matured and developed better infrastructure, the site became more adept at handling these traffic booms, but after Musk arrived and began laying waste to the employee base, those who were left became concerned that the company wouldn't have the people or know-how to deal with a major outage. As some engineers put it, Twitter was like a running car. It was chugging along for now, but if the check-engine light came on or the transmission failed, it wasn't clear if there would be mechanics around to fix it.

Despite the possibility for failure, the sixty-three games through November and December caused no such outages, and by the time Musk landed on the ground in Qatar for the World Cup Final on December 18, Twitter had had a largely charmed time dealing with the traffic. Musk flew more than fifteen hours to get to the game and arrived at the stadium just before kickoff, shaking hands, greeting fans, and snapping photos with everyone from the likes of Turkish strongman president Recep Tayyip Erdoğan; Nusret Gökçe, the influencer restaurateur better known as Salt Bae; and a Russian state television presenter and propagandist. Then Musk settled into a seat next to Jared Kushner, Donald Trump's son-in-law and close adviser, and Mansoor bin Ebrahim Al-Mahmoud, the CEO of the Qatar Investment Authority.

Musk was there on business, honoring his commitment to his Qatari backers and hoping to raise more money from investors to lessen his financial burden. The game was attended by a who's who of the rich and famous, including the former English soccer star David Beckham and Indian steel billionaire Lakshmi Mittal, and it was a perfect opportunity for Musk to get in front of people. The fundraising effort had largely been kept under wraps, overseen by Birchall and Pablo Mendoza, an associate from Vy Capital, a Dubai-based venture fund, who had been lent out to Musk's Twitter for finance help. "Over recent weeks we've received numerous inbound requests to invest in Twitter," Birchall wrote to investors in an email. "Accordingly, we are pleased to announce a follow-on equity offering for common shares at the original price and terms, targeting a year-end close."

The terms were bold, if not preposterous, as they sought to raise up to $3 billion. Musk had spent the last month and a half destroying the company's advertising revenue streams and firing employees. His chaotic approach, combined with the churning global financial markets, led some banks, which had provided $13 billion in debt financing, to try and sell debt at steep discounts. Analysts and critics believed that it was ludicrous to think anyone would pay the full $54.20 for shares in Musk's Twitter. And as he went out to try and fundraise, he would only create more chaos for himself.

As the final wrapped up, with Argentina winning on penalties, Twitter announced a new policy that sent its users into a frenzy. "We will remove accounts created solely for the purpose of promoting other social platforms and content that contains links or usernames for the following platforms: Facebook, Instagram, Mastodon, Truth Social, Tribe, Nostr and Post," the company announced on one of its official feeds. Musk had demanded the policy in an attempt to preserve traffic and stop eyeballs from going to other domains. He was particularly incensed by Dorsey, who had encouraged people to download and use Nostr, a new decentralized social network.

The policy drew backlash from all corners of Twitter and across the political spectrum. It went against all social media norms and was a direct contradiction to the idea of free speech. The move was also antithetical to the Silicon Valley ethos of competition, which posited that companies should simply build the best products to win, not force people to use them.

Among Musk's critics were whistleblower Edward Snowden; Aaron Levie, the chief executive of Box; and Balaji Srinivasan, a crypto entrepreneur and former partner at Andreessen Horowitz. But one of the loudest was Paul Graham, the founder of start-up incubator Y Combinator, who had tweeted the previous month that "it's remarkable how many people who've never run any kind of company think they know how to run a tech company better than someone who's run Tesla and SpaceX." After Twitter's new policy, he changed his tune, tweeting a link to his Mastodon account and calling Musk's move "the final straw." He was promptly banned.

"I still think Elon is a smart guy," Graham later wrote on a message board popular with start-up founders. "His work on cars and rockets speaks for itself. Nor do I think he's the villain a lot of people try to make him out to be. He's eccentric, definitely, but that should be news to no one. Plus I don't think he realizes that the techniques that work for cars and rockets don't work in social media. Those two facts are sufficient to explain most of his behavior."

Within hours, Musk posted that he would change the policy to only suspend accounts whose "*primary* purpose" was to promote competitor sites. Then he wrote, "Going forward, there will be a vote for major policy changes. My apologies. Won't happen again." That was followed by a poll that was meant to be a referendum of his leadership.

"Should I step down as head of Twitter?" he tweeted at 1:20 a.m. in Doha. "I will abide by the results of this poll." He stayed up through the night posting away before his plane took off for Austin, with a layover in London, the next day.

Musk flew out of the Middle East empty-handed, his search for new funding failed. The rejections and the policy debacle left Musk in a worse mental state than when he arrived in Qatar. But he was not prepared for the results of his poll, which ended while he was in the air.

—

>>> BY THE TIME he touched down in London, 57.5 percent of the more than 17.5 million accounts that had voted were calling for him to resign. It was a shock to Musk's system. For a man who read sci-fi and superhero fantasies growing up as a kid, the rejection felt like Gotham voting to exile Batman. Musk was shorn of the deep confidence he often had in himself and driven into further misery by Tesla's tumbling stock. The company's shares set a new two-year low upon his return stateside on December 20 and were down nearly 66 percent since the start of the year. Musk spent the day in Austin at Tesla, attempting to do damage control and prove to the market he wasn't an absentee boss. The stock continued to fall in the days after.

When he arrived back in San Francisco, he stopped responding to some emails and texts, while those who did come in contact with him reported to others in his circle that he was in the throes of a manic event. His behavior would begin a period of what Marc Andreessen, a cofounder of Twitter investor Andreessen Horowitz, would term "an episode." The venture capitalist became so concerned that he messaged Twitter staffers to see if Musk, who confined himself largely to his conference room, was in the right headspace.

In one conversation with a confidant, Musk choked up and began to doubt his ability to run the company. He wanted to be liked, and his realization that millions of people—including some of his friends and supporters—could turn against him in an instant sent him further into depression. "I'm never going to recover from this," he said.

Employees closely monitored Musk's Twitter feed, where he was still posting, for any responses. "I will resign as CEO as soon as I find someone foolish enough to take the job!" he wrote. "After that, I will just run the software & servers teams."

Some engineers, who had participated in the poll and voted for Musk to step down, were initially encouraged, hoping that his statement meant they would get a respite from the breakneck pace within the company. The brief relief, though, turned into panic when they realized that polling data was unencrypted and that those with access to Twitter's internal systems could see how individual accounts had voted. Fearing they could be outed and fired, the engineers scrambled to obscure the data and how to find it.

It was during this episode that some Twitter employees had a realization about Musk. Tesla and SpaceX were companies that were built in his image, and they had people and processes in place to manage his expectations and shifting demands. He also never spent multiple days on end fixated on one company, preferring to bounce around to address whatever needed his immediate attention, whether that be a booster design or a factory manufacturing issue. At Twitter, however, there was no buffer, just him banging his head against the wall day after day to solve a business problem he had created for himself.

Musk seemed serious about relinquishing the CEO role. Within days, he invited a CEO candidate to Twitter's offices for an interview—it was Emil Michael, an aggressive operations executive who had been Travis Kalanick's right hand at Uber before he was fired in 2017. Since his defenestration, Michael had dabbled in investing, but he was open to returning to the cutthroat world of Silicon Valley start-ups. The meeting did not result in a job offer.

—

>>> SHEEN AUSTIN, the Tesla infrastructure leader, knew people who said no to Musk tended not to last for very long in his orbit. In fact, Austin knew he'd signed a death warrant by remaining at Twitter. He had seen others be-

SELF-DOUBT

fore him, like Nelson Abramson, the Twitter vice president who led infrastructure, lose their jobs for tiny transgressions. Austin had tried to coach Abramson in how to speak to Musk and his lackeys, while managing their expectations, but it was no use. He was fired earlier that month after getting in a disagreement with Davis over budget cuts.

Fatalism consumed Austin as he got up every day for work, which was now an exercise in keeping one of the world's most trafficked websites online as its new owner did everything to break it. He hated it.

Keeping Twitter online was not cheap. Labor aside, the company spent more than $1 billion annually on server and cloud computing costs, easily making it one of the highest operating expenses at the company. Much of that went to its three physical data centers, located in Atlanta; Portland, Oregon; and Sacramento. These locations were effectively the brains of Twitter, storing every profile, tweet, like, and other data on racks of servers that spanned football fields' worth of space.

Sacramento, or SMF, named for the local airport code, was the company's largest data center. Located in the northern part of California's capital, the facility was owned and operated by Nippon Telegraph and Telephone (NTT), a Tokyo-based telecoms company from which Twitter leased space. While the company had used the facility for more than a decade, a heat wave that June knocked out all power to the data center and led some leaders to reconsider the company's presence in Sacramento. By the time Musk arrived on the scene in a cost-cutting frenzy, that decision was a no-brainer. SMF was a goner.

Austin agreed that it was the right call. But he also knew that shutting down one of the three main data centers could not be done overnight. This wasn't as simple as flipping a light switch; data would have to be ported safely to ensure that nothing was lost. Servers would have to be carefully packed and shipped. Engineers would have to reroute traffic to ensure there were no service outages. Done right, the process would take weeks, if not months. He tried to impress that idea upon his boss, who eventually agreed that they would have until early January to sort out the transition.

On December 23, one of Austin's deputies sent out an email to Twitter staff informing them that SMF would be shut down, while its Atlanta data center would also be downsized by the following month. No reasons were given, but it was clear among those who knew that it was aimed at reducing costs. Austin was torn. He saw the savings rationale behind the Sacramento closure but worried that directing more traffic to Atlanta and Portland might lead to instability and outages. Data centers rely on redundancy, or the idea that if one failed, the others could step up and handle the traffic responsibilities, keeping a site online.

That night, Musk's plane had already taken off from the private jet tarmac at Oakland International Airport with the billionaire, Davis, Andrew and James Musk, and their families in tow. Austin, who had been working all week to ready the data center transition for after the holidays, was in the middle of preparing for a celebration of two of his children's birthdays when he got a surprising call from James Musk.

The plane had been turned around, he told Austin as others laughed in the background. On the flight, Musk had become fixated on SMF and the time it would take to transition out, leading the cousins to suggest they could start removing servers themselves. Musk agreed and told his pilot to turn the plane around near Las Vegas. The SMF shutdown would begin tonight.

At Twitter, few employees knew about what was happening. Austin, who had pleaded with Musk to reconsider, with no success, informed some engineers there was an ongoing "issue," but didn't provide details about what his boss was preparing to do. As Musk, his cousins, and Davis arrived at SMF after midnight on Christmas Eve and began walking the floors of the data center, Austin logged on from home to monitor Twitter's systems and take note of any potential damage. At one point, Musk used a pocketknife to unplug one server, around the time a number of Agent Tools, or services that were used to monitor abuse, hate, or illegal activity, also went down. The failure was enough to lead Musk to stop pulling out cords.

On Twitter, however, Musk pretended that everything was going smoothly. "And yet work it does," Musk tweeted cryptically about his escapade, "even after I disconnected one of the more sensitive server racks."

Behind the scenes, Austin pleaded with Musk and Davis not to do anything else. The whole site could go down and it might take days for Twitter's infrastructure team to diagnose the issues and get it back up and running. There was also the possibility that user data could be damaged or lost, leading to a potential violation of Twitter's agreements with the FTC. His attempts to strike the fear of God into his boss were an exercise in buying himself more time so he'd be able to reroute traffic away from SMF and to the other remaining data centers.

The move worked. Musk unplugged nothing else, and most of the traffic to SMF was stopped within twelve hours as Twitter's infrastructure engineers worked around the clock to ensure nothing went down. Miraculously, the site stayed mostly online, as trucks lined up outside the data center, equipped with AirTags to track their cargo and ready to transport Twitter's server racks off the premises. Musk didn't stay long, leaving in the morning hours of Christmas Eve, having lobbed another grenade at his workers.

The following week, users around the globe began experiencing outages and were unable to use Twitter on desktop computers. In other cases, people could not view the replies to their tweets or encountered error messages while scrolling their timelines. It was exactly as Austin had warned. Days after that, Twitter stopped working for many in Australia and New Zealand, and in the coming months, the site would be stricken with half a dozen major outages. It took more than four months to install the servers from Sacramento in the Atlanta data center, with employees working constantly on the project.

Austin, however, didn't care if he was right. The SMF shutdown shattered his image of Musk. He may have built billion-dollar companies, but this was a man who chewed through people, using their loyalty to work them into the ground. Austin thought he had earned Musk's trust and respect, but he realized he was simply a pawn.

On New Year's Day, Austin and Kiko Smith, the architect who had been mistakenly fired, flew to Portland to oversee the transfer of server racks that had been driven up from Sacramento. As they ate breakfast in the dining room of an Embassy Suites hotel the following morning, Austin commiserated with his colleague.

"This is not something I can tolerate," he said of Musk's commands. "And then forcing it on other people. I don't even want it for myself. How can I force it on other people?"

Austin oversaw the SMF shutdown for a few more weeks before resigning, not just from Twitter, but also from Tesla. He was no longer a believer.

—

>>> CRAWFORD WAS ALSO losing her belief in Musk. While she had viewed him with optimism when he arrived and touted his abilities as a product visionary who could get people to move mountains, his two months at Twitter had shattered that image. The Blue launch had failed dramatically, revenue was crushed, and employee morale went through the floor. She had watched from a cautious distance as Musk passed through his episode, and at the end of the month made an attempt to correct his course.

She readied a presentation knowing that, much like with a fourth grader, she would have to deliver the bad news with praise. Then she met him in Caracara and ran him through the numbers, showing dramatic declines in year-end revenue since his takeover.

"Some of your tweets are the reason why advertisers left or paused spending," she told him, lining up the timing of some of his most controversial posts with drops in advertiser spend. "But the good news is that only you can turn that around!"

She thought the numbers, plain as day, would give him the impetus to change. But instead, he started to argue. "These are due to macroeconomic factors," he said, turning to his favorite talking points of rising interest rates and a coming recession. Musk convinced himself that he was not the problem.

"We still have too many people at the company," he said, suggesting that bankruptcy was still on the table. He repeated the line every time Crawford pointed to a new data point showing his effect on advertising.

The product manager could sense she was getting nowhere. She knew other companies in the space, like Facebook's parent company, Meta, were not experiencing the same type of impact on advertising and that Twitter's issues went far beyond macroeconomics.

As she went to get up, Musk patronized her.

"Don't worry about it, Esther. I can tell you care a lot."

Soon he'd stop texting and including her in product meetings.

42 > Red Pilled

At Twitter, the cuts continued through January. In the U.S., rolling rounds of layoffs hit engineers working on the company's ad product and trust and safety teams, while tweeps in Australia were cut. In some cases, employees were unsure if they had been exited at all.

In early January, one worker took it upon himself to figure out what was going on. He was a designer based in Europe and had resigned two months earlier. But for whatever reason, he still had access to his company email and Slack, and was still receiving paychecks. He had reached out to the company's human resources representatives, but Musk had laid many of them off, leaving it unclear if the company received his messages in the first place. With no recourse, he messaged some people who he thought were in charge on Slack.

"I'm basically the stapler guy from *Office Space*," he wrote, referencing the 1999 dark comedy that features a character who is unceremoniously moved into an office basement and forgotten. Eventually, someone figured out his predicament and properly exited him from the company.

Davis also continued to kill employee benefits, including family planning and company support for in vitro fertilization services. The moves left employees who had frozen their eggs scrambling to figure out how to pay for something they had expected their employer to cover. It all felt unnecessarily cruel and inconsistent. After all, didn't Musk want people to repopulate the earth?

Musk had recovered from his "episode" at the end of December, but his tweeting continued at an unhinged pace, revealing a mind stewing within its own filter bubble. While he had always seen himself as a libertarian centrist, he tweeted in the early hours of January 5, that "Kevin McCarthy should be speaker," pushing for the Republican representative from California who had supported Trump's claims of voter fraud in 2020 to lead the House.

Musk also leaned on Twitter's remaining content moderation team in the aftermath of the Brazilian election, second-guessing its employees when they took down tweets questioning the defeat of Bolsonaro. After Lula da Silva won the election in October, Bolsonaro's supporters claimed the election had been stolen, an echo of Trump's supporters after his 2020 defeat. Twitter's moderators knew they had to act or risk an outcome like the U.S. Capitol riot, so they took down tweets that violated the platform's rules.

When Musk discovered what was going on, he stopped them. Only tweets that explicitly called for violence or were the subject of government orders could be removed, he said, peeved by a Brazilian judge that sent regular takedown requests to the company. On January 8, people stormed Brazil's federal buildings in an attempt to keep Bolsonaro in power.

By then, Musk had become openly conservative. He regularly replied to right-wing accounts and personalities including @Catturd2, the trolling alter ego of a Trump-supporting Florida man, and Jack Posobiec, an activist who promoted Pizzagate, offering to personally look into their complaints about Twitter. He also gave a tour of the office to Dave Rubin, a conservative podcast host, allowing him to spend two days at Twitter headquarters asking employees questions about why his own account had been limited in its reach.

While he faced questions about his political tweeting, notably on a Tesla earnings call on January 25, where he was asked if his politicization would hurt the automotive company and alienate some buyers, Musk portrayed his tweeting as a net benefit. He had 127 million followers, which suggested that he was "reasonably popular."

"Twitter is actually an incredible tool for driving demand," Musk said to analysts and investors.

The following day, Musk flew to Washington, DC, to meet with McCarthy on the politician's birthday, discussing Twitter and its policies with the new speaker. He also found time for Jim Jordan, the Ohio representative and

Trump attack dog, and Kentucky representative James Comer, who had announced his intentions to use his position on the House Oversight Committee to investigate President Biden. It was an area of interest for Musk, who was facing various government inquiries across his companies and began to wonder if Biden would weaponize federal agencies against Twitter.

With the FTC investigation well underway, he now saw himself as a direct target of the White House and aligned himself with those who seemed best suited to protect his interests. But he was also eager to worm his way out of the fight with the regulator, which had subpoenaed him for a February deposition. His team wrote to Lina Khan, the FTC chairwoman, asking for an informal sit-down during his DC trip, but she rebuffed him, telling him in a letter that he should focus his efforts on responding to her investigators.

"I recommend that Twitter appropriately prioritize its legal obligations to provide the requested information," she wrote. "Once Twitter has fully complied with all FTC requests, I will be happy to consider scheduling a meeting with Mr. Musk."

In February, Comer held a hearing in the House to examine the supposed ties between Big Tech and government. As Roth, Gadde, and others answered questions about what conservative lawmakers saw as supposed anti-right-wing censorship, Republicans praised Musk. Among them was Marjorie Taylor Greene, the Georgia congresswoman and QAnon endorser whose Twitter account was suspended in early 2022 for spreading misinformation about the COVID vaccine.

"Thank God Elon Musk bought Twitter," she said.

>>> THE FAR RIGHT had plenty of reasons to celebrate Musk's acquisition. In him, they saw a leader who would loosen the site's speech rules—many of which had led to the suspension of notable figures. Greene's accounts had been reinstated in late November—after McCarthy used his influence to push Twitter's leadership to bring her back—beginning what Musk would term a "general amnesty" of previously suspended accounts.

The returnees were a who's who of the far right. Ali Alexander, an organizer of the Stop the Steal election denial movement who was banned in the aftermath of January 6, came back, as did Ron Watkins, the suspected creator behind QAnon. Musk also welcomed back Nick Fuentes, a white supremacist and incel whose hateful ideologies found a following among a group of racists known as the Groypers. Fuentes, however, didn't last long, logging in to a Twitter Space on the same night as his reinstatement to scream that he liked Hitler and was "going to war with the Jews." He was eventually re-banned, though Musk allowed him to return again in May 2024.

Musk's impact, however, went far beyond the amnesties. "People on the right should see more 'left wing' stuff and people on the left should see more 'right wing' stuff," he tweeted on January 16.

In practice, however, people began seeing the amplification of right-wing voices. While Musk had espoused the idea of "freedom of speech, not freedom of reach," Twitter's recommendation algorithms began pushing conservative accounts and posts into people's timelines. Users setting up new accounts were given recommendations to follow the likes of Republican Florida governor Ron DeSantis; Texas senator Ted Cruz; and Posobiec.

Activist groups that study online platforms, including the Center for Countering Digital Hate and the Anti-Defamation League, found that slurs against Black Americans had tripled, while antisemitic speech was up more than 60 percent since the change in ownership. Accounts showing terrorist affiliations with groups such as the Islamic State (IS) also surged in the wake of the takeover.

In mid-January, Twitter's recommendation algorithms pushed a post featuring a video of a man stabbing a woman multiple times in a bedroom and killing her as something "you may like" on users' "For You" feeds. The promotion of the video, which racked up more than 1.2 million views, 7,500 retweets, and nearly 22,000 likes, was the result of Musk's prioritization of engagement combined with the removal of the site's guardrails.

Twitter was also still crawling with child sexual exploitation material. The problem wasn't an easy one for any social media company to solve. The material is illegal to look at, which prevents companies from deputizing employees to go out and proactively scour their platforms for it. Instead, the companies largely rely on databases of hashes, digital signatures created to match abuse images that had previously been caught by law enforcement.

By searching for the hashes, tech companies could take down material they knew was illegal without ever looking at it.

Twitter used the same mechanisms, but somehow known material was still slipping through. In the chaos of layoffs and cost cutting, it seemed that the company had lost some of its ability to detect and remove the material. One of the contracts Twitter had stopped paying was its deal with Thorn, a tech company that had built a hash database for videos of child exploitation.

The gaps left up a graphic video of a boy being sexually assaulted on Twitter, where it was viewed more than 120,000 times, a *New York Times* investigation revealed. Twitter's recommendation algorithm also surfaced and suggested other accounts that posted images of child abuse.

>>> AS MUSK'S TWITTER elevated conservative voices, the billionaire was developing a budding relationship with one Republican congressman in particular. During the February 8 hearing, Representative Jordan made it clear that he believed Musk's ownership had turned the company around. He slammed the company's previous management, notably Roth and Gadde, for censoring posts and communicating with federal law enforcement agencies, which the congressman saw as violations of the First Amendment and free speech. In Jordan, Musk saw potential for a new political terrier who could argue on his behalf as he faced scrutiny from Democratic lawmakers and regulators.

Some of the heaviest scrutiny came from the FTC. By early February, the agency had sent twelve letters to the company demanding information and had interviewed former executives, including Lea Kissner, Twitter's old chief information security officer, and Damien Kieran, its former chief privacy officer.

To Musk, this was government harassment by the Biden administration, which he had come to loathe for its positions on immigration and organized labor. He authorized one of the company's top remaining lawyers, Christian Dowell, to share the FTC's requests and other documents with Jordan's office, giving the congressman ammunition for his House select subcommittee on "the weaponization of the federal government." Musk later invited staffers from Jordan's office to visit SpaceX's facilities in Texas.

He also found a sympathetic ear in the FTC's lone Republican commissioner, Christine Wilson. In February, he secured a meeting with Wilson and confided in her about the government persecution he believed he was facing. Jim Kohm, a leader in the FTC's enforcement division, sat in and listened to the complaints. Wilson agreed that the FTC had gone beyond its mandate in pursuing him personally—but there was nothing she could do. Shortly before her meeting with Musk, Wilson had tendered her resignation. He would get no protection from her. The meeting seemed to do little to spook Kohm—the demand letters from his team kept coming.

Still, Musk managed to get the agency to delay his deposition. He was not interested in answering the agency's questions under oath.

>>> AS MUSK STEWED over the external government inquiries, he also became fixated on Twitter's internal platform issues. Beyond the site's outages that were the result of the SMF shutdown and reckless tweaks to internal tools and services, Musk became obsessed with the falling engagement on his tweets.

Twitter's engineers could see the drop-off in Musk's engagement as well but didn't have any explanations for it. There had been various changes to the algorithm and in Twitter's overall product since Musk took over, and any multitude of factors could have caused it. At one point, Twitter's owner took his account private—preventing non-followers from seeing his posts—in an attempt to see if that would rectify his engagement problem. "Something fundamental is wrong," the billionaire tweeted on February 1. He stayed late into the night at Twitter's offices attempting to rectify what he believed was an existential problem.

To the engineers who had to field Musk's questions about his own tweets, there was a certain level of absurdity to the requests. On one hand, this was the world's richest person, obsessing over the dissemination of his posts, which were still getting millions and millions of views even with the supposed decline. On the other hand, they sympathized with Musk's intuition that perhaps his account was simply a canary in the coal mine, warning of potential drops in engagement for others. Tests, though, didn't show meaningful declines for other users.

On Tuesday, February 7, Musk's anger would come to a head. He had called another meeting to deal with the subject of his engagement, calling the situation "ridiculous," as Davis backed him up by pulling up old tweets and comparing their stats with newer ones to sycophantically drill home the point. "Somebody tell me what's going on," Musk said, growing more and more upset.

While the employees in the room still didn't have a reason, they put forward Yang Tang, a machine-learning engineer with almost a decade of experience at the company. Tang, an expert on the artificial intelligence used to drive the social network's personalized feeds, had little experience in talking to Musk. Those who had survived the billionaire's blast radius knew never to speak out of turn or offer guesses. They also learned to create presentations that Musk could easily digest from his phone—as he rarely, if ever, used a computer. No one told Tang.

Standing up, the engineer began riffing with a presentation from his laptop. Instead of saying he didn't know the cause of the issue, he pointed to other trends. Likes on the platform were decreasing overall, he said, before Musk cut him off.

"I'm not talking about likes," he growled. "I'm talking about view counts."

Tang pushed on, citing external factors. There were decreases in Google searches for Musk's name and he correlated that with internal engagement data. Perhaps the answer was simply that people were less interested in Twitter's new owner now that the deal had concluded. Maybe it had nothing to do with Twitter's algorithm.

Musk snapped. "You're fired! You're fired!" he bellowed at the engineer.

Tang closed his laptop and walked out of the room. Silence descended across the room, as the rest of his now former colleagues shifted their eyes away from Twitter's owner hoping they wouldn't be called on.

Eventually, after an awkward silence, a senior manager spoke up and said the team would figure out the issue and get back to Musk in twenty-four hours.

By Saturday, however, the issue still hadn't been corrected and workers stayed at the office late into the night with Musk to come up with answers. At one point, some of them decided to call a recently departed colleague, a former principal engineer with nearly a decade of experience at Twitter, who they believed was the only person with the know-how to fix the problem. After dialing him, the workers put Musk on speaker with the principal engineer, who was a bit taken aback, but still offered up some potential solutions out of pity for his former coworkers. Musk, impressed, asked him if he wanted to come back to the office that night and perhaps work for a half day to straighten everything out.

"I'll make it worth your while," the billionaire said. But the principal engineer had seen how Musk had treated his former colleagues in the months since he left. He turned him down and hung up.

>>> THE FOLLOWING DAY, Musk was sitting, despondent, in a suite at a cavernous football stadium in Glendale, Arizona. It was Super Bowl LVII between the Philadelphia Eagles and Kansas City Chiefs, and despite having one of the poshest seats as a guest of Rupert Murdoch, the media tycoon whose Fox network was broadcasting the game, Twitter's owner simply could not look away from his phone. During the first quarter of the football game, he had tweeted, "Go @Eagles!!!," a harmless show of support for the team from Philly, where Musk had spent time as an undergrad at the University of Pennsylvania.

About forty minutes after Musk's tweet, President Joe Biden used the official @POTUS Twitter account to broadcast his own Super Bowl message. "As your president, I'm not picking favorites," he wrote, including a video of the First Lady in a custom Philadelphia jersey. "But as Jill Biden's husband, fly Eagles, fly."

The president's tweet rocketed across the platform. Despite the account having less than a fourth of Musk's followers, the @POTUS post quickly surpassed Musk's own pro-Eagles proclamation, eventually racking up 29 million views to Musk's 8.4 million.

As an enthralling game played out in front of him, Musk ignored the cheers and the bone-crunching tackles, and kept comparing his tweet against Biden's, before deleting his post in anger. He fired off an email to Twitter's product engineers, demanding they immediately find the cause, and then began making arrangements for people to meet him back at Twitter's office that night. Musk left the game early and was already back on his private jet flying to San Francisco before the red and white confetti fell for the Chiefs' victory.

Dozens of people arrived at headquarters on Sunday night, some coming straight from their own Super Bowl parties to deal with Musk's five-alarm fire. At the office the billionaire addressed the engineers directly.

"I don't know whether this is due to incompetence or sabotage," Musk said, a bit calmer.

With James Musk running point on the "high urgency" issue, some eighty employees ran through the possibilities for Musk's sinking engagement through the night. One theory was that Musk had been blocked or muted by so many users in recent months that it led to fewer people seeing what he posted and caused the algorithm to penalize him when it came to distribution.

By Monday, however, several engineers had found the root of the problem. Musk's tweets were not being surfaced to his followers at the expected rate by Twitter's algorithm. In turn, a system called "out of network tweets," which recommends posts from accounts that users do not follow into their feeds as a way of potentially building new connections, was not surfacing Musk's tweets to non-followers, further depressing his engagement. The issue affected only a handful of high-profile users, the engineers determined.

While they had found the problem, the engineers lacked an elegant solution and instead went about cobbling together a fix to calm Musk. Within the recommendation algorithm, they introduced code "author_is_elon," essentially ensuring that the system would place a heavier emphasis on pushing his posts into users' personally curated feeds. In effect, Musk's tweets would have higher priority over any other post.

By Monday afternoon, some users who opened the site saw "For You" feeds that featured only Musk tweets over and over again, with little other content in between. The billionaire had finally built his very own echo chamber.

Sensing how embarrassing it was, the billionaire attempted to defuse the situation with humor. He shared a meme of a woman, labeled "Elon's tweets," force-feeding another, labeled "Twitter," at 9:35 p.m., on Monday night as engineers worked to undo the heavy emphasis.

>>> MUSK'S MISTRUST OF his own employees only grew. In an attempt to enforce his policy of working from the office, he had Davis and others monitor employees' badge swipes. Those who were not scanning into the office as much as expected were earmarked for firings and placed on "performance-improvement plans." In response, some employees traveled to the office simply to run their badges, then returned home to avoid Musk or his lackeys.

On Thursday, February 23, Twitter's Slack went offline for what employees were told was "routine maintenance." Following the dissent firings in November, chatter on the platform in large channels had slowed to a trickle, and few dared to say anything that could be remotely construed as criticism of Musk or his management. Slack, however, was still the company's main communication platform, allowing workers to contact one another quickly, track issues on the site, or monitor the deployment of new features. Without it, work ground to a halt, as employees grew suspicious. Twitter was late on paying its Slack bill, but a spokesperson for the service confirmed that it had not cut off the company or any of its accounts.

A few days earlier, a handful of managers had also been told by Davis to start creating lists of their best performers for the distribution of potential stock bonuses. At the start of February, three of the company's top engineers had been fired by Musk after they asked for improved compensation for them and their teams, as they worried low morale and long hours would lead to attrition. Perhaps the ask from Davis was a sign of compromise from management, some thought. Maybe they realized they would need to incentivize the best to stay.

On Saturday, the real reason for the lists became clear. Musk was engaging in another large layoff, cutting hundreds of jobs across the organization. One engineering manager who oversaw about twenty people and who had marked four of them as "high performers" for Davis, realized that the only people that still had jobs at Twitter were those four. The "bonuses" were the gift of their continued employment.

Esther Crawford would be locked out of her email and computer on Saturday as well. Her time serving Musk had run its course, and she was part of a group of well-paid employees who had come to Twitter through the acquisitions of their companies. These founders were contractually due large payouts in stock bonuses that vested over time, or would have to be paid in full if they were terminated early. To Musk, though, these contracts could be challenged, and he saw their compensation as another cost that could be excised. Crawford, who had come off to some as a booster of Musk's methods, tweeted nothing about her firing.

43 > Endorsements

As Musk faced heat from U.S. regulators, he began to learn of the pressures placed on the social network from around the world. In March, the European Union, which was set to roll out its new Digital Services Act, expressed concerns that Twitter did not have enough human content moderators to comply with the new law, which required that online platforms establish robust content moderation practices to deal with illegal material. While Musk typically disdained regulators, he made time to meet and engage with EU commissioner Thierry Breton, perhaps concerned by the prospect of fines that could amount to up to 6 percent of a company's annual global revenue for noncompliance.

In Turkey, Twitter was fined 0.1 percent of its gross income in the country by the nation's competition board for not seeking its permission in the approval of Musk's acquisition. While the amount was de minimis, it signaled to the company that Turkey was keeping it under its watch. Weeks earlier, following a destructive earthquake, the Turkish government blocked access to the platform in an effort to quell fears, a stark reminder that Musk's vision for "freedom of speech" was not so welcome in autocratic countries. Musk, whose SpaceX was pushing to launch its Starlink service in the country, said nothing of Turkey's government or its president, Recep Tayyip Erdoğan, over the shutdown.

Yet it was in India where Musk's commitments were tested most in his early months of leadership. Under past leadership, Twitter had faced immense pressure from Narendra Modi's government to take down accounts of dissidents and critical voices, at one point raiding its Delhi and Gurgaon offices in 2021 after the company had labeled a tweet from a Modi-aligned politician as misinformation. Musk, however, caved, blocking more than 120 accounts belonging to journalists, activists, politicians, and even a poet. In the country, users searching for these accounts received messages noting that the account was "withheld in India in response to a legal demand."

These were the tough decisions that Gadde had warned Musk about in her final meeting, and Twitter's actions under new management made it clear just how contradictory the views of the self-proclaimed "free-speech absolutist" actually were.

>>> ON MARCH 7, a rainy Tuesday in San Francisco, Musk was whisked into a swanky ballroom at the downtown Palace Hotel in order to reassure some of his investors and supporters. His early months running Twitter had been rocky, but onstage at Morgan Stanley's Tech, Media & Telecom conference, he turned on the charm. In a white shirt unbuttoned at the collar, under a blue blazer, Musk leaned forward in a small white leather chair to eagerly swing at softball questions underhanded to him by Michael Grimes, the lead Morgan Stanley banker on the deal.

"The reason I did the Twitter acquisition was not because I thought this would be some lucrative goldmine, and in fact, it has been arduous and difficult with being dumped on every day," he said, as Grimes nodded along. "That's not the most fun thing in the world. But if we do not have a strong foundation of free speech, I fear for the future of our civilization. We must have this. That's why I did it."

Grimes, who came off as more cheerleader than interviewer, failed to see any contradictions in Musk's statement. Instead, he encouraged the billionaire to paint himself as a rescuer of the business who had slashed $3 billion in operating expenses to deal with $1.5 billion in debt servicing a year and a 50 percent decline in advertising revenue. "It would have gone bankrupt in four months," Musk said. He called the advertising decline "cyclic" and "political," and blamed advertisers for believing what they read about him in news outlets.

Grimes also let him wax lyrical on his idea for an everything app, which Musk pitched as "X" and said would enable payments between Twitter accounts. Twitter had the possibility to become the "biggest financial institution in the world," Musk said, sharing few details, before he called the company "an easier problem than Tesla by a long shot."

Three weeks later, Twitter's employees received an email about a new stock compensation program. Twitter was an "inverse start-up," Musk wrote in his note, saying that the layoffs and cost cutting were all part of a rapid reshaping to avoid bankruptcy. Twitter, he said, could one day be worth $250 billion, but for now, employees' stock would be based on a company valuation of $20 billion. He didn't explain how he had arrived at that valuation, which was $24 billion less than what he had paid for the company five months earlier.

—

>>> CRAWFORD AND MANY of the employees who had launched Blue Verified were gone, but Musk continued to limp along with the subscription product. He and Morgan Stanley had pitched investors that his Twitter would be able to drive $10 billion in revenue from subscriptions within five years, but at the March conference with Grimes, little was said onstage about it.

By February, it had become clear that subscriptions were not the silver bullet Musk had been hoping for. *The Information*, a tech publication, found that less than 300,000 accounts around the world had signed up for Blue, equating to about $28 million in annual revenue. For a company that recorded about $5 billion in revenue in 2021, it was a rounding error. Still, Musk continued to push for improvements centered on Blue, vowing to eventually take away the verified check marks on the legacy accounts of celebrities, journalists, and others by April 1.

"Only verified accounts will be eligible to be in For You recommendations," Musk wrote on March 27, noting that it was the "only realistic way to address advanced AI bot swarms taking over."

Behind the scenes, though, Twitter's engineers were trying to convince him otherwise. Per his request, teams overseeing Twitter's timeline feeds had cobbled together an experiment in which the For You page showed only Blue subscribers. Testing it on less than 1 percent of users on the site, engineers found that one of the main metrics for measuring site engagement—"user active seconds"—began to drop precipitously. There were simply not enough subscribers to provide the amount of content needed to keep people staying on the site. Nor was their content particularly compelling for the average user. Musk saw the numbers and conceded.

"Forgot to mention that accounts you follow directly will also be in For You, since you have explicitly asked for them," Musk tweeted less than twenty-four hours after his original announcement.

At the same time, Musk demanded that Twitter publish its recommendation algorithm to be transparent about how the company decided which tweets would appear at the top of the For You feed. His edict set off a mad dash among engineers, who ripped out any code they thought would attract criticism. But they missed the piece of code that read "author_is_elon," leading to widespread mockery when the code was published.

As April 1 approached, several celebrities, who were fed up with the changes, voiced their opposition to the idea of paying for verification. "Welp guess my blue ✔ will be gone soon cause if you know me I ain't paying," NBA superstar LeBron James tweeted on March 31.

"I've been here for 15 years giving my ⏲ & witty thoughts all for bupkis," posted actor William Shatner. "Now you're telling me that I have to pay for something you gave me for free?"

For some Twitter employees, Shatner's criticism was spot on. For years, celebrities had been contributing their thoughts and commentary to the platform for free, driving engagement from regular folks who wanted to see what their favorite athletes or musicians or actresses had to say. Twitter was able to make money off that content by advertising. Why should these content creators now have to pay? And wasn't it in Twitter's best interest to show to users that whatever was tweeted from @WilliamShatner actually came from the verified, real Shatner?

Three weeks later, the planned badges began to be removed en masse, though the process was comical. Musk himself would pay the subscription fees for celebrities like James and Shatner so that they would keep their check marks. Ultimately, the process was so confusing that users began associating the checks with a new meaning. The past system may have been one of "lords & peasants," as Musk called it, but the new one represented something entirely different: you were with him or against him.

>>> IN EARLY APRIL, Twitter began to label the account of National Public Radio as "state-affiliated media." Under the old regime, the state-affiliated media label had been placed on accounts like Russia Today (@RT) or Xinhua News Agency (@XHNews), which belonged to outlets in countries where governments exercised control over content by political pressure and the control of financial resources, production, and distribution. The designation typically did not apply to state-financed organizations like the United Kingdom's BBC or NPR, which took government funding but maintained editorial independence.

Musk unilaterally changed Twitter's policy and began to lay out his rationale to Bobby Allyn, an NPR technology reporter who had gone back and forth with him over email, asking questions about his decision-making. "If you really think that the government has no influence on the entity they're funding then you've been marinating in the Kool-Aid for too long," Musk wrote to Allyn, who had pointed out repeatedly that less than 1 percent of NPR's operating budget came from government grants, which went to funding small stations in remote parts of the country.

On April 12, the outlet, tired of Musk's antics, announced it would no longer post to Twitter, prompting Musk to tweet a right-wing slogan. "Defund @NPR," he tweeted. And behind the scenes, the billionaire trolled Allyn. "Stop lying to the public," he wrote in an email that afternoon. A few days later, he did away with all labels for NPR and government-controlled outlets like RT. "I asked Walter Isaacson's advice and he recommended dropping labels for all media, so we did," Musk told Allyn.

Musk removed the limitations on Russian and Chinese state media accounts that previously prevented those accounts from being recommended by Twitter's algorithm. Twitter's old leaders had put the measures in place to limit the propaganda from state-controlled accounts, but as Musk declared, "All news is to some degree propaganda. Let people decide for themselves."

One media figure that Musk was content to support, however, was Tucker Carlson, the Fox News host and demagogue whose broadcasts on immigrant "invasions" and threats to white society found a captive audience among disaffected Americans. While Musk rarely watched television, he started to steadily interact with clips from Carlson's show that were posted on Twitter, including from a program in March that suggested the coverage of the January 6 Capitol riots was "a lie" and that those who stormed the Capitol were "sightseers" and "not insurrectionists."

Musk's respect for Carlson eventually led him to sit down with the Fox News star in a wide-ranging interview that aired on April 17, in which he admitted that his timing was "terrible" when he made his offer for Twitter and believed he should have paid half of what he actually did. He continued to paint Twitter's past management as corrupt, noting he had reduced the company to about 20 percent of its previous head count because he was no longer trying to "run some sort of glorified activist organization," and stating, without evidence, that government intelligence agencies had "full access" to users' direct messages. The interview was chummy, with Carlson stroking Musk's ego as they laughed together. It would be one of Carlson's last interviews for the network.

A week after the interview aired, Fox News fired its most popular host after a landmark lawsuit by Dominion Voting Systems, a company that said it had been defamed by Carlson and his network as they spread misinformation about the 2020 presidential election. While Dominion and Fox settled for more than $780 million, the day after the Musk interview, the months of discovery in the case had been embarrassing for the network, drawing an unprecedented amount of scrutiny.

Fox's loss was Musk's opportunity. In conversations with his lieutenants at Twitter, he wondered if Carlson would bring his show to Twitter, contributing to the site's shift toward video. Musk told those around him that it was simply another way to generate attention for the platform—potentially drawing subscribers—and he began communicating with Carlson about the idea.

On Monday, May 8, Carlson met with Lachlan Murdoch, Fox Corporation's executive chairman, to discuss the nature of his exit. His contract, which ran until 2025, ensured he would be paid out $20 million a year to do nothing, though a noncompete clause prevented him from starting a new show at a competing network. Still, he was intrigued by his conversations with Musk and decided to roll the dice, arguing his noncompete agreement was unenforceable. The next day, Carlson announced in a slickly produced video he would be bringing a version of his show to Twitter. "For now, we're just grateful to be here," he said.

The week after Carlson's announcement, Musk took one step further right. In recent months, he had become fixated on George Soros, the billionaire financier. At ninety-two, Soros, a Holocaust survivor, had become a frequent target of antisemitic conspiracy theories because of his foundation's funding of progressive ballot initiatives and politicians. Musk was particularly angry with Soros for backing district attorney candidates around the country who favored criminal justice reforms and, in the view of conservatives, had gone light on crime. From Musk's perspective, Soros was contributing to a woke agenda that was destroying cities like San Francisco.

"Soros reminds me of Magneto," Musk tweeted on May 15, comparing the financier to an X-Men comic book supervillain. In Marvel lore, Magneto survived the Holocaust and seeks to unite mutants to destroy the human race.

"He wants to erode the very fabric of civilization," Musk continued in a follow-up post. "Soros hates humanity."

His comments drew immediate condemnation from the likes of the Anti-Defamation League and sent further doubts through the minds of advertisers. But Musk was unrepentant.

"So many conspiracy theories have turned out to be true," Musk said in an interview with CNBC's David Faber the next day.

Faber nudged gently, asking if Tesla owners were worried about buying his cars if they saw his political views or if brands stopped advertising on Twitter because of his posts. *"Do your Tweets hurt the company?"*

Musk paused for thirteen seconds, his eyes shifting back and forth as he collected his thoughts.

"I'm reminded of a scene in *The Princess Bride* where he confronts the person who killed his father. And he says, 'Offer me money. Offer me power. I don't care.'"

"So, you just don't care. You want to share what you have to say?"

"I'll say what I want to say and if the consequence of that is losing money, so be it."

>>> STANDING IN HER New York City apartment window on June 3, more than twenty floors above the mid-Manhattan skyline, Linda Yaccarino braced herself for a new job that would begin in two days. She snapped a photo to commemorate the moment. "Current view today," she tweeted with her image of concrete and glass framed by a clear June sky. "Bay Area views coming soon!"

Three weeks earlier, Musk had named Yaccarino, a sixty-year-old advertising executive who had spent her whole career in cable television, as Twitter's chief executive. Ambitious and hungry, the former NBCUniversal ads chief, who had been nicknamed the Velvet Hammer for her negotiating skills, had waited her whole life for an opportunity like this. She had been close to the social media company, sitting on its advertiser "influence council" and attending a call with Musk in the early days of his takeover as he tried to reassure brands. So when Yaccarino was passed over for the top job at NBCUniversal in November, she began courting Musk for the top Twitter job, first approaching him around February's Super Bowl to explore an ads deal between her company and his, while also sussing out if he was serious about finding a CEO.

The two met for the first time in person at a Miami advertising conference in April. Musk was the headline speaker and Yaccarino was his moderator, and they made for an awkward pairing. The billionaire showed up unshaven and spoke off-the-cuff, while the ad exec, dressed in a lemon-custard power suit and matching pumps, had polished, pre-scripted talking points that sought to elevate her guest. It was to be Yaccarino's first test: making Musk seem palatable to a room of chief marketing officers and advertising leaders. She passed with flying colors.

"I think what we heard today was some really important and profound things," Yaccarino said to conclude her forty-minute interview. "Freedom of speech doesn't mean freedom of reach. And if freedom of speech is the bedrock of this country, I'm not sure there is anyone in this room who could disagree with that."

In Yaccarino, Musk saw a political ally. The Long Island native was a Trump supporter who had been appointed by the former president to his Council on Sports, Fitness and Nutrition, and believed that progressive "wokeism" was destroying the country. More important, however, Musk saw Yaccarino as a subservient foil, a lieutenant who would ultimately answer to him, and draw from her decades of experience and relationships to bolster Twitter's ad business.

The limits of Yaccarino's power became clear even before she was formally announced in her new role. On May 11, Musk tweeted without naming her that he had "hired a new CEO" who would "be starting in ~6 weeks." The post blindsided Yaccarino, who had yet to tell NBCUniversal of her exit. At the time, she was preparing for the Upfronts, a presentation made by major networks to advertisers to sell them airtime, but was eventually pulled from the event and quickly shown the exit door by management. NBC-Universal also made Yaccarino sign a noncompete, preventing her from doing ad deals for Twitter in her first few weeks on the job. Musk announced her new role the next day.

Yaccarino immediately became viewed as a proxy for Musk and his behavior. She was hammered in the wake of Musk's comments about Soros and ridiculed as he drove advertisers away with his comments about "losing money" in his interview with CNBC. It became apparent that if she was meant to be the adult in the room to watch the manchild, she would have little control over his impulses.

—

>>> ON THURSDAY, JUNE 1, five days before Yaccarino was set to start, Musk once again thrust his company into crisis. The previous month, Twitter had sold an ad package to the Daily Wire, a conservative media company cofounded by right-wing podcast host Ben Shapiro, to promote a transphobic, trolling documentary called "What Is a Woman?" The promotion, which cost hundreds of thousands of dollars, had been meant to coincide with Pride Month in June and would feature a dedicated page on Twitter that would be promoted to every person on the site for ten hours.

Yet after selling the Daily Wire the ad package, Twitter's brand safety team pushed back after reviewing the movie. Twitter employees found that the video violated the company's hateful conduct policies because of issues around misgendering, and as such the company revoked the promotion, eventually making Musk aware of the decision. Twitter also notified the

Daily Wire that if they were to post the film on the site organically, without paying for promotion, it would still be limited in its reach. The Daily Wire would be able to post it on their page, but posts of the video would be prevented from circulating in people's feeds. It was "freedom of speech, not reach" in action.

But after Jeremy Boreing, the Daily Wire's chief executive, complained about the treatment on Twitter and threatened to post the video anyway, Musk pulled an about-face. With right-wing accounts ramping up the pressure and shattering his image as a free speech hero, he declared that the ruling had been "a mistake by many people at Twitter" and added that the video was "definitely allowed."

"Whether or not you agree with using someone's preferred pronouns, not doing so is at most rude and certainly breaks no laws," Musk tweeted.

Musk's tweets were a slap in the face to Ella Irwin. For months she had pushed to defend Musk's company and the billionaire's ad hoc policy decisions. And while they had earlier agreed on what to do about the Daily Wire video, he had caved at the first sign of pressure from his base and called her and her team out in public. She resigned that day.

While the Daily Wire would go on to post its video Thursday evening, Musk, without a trust and safety head to push back on him, would work to remove the restrictions that prevented the film's circulation. By the next day, he had tweeted the video from his own account, drawing millions of viewers, and leading to the movie's title trending on the site. As Twitter's owner celebrated the massive reach of the spectacle, other members of Twitter's trust and safety team and A. J. Brown, the company's head of brand safety, would quit in protest. Musk's audible had alienated them all and left Yaccarino without anyone overseeing content moderation at a company where ad sales were already down almost 60 percent in the U.S.

44 > Linda

Yaccarino brushed off the criticism that she was Musk's pawn as another twist of the sexism she'd dealt with throughout her career.

"As someone used to wearing 4 inch heels, let's be crystal clear: I don't teeter," she tweeted in May, responding to the idea that she was serving as Musk's lackey. The girlboss branding didn't mesh naturally with Twitter's workforce, which had been culled to a largely male crew of young engineers. But the company still tried to embrace the new chief, displaying her tweet in one of the cafeterias in the San Francisco office. In the framed image of her post, no retweets or likes were included, which inadvertently made her message look unpopular.

But Yaccarino was on a mission to be liked. On her first day in headquarters, Thursday, June 8, she hosted a happy hour for employees and asked them to point her to Twitter's toughest problems. Making herself available went a long way with the company's employees, who had been scarred by months of living in fear and silence under Musk.

In an email to employees that day, she made her admiration for Musk clear. "From space exploration to electric vehicles, Elon knew these industries needed transformation, so he did it. More recently it has become increasingly clear that the global town square needs transformation," she wrote. "Let's dig our heels in (4 inches or flat!) and build Twitter 2.0 together."

That morning, a Syrian refugee entered a park in Annecy, France, and stabbed several children. A brutal video of the attack quickly emerged and began spreading across Twitter. The few employees who remained on the gutted trust and safety team scrambled to remove it. In the past, when violent imagery went viral on Twitter, the company used a data matching tool to detect any tweets containing it and wiped them out all at once. But the employees discovered that the tool they relied on wasn't working. When they investigated, they learned that it was one of the bits of code that engineers had torn out months earlier. The tool had mistakenly flagged an image of a SpaceX rocket launch, so Musk had ordered the entire system be killed.

It wasn't the only problem facing Yaccarino. Her arrival meant that some of Musk's employees from Tesla and SpaceX could finally go back to their day jobs. Musk had ignored sales and human resources in favor of product and engineering. For tasks he didn't enjoy, Musk simply deputized people he trusted.

But Yaccarino wanted Twitter to have some of the traditional structure of an actual company. That desire put her on a collision course with Davis, who had continued to tighten Twitter's budget. Davis pored over monthly spending reports and then called up managers at random, blindsiding them with questions about the employees who reported to them. "What does this person do?" Davis would ask. "Are they any good?"

During her first days at Twitter, several employees approached her to drop hints that they were sick of Davis's micromanaging. The constant layoffs were killing morale, and Davis's efforts weren't saving enough money to make the stress worthwhile, they argued. "Give me a few months," Yaccarino responded.

Every quarter, another vesting date rolled around, giving workers the cash bonuses they were entitled to after Musk took the company off the stock market. As each vest approached, the entire company would hold its breath, waiting to see who would take the money and then quit. And sure enough, every quarter, more people fled. The company had become so lean that everything was on the verge of falling apart. At one point, the last remaining person in charge of approving expenses went on vacation. During the employee's two-week absence, no one's expense reports were processed.

Within weeks of Yaccarino's arrival, Davis—who seemed to have been angling for the CEO role himself—had all but vanished from Twitter's offices. Some Twitter workers spotted him one last time near the end of June, when

Davis and Nicole Hollander flew to Boulder, Colorado. The company was being evicted from its office there, and the couple was responsible for clearing out any remaining valuables before the property manager changed the locks. Hollander showed up with several moving trucks and started loading up anything she thought Musk wouldn't want to lose, including servers and a few flat-screen TVs. She even packed up a massage chair that she thought Musk might want.

With the trucks piled to the brim, Hollander hosted a yard sale for employees, allowing them to buy their own desks and art from around the office. She accepted payments for the items on her personal Venmo account.

Twitter employees had some hope that Yaccarino would be a stabilizing force, counterbalancing Musk's most erratic impulses. Davis's less frequent appearances were a sign that Yaccarino had sway with the boss, having managed to push out one of his most loyal followers. But those hopes were quickly dashed.

In a fit of paranoia, Musk became convinced that an attack was being launched against Twitter by people trying to steal its data. "Several hundred organizations (maybe more) were scraping Twitter data extremely aggressively, to the point where it was affecting the real user experience," Musk tweeted on June 30.

Musk had become increasingly protective of tweets themselves, as though anything posted on the platform belonged to him. He ordered Twitter's engineers to prevent tweets from being viewed by anyone who wasn't logged in to the service, making it impossible for people to share links to popular tweets or embed them in news articles. Musk wanted to force people to sign up for an account, but the change broke one of Twitter's most fundamental uses—sharing real-time information.

And Musk wasn't done. He was fixated on the popularity of OpenAI, the artificial intelligence company that he had cofounded and then abandoned. AI models like the ones OpenAI was building relied on vast amounts of training data in order to learn how to create compelling, lifelike chatbots, and Musk was convinced that AI companies were harvesting data from Twitter to train their models.

He remained fearful that external forces were conspiring to steal data from the platform, so he imposed strict limits on how many tweets someone could view from their account. Verified accounts would be limited to reading 6,000 posts per day, unverified accounts to 600 posts, and new unverified accounts to 300 posts.

Twitter erupted in outrage. Users couldn't even read tweets on the platform for more than a few minutes without getting a warning message that claimed they had exceeded the rate limit and couldn't use Twitter for the rest of the day. Within two hours, Musk could see he was in trouble. He raised the limits to 8,000 for verified, 800 for unverified, and 400 for new unverified.

The phrase "rate limit exceeded" began trending on Twitter as users complained about the strict new limits. People began begging each other for invitation codes for Bluesky, the competing social media service maintained by Jay Graber, and letting their friends know how to find them on other social platforms.

On the night of Saturday, July 1, Musk raised the rate limits yet again, to 10,000; 1,000; and 500. But the damage was done—and one of his old nemeses saw an opening to attack him.

On July 5, Mark Zuckerberg's Meta, the parent company of Facebook, debuted its newest service: a Twitter copycat called Threads. While plenty of start-ups had tried to cannibalize Twitter's business, most of them were starting from scratch, building new apps and trying to convince Twitter's addicted users to jump ship. Zuckerberg tied Threads to Instagram, prompting Instagram users to sign up and automatically connecting them with their friends on Threads. The setup allowed Threads to leap ahead of other Twitter competitors, which required users to slowly rebuild their social networks from scratch.

"I think there should be a public conversations app with 1 billion+ people on it," Zuckerberg posted on Threads that day. "Twitter has had the opportunity to do this but hasn't nailed it. Hopefully we will."

Within hours of its debut, 10 million people signed up for Threads. It was the most threatening challenge to Musk's Twitter yet.

"It is infinitely preferable to be attacked by strangers on Twitter, than indulge in the false happiness of hide-the-pain Instagram," Musk tweeted that night. But it was clear he'd been stung.

As rumors had swirled ahead of the Threads launch, Musk had joked on Twitter that he would battle Zuckerberg in a cage match, with the winner becoming the king of short-form social media. But Zuckerberg seemed more than willing to kick Musk's ass, in the ring and on the App Store. The day after Musk cracked a joke about fighting Zuckerberg, the Facebook founder texted Dana White, the president of the Ultimate Fighting Championship.

"Is he serious about a fight?" Zuckerberg asked. During the pandemic, he had picked up Brazilian jiujitsu, becoming deeply engrossed in the sport. At first, he sparred with friends in his garage. But soon, he was taking it more seriously, hiring professional trainers and trying to apply lessons from the grappling sport, which required strategy as well as strength, to his business responsibilities.

White was flabbergasted. "I've been talking to Zuckerberg now for maybe close to two years," he said in an interview at the time. "And there's never like banter or we're joking and laughing." Zuckerberg was "dead serious all the time."

With Zuckerberg committed to the battle, White called Musk to find out if he would commit to a fight. Musk was thirteen years older than Zuckerberg, about seventy pounds heavier, and severely out of shape. After unflattering photos of him on Ari Emanuel's yacht emerged the previous year, Musk had turned to diet drugs to lose weight but didn't regularly exercise. He also struggled with neck and shoulder pain after his last impromptu fight—a battle against a sumo wrestler at his forty-second birthday party in 2013 that left him with a ruptured disk.

But Musk's ego wouldn't allow him to back down. He told White he was game. The billionaires began training for the match, and White set out to find a venue.

Musk kept throwing punches elsewhere, too. On July 13, Spiro asked a federal court to extract Twitter from its settlement with the FTC, claiming the agency was harassing Musk. The investigation "has spiraled out of control and become tainted by bias," Spiro wrote. Although the judge eventually threw out the request, it seemed to have the desired effect: the demand letters from the FTC finally tapered off.

>>> MUSK, ALWAYS DISTRACTIBLE, was soon swept up with other concerns. On the evening of Saturday, July 22, the billionaire announced that he planned to rebrand the company. "And soon we shall bid adieu to the twitter brand and, gradually, all the birds," he tweeted. Twitter's branding and ever-present bird metaphors irritated Musk, who found them twee, and he was ready to start molding Twitter into the "everything app" he'd been promising.

The company would be called X, Musk declared, hearkening back to his second start-up, X.com. Since purchasing Twitter, Musk had teased that X would make a comeback. The holding company he used to manage the acquisition was called X Holdings, and he publicly daydreamed about building an app where users would message each other, exchange payments, and even order food delivery.

In Muskian fashion, the name change was a surprise order with a harsh deadline. By Sunday, his loyalists were ripping through Twitter's headquarters in San Francisco like looters, tearing down any bird insignia they could find. They took down a ten-foot blue bird logo from the cafeteria and began projecting X's on the walls instead. In one instance, someone tried to remove a logo from the security desk but partly failed, leaving a broken bird. Conference rooms, which were all named after species of birds, were also rebranded to words with X in them, like "eXposure" and "eXult." Caracara became "s3Xy."

Musk ordered his mark to be made on Twitter, too. The bird logo vanished from its website and app, replaced with a white X. Late on Sunday night, Musk tweeted a photo of a giant white X projected on the wall of the Twitter building. "Our headquarters tonight," he wrote proudly. It was one of Musk's last official tweets—he had the "tweet" button swapped out for one that read "post" instead.

Marketers shook their heads. So many brands dreamed of the day their companies would become a verb, and few had ever come close. But "tweeting" was a word known around the world. For many longtime users, the erasure of the bird caused another round of mourning. But some former employees greeted the name change with relief. They'd been forced to watch as Musk decimated the platform they'd loved and dedicated their careers to, parading its corpse around like a trophy. They could finally move on. They had Twitter. Musk had X.

Even Dorsey weighed in. Changing the name wasn't "essential," he wrote on the platform Sunday night, "but you can make an argument for reconsideration being the best path forward. the twitter brand carries a lot of baggage. but all that matters is the utility it provides, not the name," he added.

On Monday morning, workers in neon construction vests boarded an orange cherry picker and lifted themselves above Market Street to begin ripping down the sign that marked the company's headquarters: "@twitter." It mattered little to Musk that he didn't have permission from the building's owners to alter the sign, nor from the city of San Francisco to do construction. The workers managed to rip down the first six characters from the sign before police ordered them to stop.

Musk was undeterred. He ordered workers to build a giant metal X on the roof of the building, then decked it out with strobe lights. By Friday, July 28, his vision was complete. At night, the sign turned on and began to flash searing white light into the dark sky. The neighboring buildings, many of which held apartments and condos, immediately overflowed with complaints. By Monday morning, workers were back on the roof, dismantling the sign.

Meanwhile, Musk was dilly-dallying about setting a date for his showdown against Zuckerberg. He blamed his neck injury and said he might need to get surgery before he could get into the ring with the Facebook founder.

Zuckerberg kept jabbing. Even though Musk was heavier and taller, Zuckerberg was confident that he could beat the Twitter owner in a fight—and he knew the world would relish watching him punch Musk in the face. The two billionaires started texting each other directly, egging each other on instead of using White as an intermediary.

While Musk was all bravado in public, he appeared to be attempting to weasel out of a public showdown. After floating the idea of a neck surgery that could be followed by months of recovery, he texted Zuckerberg in August: "Wanna do a practice bout at your house next week?"

Zuckerberg saw the overture for what it was: a way for Musk to privately battle with him in his garage, then go on Twitter and claim he had won. It was classic Musk spin.

"If you still want to do a real MMA fight, then you should train on your own and let me know when you're ready to compete," Zuckerberg texted back. "I don't want to keep hyping something that will never happen, so you should either decide you're going to do this and do it soon, or we should move on."

Musk insisted on coming to Zuckerberg's backyard for a casual fight. "I have not been practicing much," Musk texted on Sunday, August 13. "While I think it is very unlikely, given our size difference, perhaps you are a modern day Bruce Lee and will somehow win."

He continued goading Zuckerberg and, on Monday, August 14, threatened to show up and bang on his door. Zuckerberg replied that he wasn't home to entertain Musk's last-minute demand for a brawl, and Musk's fans pounced, accusing Zuckerberg of fleeing to avoid a confrontation. But that evening, Musk never showed. The only person outside Zuckerberg's home in Palo Alto was a lone security guard.

>>> YACCARINO HAD TO contend with the long line of other people Musk had angered—all while he continued to court controversy on the platform. X's bankers, advertisers, regulators, and users were unhappy; all of them wanted a piece of her. She set up a "listening tour," but Yaccarino did less listening than she did pleading for frustrated constituents to give her a chance.

She took a meeting with Thierry Breton, the European commissioner who had been probing X over its ability to police misinformation and hate speech. She did several sessions with the bankers who had loaned Musk money to buy the company, and had been unable to unload the debt as Musk tanked X's value. The bankers grilled her about the company's financial prospects, hammering for updates about how Yaccarino planned to move X away from its dependence on advertising. Yaccarino, who had come to X specifically to woo advertisers back, couldn't give a straight answer. Musk's plan to sell verification hadn't earned significant revenue, and his scheme to turn the social media app into an online bank was still years away as he waited on state governments to approve the necessary money transfer licenses for X to transact across the nation. She hewed to her strengths, telling the bankers that the advertising market was skittish yet cyclical, and brands would come slinking back in no time.

Musk's bankers also peppered Yaccarino with questions about content moderation and how Musk was tracking success with users. Old Twitter just measured whether people logged on or not, she said. Musk's strategy was to keep them online longer. She spoke about X's "multidimensional interface," arguing that the company would soon offer so many different kinds of content that users would stay engaged for hours, reading tweets, watching videos, and sending payments.

The next people to appease were X's public constituents, which included the advocacy groups that had objected to Musk's lax content moderation and pressured advertisers to abandon the platform immediately after his takeover. Among them were the Anti-Defamation League and the American Jewish Committee, two groups that focused on fighting antisemitism. Ted Deutch, the chief executive of the AJC, flew from Miami to New York for an uneventful sit-down with Yaccarino.

But Jonathan Greenblatt, the director of the ADL, was another matter. Musk had begun speaking with Greenblatt before his acquisition closed, after Greenblatt publicly criticized his views on content moderation. The men maintained a friendly rapport, chatting as Musk took over Twitter and Greenblatt encouraged advertisers to stop spending on the platform.

Once he had Yaccarino to manage advertiser issues, Musk passed Greenblatt off to her and encouraged the two to meet. Yaccarino invited the advocate to speak with her about his concerns and he agreed, joining a call with the new CEO on August 29 to have a frank but friendly conversation about the rise of antisemitic language on X and in the real world. Greenblatt had chastised Twitter's previous leadership about the same problem, one that had always been the company's Achilles' heel—the fact that hateful rhetoric on the platform had a way of manifesting in violence. Hate speech couldn't be ignored in Musk's pursuit of free speech at all costs, Greenblatt warned.

Yaccarino received the message well, reassuring him that she wouldn't tolerate antisemitism. After the discussion, one of her staffers encouraged Greenblatt to tweet about it, publicly indicating that X was open to hearing from its critics.

Greenblatt did just that, crediting Yaccarino with hosting a productive discussion. "@ADL will be vigilant and give her and @ElonMusk credit if the service gets better . . . and reserve the right to call them out until it does," he tweeted.

X's underbelly of rageful, paranoid, and perennially aggrieved posters seized on the message. As far as they were concerned, the ADL was the enemy, an organization that was in league with shadowy Democratic forces and would stop at nothing to censor Musk's platform. They called the meeting to Musk's attention, presuming that he hadn't arranged it and that Yaccarino was stepping out of line to meet with his foes.

By the weekend, Nick Fuentes, the white nationalist and live-streamer, had picked up on the story and encouraged his followers to campaign to get the ADL kicked off X. The hashtag #BanTheADL began trending on the platform. That weekend, men in masks marched on a Florida synagogue, chanting "Ban the ADL," the organization said.

Musk couldn't resist jumping in as his followers egged him on. "Since the acquisition, the @ADL has been trying to kill this platform by falsely accusing it & me of being anti-Semitic," Musk posted on Monday, September 4. "Our US advertising revenue is still down 60%, primarily due to pressure on advertisers by @ADL (that's what advertisers tell us), so they almost succeeded in killing X/Twitter!"

Musk's endorsement of the online campaign against the ADL propelled it into the mainstream. His acolytes piled on, accusing the organization of trying to strangle Musk. Greenblatt sent texts to Musk and Yaccarino, asking what was going on—after all, Musk had helped set up the meeting, and Yaccarino's staff had encouraged Greenblatt to post about it. Why was Musk now on the warpath?

Musk never responded. Yaccarino did, but sidestepped any criticism of her boss. Musk continued to fan the flames. "If this continues, we will have no choice but to file a defamation suit against, ironically, the 'Anti-Defamation' League," he posted.

The lawsuit never came to fruition, but Yaccarino was in a tight spot. She couldn't speak against Musk, but she also had to convince advertisers not to believe what was going on in front of their very eyes—Musk was instigating harassment against a prominent Jewish group.

>>> ON SEPTEMBER 27, Yaccarino flew to Southern California, where she was scheduled to make an appearance at CODE Conference, a bougie technology summit at the Ritz-Carlton in the coastal Orange County town of Dana Point. Musk himself had spoken at the conference years earlier, and other top executives like Apple's Tim Cook and Uber's Travis Kalanick had graced the stage. For Yaccarino, it was her coming-out party to the technologists.

She was wary, but agreed to the interview provided that the conference organizers sat her down with Julia Boorstin, a polished CNBC correspondent whom Yaccarino had viewed as a mentee while at NBCUniversal. While Boorstin was excited to score one of the first major conversations with the new CEO and prepared to grill her about the business, Yaccarino expected a gentle interview.

But the problems began before she'd even left the airport. As a representative of a Musk company, she made an effort to go everywhere in a Tesla. But the car rental company had botched the request, and no Teslas were available. Yaccarino dug in her heels, refusing to leave until the company procured a white Model Y to whisk her to the Ritz-Carlton.

She spent most of the day hiding out in the green room, preparing for her interview. But that morning, Yaccarino received word that the conference had booked a last-minute guest and scheduled his interview just before her own: Yoel Roth, the content moderation executive who Musk had driven from his home.

Yaccarino was deeply shaken. She felt like the technology journalists who Musk hated were seizing the opportunity to stab her in the back. She huddled backstage with Boorstin, debating whether she should abandon the conference altogether. It felt like a purposeful attack against her, and she was torn between lashing out and running away.

Just before Yaccarino's slot, Roth took the stage with Kara Swisher, a former admirer turned harsh critic of Musk. Swisher prodded Roth to address Yaccarino directly, asking him to offer advice to the new X executive. "If not for yourself, for your family, for your friends, for those that you love, be worried," Roth responded. "You should be worried. I wish I had been more worried."

Backstage, Boorstin was emotional. "I'm so sorry," she kept repeating to

Yaccarino. Swisher had blindsided her by inviting Roth, Boorstin said. Yaccarino was overcome with emotion, too, and suggested that Boorstin was trying to ruin her career by pushing her to go ahead with the interview.

Fifteen minutes later, the two former colleagues settled into the low-slung black leather armchairs onstage. Boorstin showed no sign of the emotion that had engulfed her moments earlier, but Yaccarino had a harder time shaking it off. When Boorstin said she would give Yaccarino a chance to respond to Roth, Yaccarino immediately interrupted.

"Chuckles," Yaccarino said through pursed lips. "Chuckling." Her eyes darted around the room, seemingly trying to gauge the crowd. "We stay strong together," she said, gesturing to Boorstin as though she were a companion rather than a journalist with a stack of questions in her lap.

But Yaccarino fell apart, unable to move past Roth's critique of the company and his caution for her in particular. As Boorstin tried to steer the interview in other directions, Roth's comments spun through Yaccarino's head, and she continued to bring them up and try to refute them. She struggled to remember key figures about X's business. At one point, she held up her phone to the audience, and some eagle-eyed observers noticed she didn't appear to have the X app downloaded on her home screen.

For her part, Boorstin asked all the basic questions that observers had been asking of Yaccarino. She gently asked what Yaccarino thought of the criticism that she was a CEO in name only, and that Musk still controlled the company's product decisions.

"Who wouldn't want Elon Musk sitting by their side running product," Yaccarino answered huffily as she stared out into the crowd. Scattered laughs erupted as some audience members raised their hands.

Yaccarino was due at another meeting in Los Angeles at 6:00 p.m., but she couldn't seem to pry herself away from the battle on the CODE stage. As the bizarre interview wound on, Yaccarino kept claiming that she needed to leave—but made no move to raise herself out of her chair. Her anger flashed on her face as Boorstin pressed her about Musk's threat of a lawsuit against the ADL.

"Everyone deserves to have that opportunity to speak their opinion, no matter who they are. Including Elon, including you, Julia," she retorted. Finally she stood up, brushed past Boorstin, and walked offstage.

As the audience buzzed over her clumsy performance, Yaccarino confronted Boorstin, saying the reporter had destroyed her career, then collected her things and raced out to the waiting Tesla. Just as the driver was ready to whisk her and her security guards away, Boorstin came sprinting out across the white cul-de-sac in front of the Ritz-Carlton, her heels clattering on the stones. With tears streaming down her face, the news anchor rushed to the car and tapped on the window, gesturing for Yaccarino to let her in. Her security guards shuffled around to make room and Boorstin piled into the back seat with Yaccarino, apologizing and thanking her for going through with the interview rather than canceling.

As onlookers stared, the Tesla pulled away to give the two women some privacy. After a few minutes, the car pulled back around to the front of the hotel and let a visibly rattled Boorstin out in front of the crowd. Then Yaccarino pulled away again.

45 > Tell It to Earth

At 7:03 in the morning on Saturday, November 18, Musk stood with his palms pressed together as if in prayer, his chin resting on his fingers, in the SpaceX launch control center in Boca Chica Village, Texas. Once a sleepy unincorporated Gulf community near the state's southernmost tip, Boca Chica had been transformed in recent years by SpaceX, which made it the home of the company's main rocket launch facility. Musk and SpaceX had brought hundreds of rocket scientists and engineers to the area, and many of them had moved for moments like this.

As the sun rose over the Gulf of Mexico, the SpaceX chief watched as massive plumes of white smoke and an earth-shaking bellow emanated from the thirty-three engines on Starship. This was the rocket he promised would one day bring humans to Mars, and on its second test launch, SpaceX was hoping to prove its vehicle could reach orbit. With the countdown finished, the 397-foot rocket strained for the blue Texas sky, eventually reaching an altitude of more than 90 miles in its eight-minute flight, before exploding. The SpaceX engineers in the control room went berserk. The test was largely a success, showing that Starship could viably take off and separate from its booster engine. They were one step closer to Mars.

While Musk celebrated with his family at the SpaceX facility, something else remained in the back of his mind. Within an hour and a half of the Starship test, Musk was already back on his phone and on X. He replied to a fan who had tweeted that advertisers on the social network didn't care about the company's past issues, and had only started to hate it for "political" reasons after his takeover. "The powers [that] be hate what this platform is doing, because it goes against their ability to control the narrative," the user wrote.

"Exactly," Musk replied.

Even in the afterglow of reaching a SpaceX milestone, Musk remained obsessed with X, stewing over a controversy of his own making that had kicked off three days earlier. Like most of his recent troubles, it all stemmed from a tweet.

>>> ON WEDNESDAY, NOVEMBER 15, Musk was flying on his Gulfstream jet from Austin to the Bay Area, when he sent off what he believed to be an innocuous post. For days, he had been raging against the Anti-Defamation League, which continued to criticize X for allowing antisemitic hate speech following the October 7 Hamas attacks in Israel. As he sat on his plane, Musk read a tweet from an anonymous, Blue-subscribed account with less than 6,000 followers.

"Jewish communities have been pushing the exact kind of dialectical hatred against whites that they claim to want people to stop using against them," the account, @breakingbaht, wrote. "I'm deeply disinterested in giving the tiniest shit now about western Jewish populations coming to the disturbing realization that those hordes of minorities that support flooding their country don't exactly like them too much."

As he read, Musk nodded in agreement. He tapped out a reply on his iPhone.

"You have said the actual truth."

The anonymous account Musk was praising had laid out the "Great Replacement Theory," a crazed white nationalist idea that Jews and the global elite encouraged the mass migration of nonwhite people into Western countries to replace Caucasian populations. Versions of the conspiracy theory had been endorsed by the likes of VDARE, the popular white nationalist publication, and Tucker Carlson, and now it had the imprimatur of the world's richest man. Musk doubled down with a follow-up tweet.

"The ADL unjustly attacks the majority of the West, despite the majority of the West supporting the Jewish people and Israel," he wrote. "This is because they cannot, by their own tenets, criticize the minority groups who are their primary threat."

His comments were celebrated on the far right. Infowars host Alex Jones called Musk's post "the truth" on his show. Nick Fuentes, the white nationalist who Twitter had reinstated before he went on to praise Hitler, called Musk's comment evidence that the fringe idea was going "mainstream."

"This is the richest man in the world, and one of the top Twitter users with 160 million followers, casually drawing a link between Jewish influence and anti-white hatred," he said on his live-streamed show.

Musk and X also faced an almost instant barrage of criticism. The following day, Media Matters, the progressive media watchdog, published a report purporting to show ads for brands like Apple and IBM appearing next to tweets from accounts celebrating Adolf Hitler and Nazism. For brands who had stayed close to X following all the controversies of the last year, Musk's tweet and his social platform's lax approach to content moderation was too much. By Friday, Apple, IBM, Disney, and more than a hundred other major brands would begin halting their advertising spend on X. It would lead to more than $75 million in losses by the end of the company's fourth quarter, typically its busiest of the year as companies pushed their products during the holiday season.

With Twitter's advertising business once again on fire, Yaccarino found herself having to defend the indefensible. She, too, had been blindsided by Musk's antisemitic tweet, but unable to criticize her boss, she sought to find another outlet for her frustration. In a meeting with the sales team the following Monday, two days after the SpaceX launch, she slammed Media Matters for its report, claiming that the organization had manipulated a Twitter feed so that ads would show up next to the white supremacist content. She also said, conspiratorially and without evidence, that the group had released its report intentionally ahead of her daughter's wedding that weekend. "I am quite sure that was not a coincidence," Yaccarino said, as she attempted to rally her troops.

"I know you have the talking points that are in front of you about what went on by a very unfortunate activist organization with bad intentions to put our company in harm's way," she said. "And there are points in your life where you make decisions based on your values. And when there are bad actors that cross the line, your integrity, your values, and—in this case—truth matters."

For some salespeople on the call, who had seen much of their business crushed in recent days, it was embarrassing. Later that day, X filed a lawsuit against Media Matters in a federal court in Texas, alleging that the organization tried to damage its relationship with advertisers. Texas's attorney general, a Musk ally, also announced an investigation into Media Matters for "potential fraudulent activity."

The lawsuit did little for Musk's and X's reputations in the eyes of advertisers, as more continued to pause their spending and affect the ads that users viewed on the site. People began seeing more advertisements for sex toys, cheap mobile games, and cam girls, as Twitter's ad inventory dwindled with the fleeing brands.

In a bind, Musk flew to Israel on Monday, November 27, to meet with Israeli prime minister Benjamin Netanyahu. Wearing a green flak jacket that was two sizes too small under a blue blazer, he walked with Netanyahu through Kfar Azza, a kibbutz that was attacked by Hamas, on an apology tour that would include a live-streamed conversation with the embattled prime minister. Musk would also meet with Israeli president Isaac Herzog and with families of some of the Israeli hostages. One father gave Musk a dog tag to wear as a reminder of his kidnapped son and the other captives.

"I think we need to fight this together, because the platforms you lead, unfortunately, have a large reservoir of hatred, hatred of Jews, anti-semitism," Herzog told the billionaire in a closed-door meeting.

Musk spent thirteen hours in the country making the requisite photo ops before heading back to the states. He landed in Austin in the early hours of Tuesday morning, and it would take less than five hours for him to be embroiled in his next controversy. That morning he tweeted a meme suggesting that the Pizzagate conspiracy theory "is real." He deleted that post a few hours later.

>>> THE MINUTE BEFORE Musk stepped out in front of a crowd of well-heeled businesspeople at the *New York Times*'s DealBook Conference on November

29 was an unexpected moment of calm. A little more than an hour before, he had landed at New Jersey's Teterboro Airport with his son X and the boy's nanny, where they met up with half a dozen security guards who'd whisk them into waiting Teslas to head toward Midtown Manhattan. By the time he arrived through a back entrance at a jazz concert hall above a high-end mall, Musk, dressed in shiny black boots and a brown leather bomber jacket with a shearling collar, was sedate. The international travel from Israel had worn him down, and he stared ahead as an eager Yaccarino, who was there to greet him backstage, fed him a stream of updates about the conference.

Jonathan Greenblatt, the ADL president, was in attendance, she told him, a subtle nudge that Musk should be on his best behavior. He nodded. Oh, and Bob Iger, the Disney CEO, had spoken onstage earlier and said his company's association with Musk and X "was not necessarily a positive one for us." The billionaire registered Yaccarino's comment, but didn't react much. Some sales employees had been hoping Musk would talk to Iger about Disney's pause on advertising but they hadn't been able to connect the pair ahead of the conference.

Musk strode out onto the DealBook stage behind Andrew Ross Sorkin, the *Times* journalist and event's host, in a jovial mood. He cackled at his own jokes as Sorkin led with a bit of charm, saying that his first impression of Musk when they met for the first time fifteen years ago was that he was the "next Steve Jobs." The billionaire laughed heartily as he settled into his seat.

Then Sorkin winded himself into position to ask what everyone wanted to know: What were you thinking when you sent that tweet?

"You know it's a real weakness to want to be liked, a real weakness," Musk said, his grin shifting to a look of annoyance. "I do not have that."

But don't you want to be at least trusted? asked Sorkin. It may be one thing not to care about being liked, but if governments are going to give you money for rockets or you want to build a financial platform on X that people are hoping to use, shouldn't they ultimately see you as a "decent and good human being"?

"Yes, I mean, I think I am, but I'm certainly not going to do some sort of tap dance to prove to people that I am," Musk responded acidly, before he bizarrely referred to Sorkin as "Jonathan" and was corrected by his host.

The billionaire let out a maniacal laugh, his mood bouncing from aggressive to affable and back again in the span of seconds.

Still Sorkin stayed on task and continued to press. There are people saying you're antisemitic, the host told him. Actual antisemites are celebrating you. "I want to know how you felt about that in that moment when you saw all of this happening."

Musk offered a half response, blaming the media for not covering his clarification of his tweet, while admitting that maybe he shouldn't have replied to that person and given a "loaded gun to those who hate me." He called himself "philosemitic" and pointed to the dog tag around his neck that he had received on his trip to Israel.

"But there's a public perception that that was part of an apology tour, if you will. That this had been said online. There was all of the criticism. There were advertisers leaving. We talked to Bob Iger today—"

Musk didn't let Sorkin finish the question. With seemingly little thought or calculation, he cocked his head back, scowled, and uttered the first impulse that came to his mind.

"I hope they stop."

"You hope—?"

"Don't advertise."

"You don't want them to advertise?"

"No."

"What do you mean?"

"If somebody is going to try to blackmail me with advertising, blackmail me with money— Go fuck yourself."

"But—"

"Go. Fuck. Yourself. Is that clear? I hope it is. Hey Bob! If you're in the audience."

There were a few stunned laughs in the auditorium, but most people in the audience, including Yaccarino, had been shocked into silence. Sorkin rubbed his nose in disbelief before trying to regain a bit of control. These were unprecedented statements for any business leader to make in public— much less the man who controlled Tesla, SpaceX, and the platform formerly known as Twitter—and it made for incredible entertainment. But on a human level, the DealBook host seemed to sense that Musk was dangerously close to falling off the rails. The lack of impulse control that typically overcame Musk late at night as he fell down Twitter rabbit holes was now on full display for those watching in the auditorium or on live streams at home or at work.

At X's office in Los Angeles, a handful of salespeople had gathered to watch Musk's remarks. He had said very little to the whole company since his "actual truth" tweet had tanked their sales, and they were hoping to see him apologize, or at least offer them some talking points that they could take to clients in an attempt to convince them to restart ad campaigns. They got the opposite.

One client partner, who helped handle the Disney account, was watching with the group. She had just joined X recently, excited by the prospects of her new job and working for a leader who had transformed industries and promised to upend online media. But as she watched Musk repeat his new expletive-laced phrase and call out Disney's Iger, she became despondent.

"That's it," she said, slowly getting up to walk back to her desk. "My job is over. My job is over."

She skipped the rest of the talk.

Onstage at the DealBook Conference, it wouldn't get any better for Musk. While he remained in tense conversation with Sorkin, his eyes darted around the conference hall as his mind cooked up his next outburst. The brands that had become wary of advertising on the platform were the forces that should be held responsible for the death of the social network, not him, he said. If Twitter was destroyed, the people would know it was the advertisers' fault.

"Tell it to Earth!"

"They're going to say Elon, you killed the company because you said these things and that they were inappropriate things and they didn't feel comfortable on the platform, right?"

"And let's see how Earth responds to that."

To him, anyone could see that X's business collapse would be attributed to the advertisers who had conspired against him. Yet with few in the audience reacting to his audacity, Musk began to riff. Tesla had gotten to its level of success with no ads at all, he said, seeming to suggest that perhaps advertising wasn't important. Sorkin raised an eyebrow.

"Tesla currently sells twice as much in terms of electric vehicles as the rest of electric car makers in the United States combined. Tesla has done more to help the environment than all other companies combined. It would be fair to say that therefore, as a leader of the company, I've done more for the environment than everyone else and any single human on Earth."

"How do you feel about that?"

"How do I feel about that?"

"I'm asking you personally how you feel about that, because this goes—we're talking about power and influence."

"I'm saying what I care about is the reality of goodness, not the perception of it. And what I see all over the place is people who care about looking good while doing evil. Fuck them."

Epilogue

On March 3, 2024, Elon Musk joined Nelson Peltz, a friend and activist investor, for breakfast at his home in Palm Beach, Florida. At the table with them were several wealthy Republican donors, and they bantered about the upcoming election. There was an air of anticipation as the men jockeyed to be the loudest and most prescriptive about what America needed ahead of the November presidential election.

Finally, Donald Trump walked in. He and Musk greeted each other with a friendly handshake—Musk had visited the White House during Trump's presidency and had briefly sat on two presidential business advisory councils before resigning in 2017 over the administration's decision to pull out of the Paris Agreement. The men had sniped at one another over the years, but they needed each other and had begun to thaw their relationship, speaking by phone about their businesses and politics. Trump was in search of funding to propel his campaign through November, while Musk saw Trump as the best alternative to another four years of Joe Biden, and hoped that he would one day come back to X.

One of Musk's first moves upon acquiring Twitter was to persuade Trump to get back on the platform. The effort culminated on August 24, 2023, the day Trump surrendered to police in Atlanta to be booked on charges of spearheading a criminal conspiracy to overturn the 2020 election. That afternoon, Trump posted his mugshot on X with the caption "NEVER SURRENDER!" Musk appeared to be aware that the post was coming, trading teasing messages with right-wing accounts on the platform before Trump's arrest.

But aside from that single update, Trump had resisted posting again. He operated his own social media service, called Truth Social, and was bound by an exclusivity contract to post there before other social platforms. Besides Trump and his surrogates, there were few prominent voices on Truth Social, and Trump seemed to realize that his platform would be in danger of collapse if he abandoned it in favor of Musk's. Weeks after their morning meeting, Trump took his company public.

The stalemate between the two men boiled down to one thing: a desire for control. Musk wanted Trump to play in his ecosystem, while Trump preferred to own his sandbox. And what Musk may not have realized was that he had long ago replaced Trump on the platform. He replied to hateful accounts highlighting Black-on-white crime, railed against "illegals," and shared conspiracies that Democrats were exacerbating the immigration crisis at the southern border of the U.S. so that they could import voters. Some of Musk's views in 2024 had become indistinguishable from those at a Trump campaign rally.

For men like them, social media was a plaything to mold in their own image. When he was banned from Twitter, Trump had essentially cloned it, stripping away the fact-checkers who disputed his claims and the outraged liberals who decried his authoritarian impulses. Meanwhile, Musk had laid waste to Twitter, chiseling away until it became X—a place where all his friends and acolytes had blue checks, his preferred pundits shot to virality, and his own posts were the most liked, most viewed, and most discussed on the platform. Four months after he took over the company, @ElonMusk became the site's most-followed account. By May 2024, he had amassed more than 184 million followers and done away with one of the last vestiges of Twitter, changing the site from twitter.com to x.com.

It was a fate that was perhaps preordained for Twitter. Its founders stumbled on a platform that was too potent for the uber wealthy to resist, and they had jockeyed to control it. Jack Dorsey, too, had schemed his way into the top job at Twitter. It was only when an activist investor tried to shake loose his grip on the company that Dorsey decided that conversation and capital should never have been intertwined.

While Musk was radicalized by Twitter, Dorsey was radicalized by its dismantling. After egging on Musk's bid and rolling over his shares to the

EPILOGUE > 433

Tesla chief's company, Dorsey came apart. He staggered from apologetic to resentful to defiant, sometimes criticizing Musk's decision-making, sometimes defending it.

In the wake of Musk's takeover, Dorsey stopped tweeting almost entirely. His absence from Twitter was meant to be symbolic. When he was ousted as the company's chief executive more than a decade earlier while remaining as a director, he refused to speak in board meetings as a silent protest. By withholding his voice, Dorsey was taking a stand.

Whenever he spoke about the acquisition of Twitter—or any other topic—Dorsey did so on Nostr, a social media protocol that resisted centralized control. Occasionally, he also posted on Bluesky, the service that had been his brainchild with Agrawal. In September 2023, he deleted his Bluesky account altogether after being criticized by its users, and the following year, he stepped down from its board. By then, founding and then deserting a social media platform had become a familiar waltz for Dorsey.

On Nostr, Dorsey shared a blog post from a dietician who signed off on his updates with a slogan associated with QAnon, the deep state conspiracy theory. He posted campaign videos for Robert F. Kennedy, Jr., a prominent anti-vaccine advocate who was running a third-party bid for the presidency. And he shared conspiratorial videos that referred to victims of the September 11 terrorist attack as "crisis actors." His online posts were the kinds of things he would have had labeled or removed while he was Twitter's CEO. His rare appearances on Twitter revolved around the promotion of Bitcoin or Kennedy.

>>> OVER TIME, the idea of Twitter itself splintered, just like the minds of the men who had tried their hands at running it. Twitter was no longer the only place where the world came together to discuss wars, news, or celebrity beefs. Instead, Twitter became a feature that one could find on nearly all social media services. On Instagram, tweets were Threads. On Substack, they were Notes, while Mastodon, the open-source platform that had inspired Musk to embark on a jealous crackdown, had toots. On Bluesky, to Graber's chagrin, the community referred to their posts as skeets, a slant rhyme on tweets that also referenced ejaculation. One no longer had to be a part of

Twitter to join the conversation—the conversation was now conversations, and they were everywhere.

The dilution of Twitter's power didn't seem to disturb Musk. All his friends and fans were verified, and their sycophantic replies to his posts were algorithmically boosted to the top of his @ mentions. Inside his own social media service, Musk was a beloved hero. After adding an audio- and video-calling feature to X, Musk said he would get rid of his phone number altogether, making it impossible for anyone who wasn't on the platform to contact him.

One year after he walked into Twitter lugging a sink, Musk handed out new stock grants to his remaining employees that valued X at $19 billion. By February 2024, the investment giant Fidelity had marked down the value of X, which was still paying off its onerous debt bill, to $11.8 billion, down more than 73 percent from its $44 billion purchase price.

As of this writing, Musk is still a billionaire, though his pockets are a bit lighter. In April 2022—the month he set out to buy Twitter—his net worth was nearing $270 billion. Two years later, he's shaved $80 billion from that amount and lost the title of the world's richest person, as X has cratered and Tesla's shares have fallen on questions about Musk's commitment to the company.

Musk found himself back in Kathaleen McCormick's court. A Tesla shareholder had sued him in 2023, arguing that his $55.8 billion compensation package was excessive and a rip-off for his investors. "This decision dares to 'boldly go where no man has gone before,' or at least where no Delaware court has tread," McCormick wrote in January 2024, as she ruled to revoke Musk's pay. Musk had exercised near-total control over Tesla, demanding whatever compensation he wanted and getting it rubber-stamped by friendly board members like Antonio Gracias, she added.

Musk, outraged by her decision, erupted. His supporters tweeted that McCormick was a Biden lackey and politically motivated to take down their hero. The billionaire reincorporated Neuralink in Nevada and SpaceX in Texas, pulling them out of Delaware's jurisdiction, and called on other companies to follow suit.

Some of Twitter's former employees took extended time off, unable to throw themselves back into work after the relentless stress and anxiety of

EPILOGUE > *435*

the takeover. Others opted for a break from fast-paced corporate life, including Vijaya Gadde and Ned Segal. As of this writing, they are still fighting Musk in court for their severance packages. But most of the company's workers—except those who were bound to X by a need for work visas, healthcare for their families, or simply a loyalty to Musk—scattered to other safe havens in Silicon Valley, landing jobs at the likes of Google, Facebook, and OpenAI. The few who joined OpenAI found themselves working under Bret Taylor once again. The former Twitter chairman was appointed chair at OpenAI after a disastrous upheaval that saw its CEO Sam Altman, removed and then reinstated.

After another corporate mess, Taylor was back in Musk's sights. In March 2024, Musk sued OpenAI, alleging the company had violated its founding charter to build AI that would benefit humanity. Meanwhile, Musk, using resources and engineers from X, raced to build his own artificial intelligence company, xAI, and branded its product as the "anti-woke" alternative to OpenAI, whatever that means.

Parag Agrawal disappeared from public view. After he left Twitter's offices for the last time, he never tweeted again. He joined the former executives' lawsuit seeking severance from Musk. In 2023, he returned to his passion for artificial intelligence and began to build a secretive start-up focused on the technology, hiring a handful of former Twitter employees to help him. While Dorsey left Bluesky in a huff, Agrawal stayed in touch with Graber and continued to advise her as she built the open social media service he and Dorsey had envisioned. While he spent long hours hunched over his computer, building his new company, he occasionally took long walks along the marshy nature preserves that smear the edges of Palo Alto into the bay. His vision for solving social media's content moderation problems once and for all fell away like the shoreline.

That left Musk to wage his war for online speech without much competition. He continued to put a premium on defending speech that questioned the experiences of transgender people, funding several lawsuits on behalf of individuals who were fired from their jobs for anti-transgender posts on the platform. Musk also welcomed back previously suspended accounts, including that of the Infowars host Alex Jones and several white nationalists. After their return, he sometimes interacted with them, giving some of the most despicable figures who had been banned by Twitter not only "freedom of speech" but also a boost in reach. Meanwhile, X continued to suspend journalists and sued critics who sought to expose the toxic stew of hate speech and misinformation that had bubbled up under Musk's ownership.

Musk had reached an unmatched level of success driving the electric vehicle revolution and pushing to make the human race extend beyond the confines of planet Earth. Of course, he was the only person who had the means and drive to buy a global internet platform to protect free speech and free expression for all.

But that was a mission Musk had given himself. He had bumbled into buying Twitter for $44 billion, overpaying to control a global internet platform where he measured his self-worth in likes and replies. A man allergic to criticism had bought himself the largest audience in the world, and hoped for praise. It wasn't an unheard-of pursuit—social media users everywhere seek affirmation from the universe. Musk succeeded in becoming Twitter's main character but struggled under the scrutiny of millions of users who loved the combative ethos of the platform.

Musk may have convinced himself he bought Twitter to protect the global town square or build the world's most important app. But the truth was much simpler. Whether or not he wanted to admit it, he had bought it for himself, and for a brief moment, he had the thing he wanted most. He owned Twitter—and then it was gone.

Acknowledgments

If this were a tweet thread, this section would be an endless string of posts tagging people who have helped us as we've reported on this wild story. We're thankful we don't have to tweet this one out.

First, we want to extend our deepest gratitude to the more than one hundred people who spoke to us about their time with Twitter and Elon Musk. Without their patience, trust, and commitment to telling this story, this book would not exist. Talking to a reporter can be unnerving and may seem to have little upside. To those people who shared their stories—at times in the face of threats from a man with seemingly endless wealth and power—and cared enough to make sure we got this right: thank you.

The New York Times brought us together and gave us the resources to report deeply on the tech industry. Its commitment to independent and authoritative journalism put us in a position to tell the complete story of Musk's takeover of Twitter from day one. It was through that collaboration that we uncovered more than could fit in a newspaper article and discovered that we wouldn't end up severely maiming one another if we embarked on writing our first book together.

At the *Times*, our editors Pui-Wing Tam, Jim Kerstetter, and Ellen Pollock were instrumental in steering our coverage of the Twitter deal and pushing us to focus on the unrivaled strangeness of the transaction. We were joined in reporting on Twitter by our incredible colleagues Mike Isaac, Lauren Hirsch, Tiffany Hsu, Kellen Browning, David McCabe, Adam Satariano, and Ben Mullin, and supported in our endeavor by our talented colleagues on the tech desk, as well as the *Times*'s peerless editing teams and research department. Cade Metz set an example for elegant, character-driven storytelling and selfless collaboration, and provided invaluable feedback on our reporting and early drafts. Kashmir Hill and Tripp Mickle also generously dedicated their time to reading our manuscript and honing our story, while Sheera Frenkel, Cecilia Kang, and Kevin Roose held our hands as we navigated the world of book publishing. We're also particularly grateful for Dai Wakabayashi's mentorship, steady spirit, and searing comebacks.

Reporting on Tesla, SpaceX, Twitter, and the wider world of Musk can be a trying endeavor. No single journalist or publication can handle it all. We're thankful for the reporters and editors who have held Musk and his companies to account, among them Dana Hull, Lora Kolodny, Jack Ewing, Kirsten Grind, Emily Glazer, Tim Higgins, Linette Lopez, Russ Mitchell, Caroline O'Donovan, Faiz Siddiqui, Will Oremus, Elizabeth Dwoskin, Joseph Menn, Drew Harwell, Donie O'Sullivan, Erin Woo, Becky Peterson, Zoë Schiffer, Kurt Wagner, Marisa Taylor, Rachael Levy, Kali Hays, Edward Niedermeyer, Will Evans, Alan Ohnsman, and so many others. While our industry thrives on competition, we owe a special thanks to Bobby Allyn and Julia Black for providing materials that helped advance our reporting.

A decade ago, neither of us thought we'd make it as reporters, much less write a book. So many people gave us opportunities to showcase our abilities and taught us along the way. For Ryan, those editors and mentors include Ann Grimes, Mark Katches, Clay Lambert, Gerry Shih, Felicity Barringer, Bob Ivry, Kerry Dolan, Luisa Kroll, Bruce Upbin, Mat Honan, John Paczkowski, and Ben Smith. For Kate, it was Jana Clark who started it all; followed by Alan Scherstuhl and Erin Sherbert, who taught her to report; A. J. Daulerio, Henry Pickavet, Kelly Bourdet, and Andrew Couts, who poured gas on the fire; and, of course, Dell Cameron, her first writing partner.

Adam Eaglin saw the potential for this book long before we did and brought a reporter's ferocity and thoroughness to every meeting, draft, and detail. He, Beniamino Ambrosi, and their team at the Cheney Agency have been instrumental to our success. Anakwa Dwamena lent his careful eyes to the fact-checking and was an important sounding board. Chris Allen designed the captivating cover.

Our editor, William Heyward, put up with a lot during this process, and his steady hand and advice, which sometimes went against our first in-

ACKNOWLEDGMENTS

stincts, were crucial. His optimism and excitement for this book carried us through. We have him to thank for the title and so much more. Natalie Coleman guided us through the madness, and we're grateful to the team at Penguin Press, including Ann Godoff, Scott Moyers, Gail Brussel, Jessie Stratton Zhou, Joy Simpkins, Chelsea Cohen, Aly D'Amato, and Darren Haggar, as well as Helen Conford at Penguin Random House UK.

Kate thanks Gillian Altman, who gave us a peaceful home to write our first (horrible) words, and many of those that followed. Thank you for your passionate heart and unrelenting support over the last decade—she'd be lost without you. Laura Jaye Cramer and Michael Belt buoyed Kate through this and much more. Sam Rogers and Emily Straley, thank you for all the nourishment, body and spirit. Deepest love and gratitude to Hanna, Lucas, and, of course, Mama—thank you for the love of words, the piles of books, and the lessons in how to argue and stand one's ground. It has all served Kate very well in journalism.

Ryan thanks Stephanie M. Lee, Ken Bensinger, Jeremy Sasson, Sivan Sasson, Lucas Manfield, Spencer Vuksic, and Rima Abouziab and the Lutefisk Lodge, all of whom opened their homes or lent a comfy couch on reporting trips and writing retreats. Rebecca Ellis and Arya Shirazi lent their eagle eyes on close reads, while Albert Samaha provided much-needed words of support. Thank you as well to Jonathan Swan, Aaron Greenspan, and Jack Sweeney for your help with reporting materials.

Ryan's mom, Hang, has been a pillar of strength. Without her, Nolan Mac, Lê Khanh "Liz" Nguyen, and Kim Mac, to whom this work is dedicated, there would be no book. Ryan loves you more than you all know.

Note on Reporting

This book is based on conversations and interviews totaling more than 150 hours with nearly 100 people. Those people included current and former employees for Twitter, X, Tesla, and SpaceX; lawyers, bankers, and other associates who worked for both sides during Elon Musk's negotiations to buy Twitter; as well as friends and acquaintances of Musk, Jack Dorsey, and other Twitter executives. Some of those interviews were conducted during the course of our years of reporting for *The New York Times* on Twitter, Musk, and his 2022 acquisition of the company. We also relied on court filings, videos, audio recordings, internal company messages, and direct messages between key figures. Of course, we also referenced countless tweets.

Quotes from those internal messages and documents appear verbatim, as do quotes from company-wide meetings at Twitter. In dialogue, quotes that are attributed to specific people are reconstructed from the memories of participants, from contemporaneous notes they made about their conversations, and sometimes from recordings of those interactions.

For the early history of Twitter, we relied in part on Nick Bilton's 2013 *Hatching Twitter: A True Story of Money, Power, Friendship, and Betrayal*, which offers a dramatic retelling of the company's beginnings. To better understand Musk, we referred to two works: Ashlee Vance's 2015 book, *Elon Musk: Tesla, SpaceX, and the Quest for a Fantastic Future*, and Walter Isaacson's 2023 authorized biography, *Elon Musk*. Isaacson shadowed Musk for two years, and his reporting helped us understand the entrepreneur's whereabouts during the deal. We also depended on clues provided by @ElonJet, the since-suspended Twitter account that diligently tracked the movements of Musk's private plane.

All of the people who appear in this book are identified by their real names. Many people spoke on the condition of anonymity because they feared retribution from Musk or one of his companies in the form of either litigation or online harassment. In the process of reporting, we attempted to contact the main figures identified in the book and offered them the opportunity to tell their stories. Musk did not answer our requests for interviews.

Notes

Introduction
1 **rioting at the U.S. Capitol**: Mac, Ryan, and Craig Silverman. "'Mark Changed the Rules': How Facebook Went Easy on Alex Jones and Other Right-Wing Figures." *BuzzFeed News*, February 21, 2021. buzzfeednews.com/article/ryanmac/mark-zuckerberg-joel-kaplan-facebook-alex-jones.

ACT I

Chapter 1: back at twttr
12 **Dorsey moved to the Bay Area in 1999**: "Tech's Best Young Entrepreneurs." *BusinessWeek*, July 11, 2014. web.archive.org/web/20140711043659/http://images.businessweek.com/ss/07/03/0326_tech_entrepreneurs/source/10.htm.
13 **service for the ferries to San Francisco's Alcatraz**: Bilton, Nick. *Hatching Twitter*. New York: Portfolio, 2013.
13 **offered a freelance coding gig**: Bilton, *Hatching Twitter*.
13 **sketched the idea in a legal pad**: Dorsey, Jack. "Twttr Sketch." Flickr, March 24, 2006. flickr.com/photos/jackdorsey/182613360/in/photostream.
14 **He thumbed through tw- words**: Bilton, *Hatching Twitter*.
14 **Williams retained a 70 percent ownership stake**: Bilton, *Hatching Twitter*.
16 **"We were just hanging on"**: Miller, Claire Cain. "Why Twitter's C.E.O. Demoted Himself." *New York Times*, October 30, 2010, sec. Technology. nytimes.com/2010/10/31/technology/31ev.html.
17 **Williams's ownership had been whittled down**: Twitter, Inc. SEC "Form S-1," October 3, 2013. sec.gov/Archives/edgar/data/1418091/000119312513390321/d564001ds1.htm.

Chapter 2: #StayWoke
19 **handed out blooming red roses**: Hu, Elise (@elise who). "STL native and @twitter co-founder @jack passing out roses to demonstrators on w florissant #Ferguson." Twitter, August 19, 2014, 4:01 p.m. twitter.com/elisewho/status/501866614284091392.
20 **"Freedom of expression means little"**: Gadde, Vijaya. "Twitter Executive: 'Here's How We're Trying to Stop Abuse While Preserving Free Speech.'" *Washington Post*, April 16, 2015. washingtonpost.com/posteverything/wp/2015/04/16/twitter-executive-heres-how-were-trying-to-stop-abuse-while-preserving-free-speech.
20 **"We suck at dealing with abuse and trolls"**: Tiku, Nitasha, and Casey Newton. "Twitter CEO: 'We Suck at Dealing with Abuse.'" *The Verge*, February 5, 2015. theverge.com/2015/2/4/7982099/twitter-ceo-sent-memo-taking-personal-responsibility-for-the.
21 **"looking out your window"**: Goel, Vindu. "Twitter Revenue up 61%, but User Growth Lags." *New York Times*, July 28, 2015. nytimes.com/2015/07/29/technology/twitter-quarterly-earnings.html.
21 **Trump credited Twitter**: Donald J. Trump interview. "I doubt I'd be here if it weren't for social media . . . because there is a fake media out there." Fox News, October 22, 2017. facebook.com/watch/?v=10156176791126336.

Chapter 3: "This is actually me"
27 **He was a devout member of the Church**: Copeland, Rob. "Elon Musk's Inner Circle Rocked by Fight over His $230 Billion Fortune." *Wall Street Journal*, July 16, 2022. wsj.com/articles/elon-musk-fortune-fight-jared-birchall-igor-kurganov-11657308426.
27 **He paid $52,000**: Mac, Ryan. "I Went to Elon Musk's 'Pedo Guy' Trial, but I Wasn't Ready to Become a Part of It." *BuzzFeed News*, January 30, 2020. buzzfeednews.com/article/ryanmac/elon-musk-cant-lose.
29 **subsisting on Jack in the Box burgers**: Isaacson, Walter. *Elon Musk*. New York: Simon & Schuster, 2023.
29 **He personally made $22 million**: Isaacson, *Elon Musk*.
30 **spending a small fortune on the web domain**: Chafkin, Max. *The Contrarian*. New York: Penguin Books, 2021.
33 **"Please ignore prior tweets"**: Musk, Elon (@elonmusk). "Please ignore prior tweets, as that was someone pretending to be me :) This is actually me." Twitter, June 4, 2010, 11:31 a.m. twitter.com/elonmusk/status/15434727182.
38 **Some of Tesla's own board members**: Gelles, David, James B. Stewart, Jessica Silver-Greenberg, and Kate Kelly. "Elon Musk Details 'Excruciating' Personal Toll of Tesla Turmoil." *New York Times*, August 17, 2018. nytimes.com/2018/08/16/business/elon-musk-interview-tesla.html.
39 **He denied using marijuana**: Gelles et al., "Elon Musk Details."

Chapter 4: OneTeam
43 **he would sometimes sit, take work calls**: Hiatt, Brian. "Twitter CEO Jack Dorsey: The Rolling Stone Interview." *Rolling Stone*, January 23, 2019. rollingstone.com/culture/culture-features/twitter-ceo-jack-dorsey-rolling-stone-interview-782298.

Chapter 5: An Invasion
50 **a 2 percent stake worth about $531 million**: Merced, Michael J. de la, and Kate Conger. "Hedge Fund May Push for Ouster of Jack Dorsey as Twitter's C.E.O." *New York Times*, February 29, 2020, sec. Business. nytimes.com/2020/02/29/business/dealbook/elliott-twitter-jack-dorsey.html.

Chapter 6: A Polynesian Vacation
63 **Musk sent out a company-wide email**: Mac, Ryan. "Elon Musk Told Workers They're More Likely to Die in a Car Crash Than from Coronavirus." *BuzzFeed News*, March 13, 2020. buzzfeednews.com/article/ryanmac/elon-musk-spacex-employees-car-crash-coronavirus.
68 **labeled some 300,000 tweets**: Fung, Brian. "Twitter Says It Labeled 300,000 Tweets around the Election." CNN, November 12, 2020. cnn.com/2020/11/12/tech/twitter-election-labels-misinformation/index.html.
71 **"Mike Pence didn't have the courage"**: Restuccia, Andrew, and Siobhan Hughes. "Trump's Tweet about Pence Seen as Critical Moment during Riot." *Wall Street Journal*, July 21, 2022. wsj.com/livecoverage/jan-6-hearing-today-trump/card/trump-s-tweet-about-pence-seen-as-critical-moment-during-riot-fmPxoFkeoTKxi0NqPLCL.
72 **"We must learn from our mistakes"**: Tiku, Nitasha, Tony Romm, and Craig Timberg. "Twitter Bans Trump's Account, Citing Risk of Further Violence." *Washington Post*, January 8, 2021. washingtonpost.com/technology/2021/01/08/twitter-trump-dorsey.

Chapter 7: Resource Plan
76 **doubling its annual revenue to $7.5 billion**: Feiner, Lauren. "Twitter Shares Soar after Company Announces Plan to Double Revenue by End of 2023." CNBC, February 25, 2021. cnbc.com/2021/02/25/twitter-sets-goals-to-double-revenue-reach-315-million-users-by-end-of-2023.html.

Chapter 8: Parag
82 **"The Board believes that you will add tremendous value"**: Twitter, Inc. SEC "EX-10.1." November 29, 2021. sec.gov/Archives/edgar/data/1418091/000119312521342255/d401229dex101.htm.

Chapter 9: Bluesky
89 **"What we see is"**: Hanamura, Wendy. "Our Social Media Is Broken. Is Decentralization the Fix?" Internet Archive Blog, January 30, 2020. blog.archive.org/2020/01/30/our-social-media-is-broken-is-decentralization-the-fix.

Chapter 10: Twitter in Trouble
94 **"clear decision-making, increased accountability"**: Oremus, Will, and Elizabeth Dwoskin. "Twitter's New CEO Announces Major Reorganization of the Social Networking Company." *Washington Post*, December 3, 2021. washingtonpost.com/technology/2021/12/03/twitter-agrawal-restructuring.

Chapter 11: Musk's Shopping Spree
99 **He had a reputation for using drugs**: Glazer, Emily, and Kirsten Grind. "Elon Musk Has Used Illegal Drugs, Worrying Leaders at Tesla and SpaceX." *Wall Street Journal*, January 7, 2024, www.wsj.com/business/elon-musk-illegal-drugs-e826a9e1.
101 **"as collateral to secure certain personal indebtedness"**: Tesla, Inc. SEC "Form 10-K." February 7, 2022. sec.gov/ixviewer/ix.html?doc=/Archives/edgar/data/0001318605/000156459022016871/tsla-10ka_20211231.htm.
101 **"So I asked myself what product I liked"**: Isaacson, Walter. *Elon Musk*. Simon & Schuster, 2023.
102 **By March 14, Musk had accumulated 9.2 percent**: Isaac, Mike, and Lauren Hirsch. "Elon Musk Becomes Twitter's Largest Shareholder." *New York Times*, April 4, 2022, sec. Technology. nytimes.com/2022/04/04/technology/elon-musk-twitter.html.
103 **"I saw your tweet re free speech"**: Twitter, Inc. v. Elon R. Musk, et al. 2022-0613-KSJM (Delaware Court of Chancery 2022).
105 **Tesla chief's phone contacts as "jack jack"**: Twitter, Inc. v. Elon R. Musk, et al.
109 **"Not today"**: BerghAnon. "The Elon Musk Rejection a Few Weeks Back." Reddit, April 22, 2022. reddit.com/r/berghain/comments/u9ahbv/the_elon_musk_rejection_a_few_weeks_back.
110 **Birchall rejected the document on his behalf**: Isaacson, *Elon Musk*.
111 **Lane Fox offered a revised agreement**: Twitter, Inc. v. Elon R. Musk, et al.
111 **The stock ownership limit remained**: Twitter, Inc. xX-10.1." sec.gov. SEC. April 4, 2022. sec.gov/Archives/edgar/data/1418091/000119312522095651/d342257dex101.htm.

Chapter 12: An Offer
114 **"We know that he has caused harm to workers"**: Nix, Naomi, Nitasha Tiku, Will Oremus, and Faiz Siddiqui. "Elon Musk's Twitter Bid Frustrates Employees. That's a Risk for Him." *Washington Post*, April 15, 2022. washingtonpost.com/technology/2022/04/14/twitter-employees-elon-musk.
114 **"Are board members held to the same standard?"**: Dwoskin, Elizabeth. "Elon Musk to Address Twitter Staff after Internal Outcry." *Washington Post*, April 7, 2022. washingtonpost.com/technology/2022/04/07/musk-twitter-employee-outcry.
117 **the Tesla chief flew to Lanai**: Isaacson, Walter. *Elon Musk*. New York: Simon & Schuster, 2023.
117 **He had a call with Agrawal to discuss engineering**: Twitter, Inc. v. Elon R. Musk, et al. 2022-0613-KSJM (Delaware Court of Chancery 2022).
120 **"convert Twitter SF HQ"**: Musk, Elon (@elonmusk). "Convert Twitter SF HQ to homeless shelter since no one shows up anyway." Twitter, April 10, 2022. polititweet.org/tweet?account=44196397&tweet=1512966135423066116. (This tweet was deleted. Note, that the time zone is UTC, so seven hours ahead of PT. It was at 1:30:10 UTC on 4/10/22, so this was at 18:30:10 PT on 4/9/22 in SF.)
121 **Birchall contacted some former colleagues at Morgan Stanley**: Isaacson, *Elon Musk*.

Chapter 13: Poison Pill
124 **It was a single emoji of a blushing face**: Musk, Elon (@elonmusk). "😊." Twitter, April 11, 2022. polititweet.org/tweet?account=44196397&tweet=1513373170333487104.
124 **"ideas worth spreading"**: Isaacson, Walter. *Elon Musk*. New York: Simon & Schuster, 2023.

Act II

Chapter 14: "Bring the cattle"
134 **"not like the other kids in the class"**: Twitter, Inc. v. Elon R. Musk, et al. 2022-0613-KSJM (Delaware Court of Chancery 2022).
136 **Musk sometimes stayed at Kives's Beverly Hills**: Bernstein, Joseph. "Elon Musk Has the World's Strangest Social Calendar." *New York Times*, October 11, 2022, sec. Style. nytimes.com/2022/10/11/style/elon-musk-social-calendar.html.
136 **earned $125 million after the crypto mogul**: Yaffe-Bellany, David, and Erin Griffith. "The Super Connector Who Built Sam Bankman-Fried's Celebrity World." *New York Times*, June 23, 2023, sec. Technology. nytimes.com/2023/06/23/technology/sam-bankman-fried-celebrity-friends.html.
137 **Morgan Stanley prepared a pitch deck**: Isaac, Mike, Lauren Hirsch, and Anupreeta Das. "Inside Elon Musk's Big Plans for Twitter." *New York Times*, May 6, 2022, sec. Technology. nytimes.com/2022/05/06/technology/elon-musk-twitter-pitch-deck.html.

NOTES > 445

138 **The directors unanimously agreed:** Hirsch, Lauren, and Kate Conger. "Twitter Counters a Musk Takeover with a Time-Tested Barrier." *New York Times*, April 15, 2022, sec. Business. nytimes.com/2022/04/15/business/twitter-poison-pill-elon-musk.html.
141 **That weekend in Austin, some of Musk's friends:** Isaacson, Walter. *Elon Musk*. New York: Simon & Schuster, 2023.
141 **American investment adviser that held a 9.2 percent stake:** Twitter, Inc. "Schedule 14A." sec.gov. SEC. April 12, 2022. sec.gov/Archives/edgar/data/1418091/000114036122014049/ny20001921x3_def14a.htm.
144 **He flew back to Austin on his Gulfstream jet:** ElonJet (@ElonJet). web.archive.org/web/20220428031935/twitter.com/elonjet.

Chapter 15: Parag's Last Stand
149 **Taylor and Pichette met with shareholders:** Twitter, Inc. n.d. "DEFM14A." sec.gov. SEC. sec.gov/Archives/edgar/data/1418091/000119312522202163/d283119ddefm14a.htm.
152 **offered him $13 billion in debt financing:** Hirsch, Lauren. "Elon Musk Details His Plan to Pay for a $46.5 Billion Takeover of Twitter." *New York Times*, April 21, 2022, sec. Business. nytimes.com/2022/04/21/business/elon-musk-twitter-funding.html.

Chapter 16: Just Say Yes
164 **Just before 11:45 a.m. in San Francisco:** Chang, Emily (@emilychangtv). "$TWTR stock halted, news pending. . . ." Twitter, April 25, 2022, 11:44 a.m. twitter.com/emilychangtv/status/1518662186675044353?lang=en
167 **"Once the deal closes, we don't know":** Isaac, Mike, and Adam Satariano. "Twitter Reports Growth in Revenue and Users as Elon Musk Prepares to Take Over." *New York Times*, April 28, 2022, sec. Technology. nytimes.com/2022/04/28/technology/twitter-first-quarter-earnings-elon-musk.html.

Chapter 17: Golden, Golden
173 **She had blocked the CIA from accessing Twitter's:** Singh, Kanishka. "U.S. Judge Block's Twitter's Bid to Reveal Government Surveillance Requests." Twitter, April 18, 2020, 11:27 p.m. EDT. reuters.com/article/us-usa-twitter-lawsuit/us-judge-blocks-twitters-bid-to-reveal-government-surveillance-requests-idUSKBN2200C5.
175 **And Musk had also secured a $12.5 billion personal loan:** Hirsch, Lauren. "Elon Musk Details His Plan to Pay for a $46.5 Billion Takeover of Twitter." *New York Times*, April 21, 2022, sec. Business. nytimes.com/2022/04/21/business/elon-musk-twitter-funding.html.
176 **"Does Sam actually have $3B liquid?":** Twitter, Inc. v. Elon R. Musk, et al. 2022-0613-KSJM (Delaware Court of Chancery 2022).
177 **he'd go "to the fucking mattresses no matter what":** Peterson, Becky. "How Antonio Gracias Became the Most Hardcore of Elon Musk's Loyalists." *The Information*, August 14, 2023. theinformation.com/articles/how-antonio-gracias-became-the-most-hardcore-of-elon-musks-loyalists?rc=620356.

Chapter 18: 😊
184 **"Putin's speech tomorrow is extremely important":** Twitter, Inc. v. Elon R. Musk, et al. 2022-0613-KSJM (Delaware Court of Chancery 2022).
184 **The more he stewed, the angrier he became:** Isaacson, Walter. *Elon Musk*. New York: Simon & Schuster, 2023.
186 **"It's critical to have the right leaders":** Isaac, Mike. "Twitter Fires Two Executives and Freezes Most Hiring After Musk's Deal to Buy the Company." *New York Times*, May 12, 2020. nytimes.com/2022/05/12/technology/twitter-elon-musk.html.
189 **"which are partially secured by pledges":** Tesla, Inc. "10-K." sec.gov. SEC. 2022. sec.gov/Archives/edgar/data/1318605/000095017022000796/tsla-20211231.htm.
190 **he was unsure if he should push through with his acquisition:** Isaacson, *Elon Musk*, 463.
190 **The reams and reams of data:** Twitter, Inc. "Schedule 14A." sec.gov. SEC. 2022. sec.gov/Archives/edgar/data/1418091/000114036122014049/ny20001921x3_def14a.htm.
191 **Spiro and Birchall frantically texted:** Isaacson, *Elon Musk*.
191 **Twitter's stock dropped nearly 10 percent:** Roumeliotis, Greg, and Sheila Dang. "Musk Says $44 Billion Twitter Deal on Hold over Fake Account Data." Reuters, May 16, 2022, sec. Technology. reuters.com/technology/musk-says-44-billion-twitter-deal-hold-2022-05-13.

Chapter 19: Bots and Horses
194 **On May 27, Musk's private jet took off:** ElonJet (@elonjet). Twitter. web.archive.org/web/20220530091209/twitter.com/elonjet.
194 **sparred verbally with Durban:** Isaacson, Walter. *Elon Musk*. New York: Simon & Schuster, 2023.
198 **He couldn't get his laptop to work:** Isaacson, *Elon Musk*.
199 **"I want Twitter to contribute to a better":** Isaac, Mike. "Elon Musk Tells Twitter's Employees He Wants the Service to 'Contribute to a Better, Long-Lasting Civilization.'" *New York Times*, June 16, 2022, sec. Technology. nytimes.com/2022/06/16/technology/elon-musk-twitter-employees-meeting.html.
200 **Business Insider had dropped a bombshell story:** McHugh, Rich. "A SpaceX Flight Attendant Said Elon Musk Exposed Himself and Propositioned Her for Sex, Documents Show. The Company Paid $250,000 for Her Silence." *Business Insider*, May 19, 2022. businessinsider.com/spacex-paid-250000-to-a-flight-attendant-who-accused-elon-musk-of-sexual-misconduct-2022-5.
201 **"Elon is seen as the face of SpaceX":** Scheiber, Noam, and Ryan Mac. "SpaceX Employees Say They Were Fired for Speaking Up about Elon Musk." *New York Times*, November 17, 2022, sec. Business. nytimes.com/2022/11/17/business/spacex-workers-elon-musk.html.
201 **Gwynne Shotwell labeled "overreaching activism":** Mac, Ryan. "SpaceX Said to Fire Employees Involved in Letter Rebuking Elon Musk." *New York Times*, June 17, 2022, sec. Technology. nytimes.com/2022/06/17/technology/spacex-employees-fired-musk-letter.html.
201 **"SpaceX is Elon and Elon is SpaceX":** Scheiber and Mac, "SpaceX Employees Say."
201 **"Mr. Musk made clear that Twitter's continuing failure":** Twitter, Inc. "Schedule 14A." sec.gov. SEC. 2022. sec.gov/Archives/edgar/data/1418091/000114036122014049/ny20001921x3_def14a.htm.

Chapter 20: Sun Valley
203 **Musk agreed to an in-person meeting on June 21:** Twitter, Inc. v. Elon R. Musk, et al. 2022-0613-KSJM (Delaware Court of Chancery 2022). documentcloud.org/documents/22084453-twittermuskcomplaint.
205 **according to court filings obtained by Business Insider:** Black, Julia. "Elon Musk Had Twins Last Year with One of His Top Executives." *Business Insider*, July 6, 2022. businessinsider.com/elon-musk-shivon-zilis-secret-twins-neuralink-tesla.
207 **"You are not presently devoting sufficient resources":** Twitter, Inc. "Schedule 14A." sec.gov. SEC. 2022. sec.gov/Archives/edgar/data/1418091/000114036122014049/ny20001921x3_def14a.htm.
207 **"That needs to stop":** Twitter, Inc. v. Elon R. Musk, et al.

208 **On July 5, she directed Twitter to sue:** Singh, Karan Deep, and Kate Conger. "Twitter, Challenging Orders to Remove Content, Sues India's Government." *New York Times*, July 5, 2022, sec. Business. nytimes.com/2022/07/05/business/twitter-india-lawsuit.html.
209 **letter claimed the company had broken several:** Twitter, Inc., "Schedule 14A," 2022.
210 **The ruling proved that Delaware was willing to force a merger:** Winter, Greg. "Judge Rules That Tyson Must Take Over IBP." *New York Times*, June 16, 2001. nytimes.com/2001/06/16/business/judge-rules-that-tyson-must-complete-takeover-of-ibp.html.
212 **Agrawal, in a black T-shirt with an ultrasoft cashmere:** Maas, Jennifer. "Elon Musk Makes Long-Awaited Arrival at Sun Valley Conference." *Variety*, July 8, 2022. variety.com/2022/tv/news/elon-musk-sun-valley-twitter-1235311152.

Chapter 21: Court of Chancery
215 **Savitt introduced himself to Musk's lawyers:** Twitter, Inc. "Schedule 14A." sec.gov. SEC. 2022. sec.gov/Archives/edgar/data/1418091/000119312522202163/d283119ddefm14a.htm#toc283119_14.
215 **"Musk apparently believes that he":** Conger, Kate, and Lauren Hirsch. "Twitter Sues Musk after He Tries Backing Out of $44 Billion Deal." *New York Times*, July 12, 2022, sec. Technology. nytimes.com/2022/07/12/technology/twitter-lawsuit-musk-acquisition.html.
216 **"Musk said he needed to take the company private because":** Twitter, Inc. v. Elon R. Musk, et al. 2022-0613-KSJM (Delaware Court of Chancery 2022). documentcloud.org/documents/22084453-twittermuskcomplaint.

Chapter 23: Mudge
226 **On August 29, a week after Zatko's claims were published:** Skadden Arps. "Exhibit Q." sec.gov. SEC. August 29, 2022. sec.gov/Archives/edgar/data/1418091/000110465922095765/tm2224790d1_ex99-q.htm.
226 **"Stunning events over the last week, however":** Twitter, Inc. v. Elon R. Musk, et al. 2022-0613-KSJM (Delaware Court of Chancery 2022). documentcloud.org/documents/22416599-public-version-of-amended-musk-counterclaims-twitter-v-musk.

Chapter 24: An Accelerant to X
228 **asked for a 30 percent discount:** Conger, Kate, and Michael S. Schmidt. "Elon Musk Offered to Buy Twitter at a Lower Price in Recent Talks." *New York Times*, October 6, 2022, sec. Technology. nytimes.com/2022/10/05/technology/elon-musk-twitter-discount.html.

Chapter 25: "Not owned by a fucking moron"
232 **In court, Twitter's lawyers were fighting:** Twitter, Inc. v. Elon R. Musk, et al. 2022-0613-KSJM (Delaware Court of Chancery 2022). documentcloud.org/documents/23119236-letter-decision-resolving-plaintiffs-seventh-discovery-motion-twitter-v-musk.
232 **when Musk was arranging:** Twitter, Inc. "Schedule 14A." sec.gov. SEC. 2022. sec.gov/Archives/edgar/data/1418091/000114036122014049/ny20001921x3_def14a.htm.
232 **she told Twitter it could have a November trial date:** Twitter, Inc. v. Elon R. Musk, et al. 2022-0613-KSJM (Delaware Court of Chancery 2022). documentcloud.org/documents/23126733-letter-decision-granting-stay.
233 **he'd pay Dorsey the full price of $54.20 per share:** Isaacson, Walter. *Elon Musk*. New York: Simon and Schuster, 2023.
234 **The Wall Street Journal reported that MBS sought:** Stancati, Margherita, Benoit Faucon, and Summer Said. n.d. "The Price of Freedom for Saudi Arabia's Richest Man: $6 Billion." *Wall Street Journal*. Accessed February 21, 2024. wsj.com/articles/the-price-of-freedom-for-saudi-arabias-richest-man-6-billion-1513981887.
235 **Agrawal would earn his salary and unvested stock options:** Twitter, Inc. "Schedule 14A." Sec.gov. SEC. 2022. sec.gov/Archives/edgar/data/1418091/000119312522202163/d283119ddefm14a.htm#toc.

Chapter 26: Let That Sink In
241 **The employees had all read a *Washington Post* report:** Dwoskin, Elizabeth, Faiz Siddiqui, Gerrit De Vynck, and Jeremy B. Merrill. 2022. "Documents Detail Plans to Gut Twitter's Workforce." *Washington Post*, October 22, 2022. washingtonpost.com/technology/2022/10/20/musk-twitter-acquisition-staff-cuts.

Chapter 27: Trick or Tweet
253 **The Wall Street Journal later reported that Musk borrowed:** Maidenberg, Micah, and Tim Higgins. "Elon Musk Borrowed $1 Billion from SpaceX in Same Month of Twitter Acquisition." *Wall Street Journal*, September 5, 2023. wsj.com/business/elon-musk-spacex-loan-269a2168.
255 **decorative pumpkins and black paper cutouts of bats:** Rebecca (@rebeccaw). "I hope @TwitterUK like 🎃 soup #TrickOrTweet." Twitter, October 27, 2022. twitter.com/RebeccaW/status/1585692866893971457.
255 **draped over giant hashtags and @-sign statues:** Lozic, Jenn (@jennifernatalie). "Bluey and Bingo visit Twitter NYC #TrickorTweet." Twitter, October 27, 2022. twitter.com/JenniferNatalie/status/1585701638739791874.
255 **In Mexico City, workers prepped for a spooky art session:** Rosales, Denisse. "Tis the Season 🎃 #TrickOrTweet 🎃🎃 @emmanuelromero7 Pic.Twitter.Com/2yhiyoriui." Twitter, October 28, 2022. twitter.com/its_deniiiisse/status/1585799565592363008.
255 **At noon in San Francisco, workers were adding the finishing touches:** Varon, Samuel. "#trickortweet Sucess 🎃 Pic.Twitter.Com/1xzqofmrqb." Twitter, October 27, 2022. twitter.com/samuelvaronn/status/1585766618713362433.
256 **running up a tab of $6,397:** Canary, LLC dba Canary Marketing v. Twitter, Inc., CGC-23-603842 (CA Superior Court, County of San Francisco 2023).

Chapter 28: "The bird is freed"
265 **Children dressed as Marvel superheroes:** Harrigan, Justine. "#trickortweet Back at the Tweet HQ! Pic.Twitter.Com/Tqsmagjspu." Twitter, October 27, 2022. twitter.com/justinelevi/status/1585780568469299200.
265 **Several workers crowded in front of a portrait station:** S, Parissa. "Gettin Spooky! 🎃 #TrickOrTweet Pic.Twitter.Com/S8Mkr1G7P8." Twitter, October 27, 2022. twitter.com/Parissa_S/status/1585773993583206401.
266 **the deal's closing fees had made Morgan Stanley millions:** Isaacson, Walter. *Elon Musk*. New York: Simon & Schuster, 2023.

Act III

Chapter 29: Code Reviews
272 **It was fifteen times smaller than Twitter's engineering staff:** Isaacson, Walter. *Elon Musk*. New York: Simon & Schuster, 2023.
272 **"Print out 50 pages of code you've done in the last 30 days":** Schiffer, Zoë, Casey Newton, and Alex Heath. "Inside Elon's 'Extremely Hardcore' Twitter." *The Verge*, January 17, 2023. theverge.com/23551060/elon-musk-twitter-takeover-layoffs-workplace-salute-emoji.

NOTES   > 447

Chapter 30: Lords and Peasants

283 **compliance department was told to eliminate 27 percent:** Peters, Jay. "Twitter Cut 15 Percent of Its Trust and Safety Staff but Says It Won't Impact Moderation." *The Verge*, November 5, 2022. theverge.com /2022/11/4/23441404/twitter-trust-safety-staff -layoffs-content-moderation.

285 **projected it would reach $10 billion by 2028:** Isaac, Mike, and Ryan Mac. "Elon Musk, under Financial Pressure, Pushes to Make Money from Twitter." *New York Times*, November 3, 2022, sec. Technology. www .nytimes.com/2022/11/03/technology/elon-musk -twitter-money-finances.html.

287 **evening of October 28, Musk was retiring:** Isaacson, Walter. *Elon Musk*. New York: Simon & Schuster, 2023.

Chapter 31: "Educate me"

290 **"'Is he going to get Donald Trump back on the platform?'":** Wagner, Kurt, Sarah Frier, and Brad Stone. "Elon Musk's Twitter Is a Shakespearean Psychodrama Set in Silicon Valley." Bloomberg.com, December 14, 2022. bloomberg.com/news/features/2022 -12-14/elon-musk-twitter-ownership-full-of -firings-ad-cuts-chaos.

292 **The next day, tabloids would circulate photos of Musk:** Silverman, Sam. "Elon Musk Parties in $7.5k Halloween Costume with Mom Maye." *Entrepreneur*, November 1, 2022. entrepreneur.com/business -news/elon-musk-parties-in-75k-halloween -costume-with-mom-maye/438282.

Chapter 32: A Blue Heart

294 **Interpublic Group, or IPG, a large advertising:** Conger, Kate, Tiffany Hsu, and Ryan Mac. "Elon Musk's Twitter Faces Exodus of Advertisers and Executives." *New York Times*, November 1, 2022, sec. Technology. nytimes.com/2022/11/01/technology /elon-musk-twitter-advertisers.html.

294 **Musk would later push to ban posts:** Isaacson, Walter. *Elon Musk*. New York: Simon & Schuster, 2023.

297 **encrypted direct messages were unveiled months later:** Isaac, Mike, and Ryan Mac. "Elon Musk, under Financial Pressure, Pushes to Make Money from Twitter." *New York Times*, November 3, 2022, sec. Technology. nytimes.com/2022/11/03/technology /elon-musk-twitter-money-finances.html.

298 **They also began circulating "A Layoff Guide":** "A Layoff Guide for Tweeps." Collective Action in Tech, November 2, 2022. collectiveaction.tech/2022/a -layoff-guide-for-tweeps.

299 **"The severance calculations are updated based":** Conger, Kate, Mike Isaac, Ryan Mac, and Tiffany Hsu. "Two Weeks of Chaos: Inside Elon Musk's Takeover of Twitter." *New York Times*, November 11, 2022, sec. Technology. nytimes.com/2022/11/11 /technology/twitter-elon-musk-takeover.html.

Chapter 33: The Snap

301 **The group held occasional meetings to give guidance:** O'Reilly, Lara, and Lindsay Rittenhouse. "In Closed-Door Meeting, Elon Musk Tells 100 Top Ad Execs That He Will Improve Brand Safety on Twitter and That He Will Personally Oversee Its New Video Product." *Business Insider*, November 23, 2022. businessinsider .com/elon-musk-meets-to-reassure-twitter -advertisers-on-brand-safety-video-2022-11.

303 **days of rest, the monthly no-meeting, recharge days:** Conger, Kate, and Ryan Mac. "Elon Musk Begins Layoffs at Twitter." *New York Times*, November 4, 2022, sec. Technology. nytimes.com/2022/11/03 /technology/twitter-layoffs-elon-musk.html.

303 **"Has the red wedding started? 👀":** Conger and Mac, "Elon Musk Begins Layoffs."

306 **She included a broken heart emoji and a link:** Conger, Kate, Ryan Mac, and Mike Isaac. "Confusion and Frustration Reign as Elon Musk Cuts Half of Twitter's Staff." *New York Times*, November 4, 2022, sec. Technology. nytimes.com/2022/11/04/technol ogy/elon-musk-twitter-layoffs.html.

Chapter 34: The Aftermath

307 **known internally as Redbird, lost about 80 percent:** Conger, Kate, Mike Isaac, Ryan Mac, and Tiffany Hsu. "Two Weeks of Chaos: Inside Elon Musk's Takeover of Twitter." *New York Times*, November 11, 2022, sec. Technology. nytimes.com/2022/11/11/technology /elon-musk-twitter-takeover.html.

307 **Dozens of product managers were laid off across the company:** Conger, Kate, Ryan Mac, and Mike Isaac. "Confusion and Frustration Reign as Elon Musk Cuts Half of Twitter's Staff." *New York Times*, November 4, 2022, sec. Technology. nytimes.com /2022/11/04/technology/elon-musk-twitter-layoffs .html.

308 **Dorsey's message came after about twenty employees:** Ogbonna, Nkechi. "Twitter Lays Off Staff at Its Only Africa Office in Ghana." BBC News, November 9, 2022, sec. Africa. bbc.com/news/world -africa-63569525.

Chapter 35: Verified or Not

311 **The fake tweets were "derogatory and demeaning":** Matyszczyk, Chris. "Tony La Russa Sues Twitter over Alleged Fake Tweets." CNET, June 4, 2009. cnet.com /culture/tony-la-russa-sues-twitter-over-alleged -fake-tweets.

314 **Balajadia also made some public appearances on behalf of the Boring Company:** Marshall, Aarian. "Elon Musk Now Wants to Dig Another Tunnel under LA." *Wired*, January 23, 2018. wired.com/story/elon -musk-boring-company-culver-city.

Chapter 36: Elections

321 **connections to Big Oil and climate change denial:** Brooks, Emily. "Elon Musk Featured at Kevin McCarthy's GOP Retreat in Wyoming." *The Hill*, August 17, 2022. thehill.com/blogs/blog-briefing-room/news /3605110-elon-musk-featured-at-kevin-mccarthys -gop-retreat-in-wyoming.

325 **In 2021, the company's general and administrative business costs:** "Twtr-20211231." n.d. Securities and Exchange Commission. sec.gov/Archives/edgar/data /1418091/000141809122000029/twtr-20211231.htm.

326 **Musk disclosed that he sold 19.5 million:** Kolodny, Lora. "Elon Musk Sells at Least $3.95 Billion Worth of Tesla Shares after Twitter Deal." CNBC, November 8, 2022. cnbc.com/2022/11/08/elon-musk-sells-at-least -3point95-billion-worth-of-tesla-shares.html.

Chapter 37: Zombie Attack

329 **In October, a former security executive at Uber:** Metz, Cade. "Former Uber Security Chief Found Guilty of Hiding Hack from Authorities." *New York Times*, October 5, 2022, sec. Technology. nytimes .com/2022/10/05/technology/uber-security -chief-joe-sullivan-verdict.html.

330 **news outlets began to take notice of the new Blue:** Peters, Jay. "Elon Musk's Twitter Blue with Verification Is Now Live." *The Verge*, November 9, 2022. theverge.com/2022/11/9/23448317/elon-musk -twitter-blue-verification-live-ios.

332 **Russia's Internet Research Agency used credit cards:** Parks, Miles. "Fact Check: Russian Interference Went Far Beyond 'Facebook Ads' Kushner Described." NPR, April 24, 2019. npr.org/2019/04/24 /716374421/fact-check-russian-interference-went -far-beyond-facebook-ads-kushner-described.

336 **"Elon has shown that he cares only about recouping":** Conger, Kate, Ryan Mac, and Mike Isaac. "'Economic Picture Ahead Is Dire,' Elon Musk Tells Twitter Employees." *New York Times*, November

10, 2022. nytimes.com/2022/11/10/technology/elon-musk-twitter-employees.html.
336 **"There is no way to sugar coat this message":** Wagner, Kurt. "Musk's First Email to Twitter Staff Ends Remote Work." Bloomberg.com, November 10, 2022. bloomberg.com/news/articles/2022-11-10/musk-s-first-email-to-twitter-staff-ends-remote-work.
336 **Musk was known for late-night email blasts at Tesla:** Tabahriti, Sam. "Read the Email Elon Musk Sent to Tesla Employees about Returning to the Office before Saying Headcount Will Increase." *Business Insider*, June 5, 2022. businessinsider.com/read-elon-musk-email-tesla-employees-return-office-2022-6.
338 **the billionaire painted a bleak picture:** Conger et al., "'Economic Picture Ahead.'"
338 **"We just definitely need to bring in more cash":** Heath, Alex. "Inside Elon Musk's First Meeting with Twitter Employees." *The Verge*, November 11, 2022. theverge.com/2022/11/10/23452196/elon-musk-twitter-employee-meeting-q-and-a.
338 **The post, which garnered more than 3,000 retweets:** Harwell, Drew. "The fake Eli Lilly free-insulin tweet has now been online for six hours. 3,000 retweets. The (actual) company responded three hours ago. But, hey, at least there's a 'Community Note' now. And all this for a crisp $8. Enjoy it, @elonmusk!" Twitter, November 11, 2022. twitter.com/drewharwell/status/1590870708824920064.
339 **"It is with heavy heart that I must concede":** McCarthy, Bill. "An impostor account using a paid-for blue verification checkmark to impersonate Kari Lake, the Republican candidate for governor in Arizona, posted a fake concession last night. The post has been up for more than 12 hours." Twitter, November 11, 2022. twitter.com/billdmccarthy/status/1591066507152609281.

Chapter 38: Fired for Shitposting
344 **He flew home that weekend to Austin to brood:** Isaacson, Walter. *Elon Musk*. Simon & Schuster.
345 **flashing a thumbs-up in a selfie with Musk:** Brynaert, Ron. "On October 26, Luke Simon posted this picture with @elonmusk on Twitter, but now he locked his account." Twitter, November 15, 2022. twitter.com/ronbryn/status/1591582973346074624.
346 **After tweeting, Frohnhoefer left home to go to Starbucks:** Newton, Casey. "Elon's Paranoid Purge." Platformer, November 16, 2022. platformer.news/elons-paranoid-purge.
349 **Smith was a tough woman and was known for championing female:** "Kiko Smith: I Am a Wife, a Mother, an Architect and a Proud Member of Our TechWomen Sisterhood." TechWomen, March 8, 2017. techwomen.org/techwomen-delegation/kiko-smith-i-am-a-wife-a-mother-an-architect-and-a-proud-member-of-our-techwomen-sisterhood.
350 **a messaging service with 800 million active users:** Singh, Manish. "Telegram Raises $210 Million through Bond Sales." *TechCrunch*, July 18, 2023. techcrunch.com/2023/07/18/telegram-raises-210-million-through-bond-sales.
351 **"the link provided is not a phishing attempt":** Mac, Ryan, Mike Isaac, and David McCabe. "Resignations Roil Twitter as Elon Musk Tries Persuading Some Workers to Stay." *New York Times*, November 17, 2022. nytimes.com/2022/11/17/technology/twitter-elon-musk-ftc.html.
352 **"I know how to win," he said, alluding:** Ryan et al., "Resignations Roil Twitter."
354 **threatened his "thermonuclear name & shame":** Mac, Ryan, Mike Isaac, and Kate Conger. "Twitter Keeps Missing Its Advertising Targets as Woes Mount." *New York Times*, December 2, 2022, sec. Technology. nytimes.com/2022/12/02/technology/twitter-advertising-targets-missed.html.

355 **the Daily Stormer, who was suspended in 2013:** Ramirez, Nikki McCann. "Elon Brings One of America's Most Prominent Nazis Back to Twitter." *Rolling Stone*, December 2, 2022. rollingstone.com/politics/politics-news/elon-musk-twitter-reinstates-neo-nazi-andrew-anglin-account-1234640390.
355 **"death con 3 on JEWISH PEOPLE":** Li, David K., and Colin Sheeley. "Ye Locked out of Twitter after Backlash for Antisemitic Posts." NBC News, October 10, 2022. nbcnews.com/news/us-news/ye-locked-twitter-violation-platform-policy-rcna51505.
356 **Trump joined Truth Social, the social network he partly owned:** Singman, Brooke. "Trump Joins TRUTH Social: 'I'M BACK! #COVFEFE.'" Fox News, April 28, 2022. foxnews.com/politics/trump-joins-truth-social-im-back-covfefe.

Chapter 39: Zero-Based Budgeting
357 **Twitter's revenue for the last three months of 2022:** Woo, Erin. "Musk's Twitter Saw Revenue Drop 35% in Q4, Sharply below Projections." *The Information*, January 18, 2023. theinformation.com/articles/musks-twitter-saw-revenue-drop-35-in-q4-sharply-below-projections.
357 **balloon to more than $1.5 billion a year:** Mac, Ryan. "Elon Musk Says Twitter's Finances Are Improving after Big Cuts." *New York Times*, March 7, 2023, sec. Business. nytimes.com/2023/03/07/business/elon-musk-twitter-finances.html.
357 **The company was incurring more than $3 million:** Jin, Berber, and Alexander Saeedy. "WSJ News Exclusive: Elon Musk Explores Raising up to $3 Billion to Help Pay Off Twitter Debt." *Wall Street Journal*, January 25, 2023. wsj.com/articles/elon-musk-explores-raising-up-to-3-billion-to-pay-off-twitter-debt-11674669412.
357 **using his Tesla shares to back new margin loans:** Scigliuzzo, Davide, Sonali Basak, and Paula Seligson. "Elon Musk's Bankers Consider Tesla Margin Loans to Cut Risky Twitter Debt." Bloomberg.com, December 8, 2022. bloomberg.com/news/articles/2022-12-08/musk-bankers-mull-tesla-margin-loans-to-cut-risky-twitter-debt.
358 **He moved his family—including the newborn:** Peterson, Becky, and Erin Woo. "Musk May Have Found a Hardcore Leader for Twitter." *The Information*, December 23, 2022. theinformation.com/articles/musk-may-have-found-a-hardcore-leader-for-twitter.
358 **With fifty offices in more than thirty cities around the world:** Wolfram Arnold, Eric Froese, Tracy Hawkins, Joseph Killian, Laura Chan Pytlarz, and Andrew Schlaikjer, Plaintiffs, v. X Corp. f/k/a Twitter, Inc., X Holdings Corp. f/k/a X Holdings I, Inc. and Elon Musk. Filed May 16, 2023. United States District Court for the District of Delaware. dockets.justia.com/docket/delaware/dedce/1:2023cv00528/82425.
359 **"We just won't pay landlords":** Arnold et al. v. X Corp. et al.
361 **Davis and Musk ended relationships with a cleaning company:** Sainato, Michael. "'It's a Nightmare': Twitter's New York City Janitors Protest over Sudden Layoffs." *The Guardian*, January 12, 2023, sec. Technology. theguardian.com/technology/2023/jan/12/twitter-janitors-new-york-city-protest-layoffs-elon-musk.
362 **the stench of the bathrooms overwhelmed:** Hays, Kali. "A Malodorous Musk: Twitter Employees Beg for Toilet Paper and Report a Wafting Stench on Slack as Elon Musk Cuts Back on Office Facilities Staff." *Business Insider*, January 5, 2023. businessinsider.com/elon-musk-twitter-layoffs-employees-beg-toilet-paper-slack-2023-1.
362 **shirts had been created after the Ferguson protests by Blackbirds:** Sullivan, Mark. "What's Jack Dorsey's

NOTES   > 449

#StayWoke T-Shirt Mean?," *Fast Company*, June 1, 2016. fastcompany.com/4009077/whats-jack-dor seys-staywoke-t-shirt-mean.

364 **checks to landlords "over my dead body"**: Arnold et al. v. X Corp. et al.

364 **Twitter skipped out on the $6.8 million owed**: Li, Roland. "Twitter Sued by Landlord at S.F. HQ after Alleged $6.8 Million in Missed Rent Payments." *San Francisco Chronicle*, January 24, 2024. sfchronicle.com/tech/article/twitter-sued-by-landlord-at-s-f-hq-after-alleged-17737102.php.

364 **In London, the Crown Estate, which oversees property**: Rhoden-Paul, Andre. "Twitter Sued by Crown Estate over Alleged Unpaid Rent at UK HQ." BBC.com, January 23, 2023. bbc.com/news/uk-64381582.

364 **In Singapore, workers were walked out by building management**: Newton, Casey. "I'm told Twitter employees were just walked out of its Singapore office—its Asia-Pacific headquarters—over nonpayment of rent. . . ." Twitter, January 11, 2023. twitter.com/CaseyNewton/status/1613303513240702976.

365 **$30,000 Twitter bird statue and a $25,000 espresso**: Ortutay, Barbara. "Twitter Auctions Off Blue Bird Memorabilia, Pricey Furniture." AP News, January 18, 2023. apnews.com/article/elon-musk-twitter-inc-technology-san-francisco-business-204237cbfdc7a2bb11d3457f6de6c19e.

365 **"overtly sexual or pornographic material"**: Apple.com, Developer. "App Review Guidelines." developer.apple.com/app-store/review/guidelines.

366 **Musk showed up at the $5 billion Apple Park**: Isaacson, Walter. *Elon Musk*. New York: Simon & Schuster, 2023.

Chapter 40: "I'm Rich, Bitch!"

372 **During the first month of Musk's ownership, it suspended nearly 300,000 accounts**: Safety (@Safety). "We've been improving our detection and enforcement methods and expanding our partnerships . . ." Twitter, December 9, 2022. twitter.com/safety/status/1601439984292360193.

373 **Soon, the *Daily Mail* picked up on the story**: Gordon, James. "Ex-Twitter Censor Yoel Roth and His Boyfriend Are Forced to FLEE Their $1.1m Home." *Daily Mail* online, December 13, 2022. dailymail.co.uk/news/article-11531441/Ex-Twitter-censor-Yoel-Roth-boyfriend-forced-FLEE-t-1-1m-home-Elon-Musk-shared-thesis.html.

375 **"Ladies and gentlemen, make some noise for the richest"**: Novak, Matt. "Elon Musk Gets Viciously Booed by Stadium Crowd at Dave Chappelle Show." *Gizmodo*, December 12, 2022. gizmodo.com/elon-musk-booed-stadium-crowd-dave-chappelle-sf-boo-1849881192.

375 **"I'm rich, bitch!" Musk screamed**: Novak, Matt. "Elon Musk Gets Booed by the Crowd at Dave Chappelle's San Francisco Show (Part 3 of 4)." Youtube, December 12, 2022. youtube.com/watch?v=u1cl8U0UCMQ.

375 **He blamed the incident on**: Lovelyti. "Elon Musk Was Drowned Out by Boos When He Joined Dave Chappelle Onstage at a Gig in San Francisco . . ." Twitter, December 13, 2022. twitter.com/lovelyti/status/1602681687946809348.

376 **Tesla's stock fell more than 6 percent during trading**: Bobrowsky, Meghan. "Tesla Investors Voice Concern over Elon Musk's Focus on Twitter." *Wall Street Journal*, December 13, 2022. wsj.com/articles/tesla-investors-voice-concern-over-elon-musks-focus-on-twitter-11670948786.

377 **Jenna, who was seen by her father as a neo-Marxist**: *Financial Times* staff. "Elon Musk: 'Aren't You Entertained?'" *Financial Times*, October 7, 2022. ft.com/content/5ef14997-982e-4f03-8548-b5d67202623a.

379 **"There is not going to be any distinction in the future between journalists"**: ElectricTonde. "Elon Musk on Twitter Spaces @Katienotopoulos' Dec 16, 2022 with Banned Journalists." Youtube, December 17, 2022. youtube.com/watch?v=SZgC6wvBsZ4.

Chapter 41: Self-Doubt

380 **In 2010, the first World Cup with a major Twitter presence**: Popkin, Helen. "Twitter down Again? Blame Justin Bieber!" NBC News, June 15, 2010. nbcnews.com/id/wbna37711973.

380 **when the Netherlands national team upset Brazil**: Popkin, Helen. "Did World Cup Fans Crash Twitter . . . Again?" NBC News, July 2, 2010. nbcnews.com/id/wbna38060788.

380 **The 2014 event in Brazil led to some 672 million tweets**: "Insights into the #WorldCup Conversation on Twitter." *Twitter Blog*, July 14, 2014. blog.twitter.com/en_us/a/2014/insights-into-the-worldcup-conversation-on-twitter.

381 **Musk flew more than fifteen hours to get to the game**: elonmusksjet. Instagram, December 18, 2022. instagram.com/p/CmT1fxLLfXj.

381 **snapping photos with everyone from the likes of Turkish strongman**: Bennett, Dalton, Samuel Oakford, Gerrit De Vynck, and Monique Woo. "From Jared Kushner to Salt Bae: Here's Who Elon Musk Was Seen with at the World Cup." *Washington Post*, December 21, 2022. washingtonpost.com/investigations/2022/12/20/elon-musk-spotted-world-cup-final.

381 **"Over recent weeks we've received numerous inbound"**: Hoffman, Liz, and Reed Albergotti. "Elon Musk's Team Is Seeking New Investors for Twitter." Semafor, December 16, 2022. semafor.com/article/12/16/2022/elon-musks-team-is-seeking-new-investors-for-twitter.

381 **The terms were bold, if not preposterous**: Jin, Berber, and Alexander Saeedy. "Elon Musk Explores Raising up to $3 Billion to Help Pay Off Twitter Debt." *Wall Street Journal*, January 25, 2023. wsj.com/articles/elon-musk-explores-raising-up-to-3-billion-to-pay-off-twitter-debt-11674669412.

382 **"I still think Elon is a smart guy"**: Graham, Paul. "Paul Graham Is Leaving Twitter for Now." ycombinator.com, December 18, 2022. news.ycombinator.com/item?id=34041985.

383 **The company's shares set a new two-year low**: Sean, O'Kane. "Musk Lashes Out at Unhappy Investor as Tesla Shares Retreat." Bloomberg.com, December 20, 2022. bloomberg.com/news/articles/2022-12-20/tesla-tsla-share-retreat-prompts-criticism-of-musk-by-longtime-investors.

385 **On December 23, one of Austin's deputies sent out an email**: Schiffer, Zoë. "NEW: Twitter is shutting down its data center in Sacramento and downsizing another in Atlanta by early Jan, likely as a cost saving measure. . . ." Twitter, December 23, 2022. twitter.com/ZoeSchiffer/status/1606408842417512455.

387 **In other cases, people could not view the replies**: Conger, Kate. "Twitter Users Report Widespread Service Interruptions." *New York Times*, December 29, 2022, sec. Technology. nytimes.com/2022/12/28/technology/twitter-outages.html.

387 **the site would be stricken with half a dozen major outages**: Mac, Ryan, Mike Isaac, and Kate Conger. "'Sometimes Things Break': Twitter Outages Are on the Rise." *New York Times*, February 28, 2023, sec. Technology. nytimes.com/2023/02/28/technology/twitter-outages-elon-musk.html.

Chapter 42: Red Pilled

390 **He also gave a tour of the office to Dave Rubin**: Rubin, Dave. "Spent last two days at Twitter in SF talking to engineers, product managers and yes, @elonmusk. . . ." Twitter, January 26, 2023. twitter.com/RubinReport/status/1618667912377810945.

390 **suggested that he was "reasonably popular"**:

Siddiqui, Faiz, and Jeremy B. Merrill. "Elon Musk Reinvents Twitter for the Benefit of a Power User: Himself." *Washington Post*, February 16, 2023. washingtonpost.com/technology/2023/02/16/elon-musk-twitter.
390 **"Twitter is actually an incredible tool"**: Ewing, Jack. "Tesla's Profit Jumped 12% in Fourth Quarter." *New York Times*, January 25, 2023, sec. Business. nytimes.com/2023/01/25/business/tesla-earnings-fourth-quarter-2022.html.
391 **Musk, who was facing various government inquiries**: Musk, Elon. "Will be interesting to see how the Biden administration reacts to this. They may try to weaponize federal agencies against Twitter." Twitter, January 18, 2023. twitter.com/elonmusk/status/1615775765702004737.
391 **His team wrote to Lina Khan, the FTC chairwoman**: McCabe, David, and Kate Conger. "Elon Musk Tried to Meet with F.T.C. Chair about Twitter but Was Rebuffed." *New York Times*, March 30, 2023, sec. Technology. nytimes.com/2023/03/30/technology/elon-musk-ftc-chair.html.
391 **Among them was Marjorie Taylor Greene, the Georgia congresswoman**: Alba, Davey. "Twitter Permanently Suspends Marjorie Taylor Greene's Account." *New York Times*, January 2, 2022, sec. Technology. nytimes.com/2022/01/02/technology/marjorie-taylor-greene-twitter.html.
391 **after McCarthy used his influence to push Twitter's leadership**: Swan, Jonathan, and Catie Edmondson. "How Kevin McCarthy Forged an Ironclad Bond with Marjorie Taylor Greene." *New York Times*, January 23, 2023, sec. U.S. nytimes.com/2023/01/23/us/politics/kevin-mccarthy-marjorie-taylor-greene.html.
392 **organizer of the Stop the Steal election denial movement**: Thalen, Mikael. "Ali Alexander Returns to Twitter, Boasts He Could Have Destroyed the Capitol on Jan. 6 If He Wanted To." *Daily Dot*, January 10, 2023. dailydot.com/debug/ali-alexander-twitter-capitol-riot.
392 **Ron Watkins, the suspected creator behind QAnon**: Sommer, Will. "Alleged QAnon creator Ron Watkins has returned from Twitter, after receiving a lifetime ban in the pre-Elon era." Twitter, January 10, 2023. twitter.com/willsommer/status/1612843340118872064.
392 **"going to war with the Jews"**: Hoff, Gabe. "Fascist Jew hater Nick Fuentes . . . Claimed He Is 'Going to War with the Jews.'" Twitter, January 25, 2023. twitter.com/GabeHoff/status/1618121530311049218.
392 **He was eventually re-banned**: Singh, Kanishka. "Twitter Suspends Account of White Supremacist Nick Fuentes a Day after Restoration." Reuters, January 25, 2023. Accessed February 21, 2024. reuters.com/world/us/twitter-suspends-account-white-supremacist-nick-fuentes-day-after-restoration-2023-01-25.
392 **antisemitic speech was up more than 60 percent**: Frenkel, Sheera, and Kate Conger. "Hate Speech's Rise on Twitter Is Unprecedented, Researchers Find." *New York Times*, December 2, 2022, sec. Technology. nytimes.com/2022/12/02/technology/twitter-hate-speech.html.
392 **Accounts showing terrorist affiliations with groups**: Ayad, Moustafa. "Islamic State Supporters on Twitter: How Is 'New' Twitter Handling an Old Problem?" GNET, November 18, 2022. gnet-research.org/2022/11/18/islamic-state-supporters-on-twitter-how-is-new-twitter-handling-an-old-problem.
393 **Twitter's recommendation algorithm also surfaced and suggested**: Keller, Michael H., and Kate Conger. "Musk Pledged to Cleanse Twitter of Child Abuse Content. It's Been Rough Going." *New York Times*, February 6, 2023, sec. Technology. nytimes.com/2023/02/06/technology/twitter-child-sex-abuse.html.
393 **By early February, the agency had sent twelve letters**: Tracy, Ryan. "FTC Twitter Investigation Sought Elon Musk's Internal Communications, Journalist Names." *Wall Street Journal*, March 7, 2023, sec. Tech. wsj.com/articles/twitter-investigation-ftc-musk-documents-db6b179e.
393 **Kissner, Twitter's old chief information security officer**: Conger, Kate, Ryan Mac, and David McCabe. "F.T.C. Intensifies Investigation of Twitter's Privacy Practices." *New York Times*, March 8, 2023, sec. Technology. nytimes.com/2023/03/07/technology/ftc-twitter-investigation-privacy.html.
394 **"Something fundamental is wrong"**: Musk, Elon. "Something fundamental is wrong." Twitter, February 1, 2023. twitter.com/elonmusk/status/1620870276279042048.
396 **he had tweeted, "Go @Eagles!!!"**: Dator, James. "Why Did Elon Musk Delete His 'Go Eagles' Tweet Right after Philly Lost the Super Bowl?" SBNation, February 14, 2023. sbnation.com/2023/2/14/23599418/elon-musk-delete-go-eagles-tweet-super-bowl.
397 **Musk running point on the "high urgency" issue**: Newton, Casey, and Zoë Schiffer. "Yes, Elon Musk Created a Special System for Showing You All His Tweets First." Platformer, February 15, 2023. platformer.news/yes-elon-musk-created-a-special-system.
398 **Twitter was late on paying its Slack bill**: Woo, Erin. "Twitter Lays Off at Least 50 in Relentless Cost Cuts." *The Information*, February 25, 2023. theinformation.com/articles/twitter-lays-off-at-least-50-in-relentless-cost-cuts.

Chapter 43: Endorsements

400 **the European Union, which was set to roll out**: *Financial Times* staff. "EU Tells Elon Musk to Hire More Staff to Moderate Twitter." *Financial Times*, March 7, 2023. ft.com/content/20141fb1-d8f7-4c9e-a0d0-ded1ac8c7947.
400 **Twitter was fined 0.1 percent of its gross income**: "Turkish Competition Board Says Fines Elon Musk over Twitter Takeover." Reuters, March 6, 2023. Accessed February 21, 2024. reuters.com/technology/turkish-competition-board-says-fines-elon-musk-over-twitter-takeover-2023-03-06.
400 **the Turkish government blocked access to the platform**: Satariano, Adam. "Twitter Was Blocked in Turkey, Internet-Monitoring Group Says." *New York Times*, February 8, 2023, sec. World. nytimes.com/2023/02/08/world/europe/turkey-earthquake-twitter-blocked.html.
400 **Musk, whose SpaceX was pushing to launch its Starlink**: Fernholz, Tim. "How Turkey Is Using Starlink to Win a Tesla Factory." Quartz, September 18, 2023. qz.com/turkey-erdogan-elon-musk-starlink-spacex-tesla-1850849958.
400 **Twitter had faced immense pressure from Narendra**: Lyons, Kim. "Police in India Raid Twitter Offices in Probe of Tweets with 'Manipulated Media' Label." *The Verge*, May 24, 2021. theverge.com/2021/5/24/22451271/police-india-raid-twitter-tweets-government-manipulated-media.
401 **Musk, however, caved, blocking more than 120 accounts**: Sakunia, Samriddhi. "Twitter Blocked 122 Accounts in India at the Government's Request." Rest of World, March 24, 2023. restofworld.org/2023/twitter-blocked-access-punjab-amritpal-singh-sandhu.
401 **"The reason I did the Twitter acquisition was not because"**: Alfar, Gail. "Elon Musk's Talk at Morgan Stanley TMT 2023 on Twitter, X.com, Tesla and Starship." What's up Tesla, March 13, 2023. whatsuptesla.com/2023/03/12/x-4.
402 **Three weeks later, Twitter's employees received an email**: Conger, Kate, and Ryan Mac. "Elon Musk Values Twitter at $20 Billion." *New York Times*, March 26, 2023, sec. Technology. nytimes.com/2023/03/26/technology/elon-musk-twitter-value.html.

NOTES > 451

402 *The Information*, a tech publication, found that less than 300,000: Woo, Erin. "Musk's Twitter Has Just 180,000 U.S. Subscribers, Two Months after Launch." *The Information*, February 6, 2023. theinformation.com/articles/musks-twitter-has-just-180-000-u-s-subscribers-two-months-after-launch.

402 vowing to eventually take away the verified check marks: O'Kane, Caitlin. "Twitter Is Officially Ending Its Old Verification Process on April 1. To Get a Blue Check Mark, You'll Have to Pay." cbsnews.com, March 24, 2023. cbsnews.com/news/twitter-blue-check-verification-ending-new-subscription-april-1-elon-musk.

404 A few days later, he did away with all labels for NPR: Yang, Mary. "Twitter Removes All Labels about Government Ties from NPR and Other Outlets." NPR, April 21, 2023. npr.org/2023/04/21/1171236695/twitter-strips-state-affiliated-government-funded-labels-from-npr-rt-china.

405 noting he had reduced the company to about 20 percent: Oshin, Olafimihan. "Elon Musk Claims the US Government Had 'Full Access' to Private Twitter DMs." *The Hill*, April 16, 2023. thehill.com/homenews/3953995-elon-musk-claims-the-us-government-had-full-access-to-private-twitter-dms.

405 Dominion Voting Systems, a company that said it had been defamed: Peters, Jeremy W., Katie Robertson, and Michael M. Grynbaum. "Tucker Carlson, a Source of Repeated Controversies, Is Out at Fox News." *New York Times*, April 24, 2023, sec. Business. nytimes.com/2023/04/24/business/media/tucker-carlson-fox-news-dismissal.html.

405 While Dominion and Fox settled for more than $780 million: Peters, Jeremy W., and Katie Robertson. "Fox Will Pay $787.5 Million to Settle Defamation Suit." *New York Times*, April 18, 2023, sec. Business. nytimes.com/2023/04/18/business/media/fox-dominion-defamation-settle.html.

405 On Monday, May 8, Carlson met with Lachlan Murdoch: Peters, Jeremy W., and Benjamin Mullin. "Carlson, Still under Contract at Fox, Says He Will Start New Show on Twitter." *New York Times*, May 9, 2023, sec. Business. nytimes.com/2023/05/09/business/media/tucker-carlson-twitter-show.html.

405 His contract, which ran until 2025, ensured: Sorkin, Andrew Ross, Ravi Mattu, Sarah Kessler, Michael J. De La Merced, Lauren Hirsch, and Ephrat Livni. "The Calculus behind Firing Tucker Carlson." *New York Times*, April 25, 2023. nytimes.com/2023/04/25/business/dealbook/the-calculus-behind-firing-tucker-carlson.html.

406 "So many conspiracy theories have turned out to be true": Calia, Mike. "Elon Musk: 'I'll Say What I Want, and If the Consequence of That Is Losing Money, So Be It.'" CNBC, May 16, 2023. cnbc.com/2023/05/16/elon-musk-defends-inflammatory-tweets-ill-say-what-i-want.html.

407 So when Yaccarino was passed over for the top job: Hsu, Tiffany, Sapna Maheshwari, Benjamin Mullin, and Ryan Mac. "Elon Musk Appoints Linda Yaccarino Twitter's New Chief." *New York Times*, May 12, 2023, sec. Technology. nytimes.com/2023/05/12/technology/yaccarino-twitter-ceo-musk.html.

407 The billionaire showed up unshaven and spoke off-the-cuff: "Linda Yaccarino Interviews Elon Musk— April 18, 2023." youtube.com, April 20, 2023. youtube.com/watch?v=ypZNWjPpOuI.

408 At the time, she was preparing for the Upfronts: Mullin, Benjamin. "NBCUniversal's New Leader, a Hollywood Outsider, Steps into the Spotlight." *New York Times*, May 22, 2023, sec. Business. nytimes.com/2023/05/22/business/media/mike-cavanagh-nbcuniversal.html.

408 preventing her from doing ad deals for Twitter: Mac, Ryan, Tiffany Hsu, and Benjamin Mullin. "Twitter's New Chief Eases into the Hot Seat." *New York Times*, June 29, 2023, sec. Technology. nytimes.com/2023/06/29/technology/twitter-ceo-linda-yaccarino.html.

408 "What Is a Woman?" The promotion, which cost: Boreing, Jeremy. "Twitter canceled a deal with @realdailywire to premiere . . ." Twitter, June 1, 2023. twitter.com/JeremyDBoreing/status/1664255321630552065.

409 While the Daily Wire would go on to post its video: Binder, Matt. "Right-Wing Musk Fans Win Twitter CEO's Loyalty. Staff Loses." *Mashable*, June 2, 2023. mashable.com/article/elon-musk-the-daily-wire-twitter-resignations-explained.

409 Musk's audible had alienated them all: Mac, Ryan, and Tiffany Hsu. "Twitter's U.S. Ad Sales Plunge 59% as Woes Continue." *New York Times*, June 5, 2023, sec. Technology. nytimes.com/2023/06/05/technology/twitter-ad-sales-musk.html.

Chapter 44: Linda

413 getting a warning message that claimed they had exceeded the rate limit: Medina, Eduardo, and Ryan Mac. "Musk Says Twitter Is Limiting Number of Posts Users Can Read." *New York Times*, July 1, 2023, sec. Business. nytimes.com/2023/07/01/business/twitter-rate-limit-elon-musk.html.

414 a battle against a sumo wrestler at his forty-second birthday party in 2013: Zilber, Ariel. "Elon Musk Blew Out Disc in Neck after Fighting 350-Pound Sumo Wrestler." *New York Post*, September 14, 2023. nypost.com/2023/09/14/elon-musk-blew-out-disc-in-neck-after-fighting-350-pound-sumo-wrestler.

415 someone tried to remove a logo from the security desk: Mac, Ryan. "Another bird update: Some bird logos are not coming off . . ." Twitter, July 24, 2023. twitter.com/RMac18/status/1683599649318535168.

415 "eXposure" and "eXult": Mac, Ryan, and Tiffany Hsu. "From Twitter to X: Elon Musk Begins Erasing an Iconic Internet Brand." *New York Times*, July 24, 2023, sec. Technology. nytimes.com/2023/07/24/technology/twitter-x-elon-musk.html.

416 "Wanna do a practice bout at your house": Musk, Elon. "This is the full message: If you still want to do a real MMA fight, . . ." Twitter, August 13, 2023. twitter.com/elonmusk/status/1690747345674137600?s=20/Pic.Twitter.Com/Uzbkoikfoc.

419 The hashtag #BanTheADL began trending: Graziosi, Graig. "White Nationalist Nick Fuentes Appears to Admit He Skirted Twitter Ban." *The Independent*, September 12, 2023. independent.co.uk/news/world/americas/nick-fuentes-white-nationalist-twitter-ban-b2410119.html.

Chapter 45: Tell It to Earth

423 At 7:03 in the morning on Saturday, November 18: Chang, Kenneth. "SpaceX Starship Launch: Highlights from the 2nd Flight of Elon Musk's Moon and Mars Rocket." *New York Times*, November 19, 2023. nytimes.com/live/2023/11/18/science/spacex-starship-launch-elon-musk.

423 Musk stood with his palms pressed together as if in prayer: Musk, Kimbal. "Focused." Twitter, November 18, 2023. https://twitter.com/kimbal/status/1725890951804276896; "Congratulations bro. I can't think of a more deserving person ever. . . ." Twitter, December 13, 2021. Accessed February 21, 2024. twitter.com/kimbal/status/1470415129799905280.

423 The SpaceX chief watched as massive plumes of white smoke: Musk, Elon. "Magnificent Machine with a 1000 ft plume. Pic.Twitter.Com/Wsyxjqjr3v." Twitter, November 18, 2023. twitter.com/elonmusk/status/1725926972423852296.

423 With the countdown finished, the 397-foot rocket: Chang, "SpaceX Starship Launch."

425 "This is the richest man in the world": Kaplan, Alex.

"Far-Right Figures and White Nationalists Celebrate Elon Musk's Antisemitic Post: 'What We Were Saying in Charlottesville.'" Media Matters for America, November 16, 2023. mediamatters.org/elon-musk/far-right-figures-and-white-nationalists-celebrate-elon-musks-antisemitic-post-what-we.

425 **The following day, Media Matters, the progressive media watchdog**: Hananoki, Eric. "As Musk Endorses Antisemitic Conspiracy Theory, X Has Been Placing Ads for Apple, Bravo, IBM, Oracle, and Xfinity next to Pro-Nazi Content." Media Matters for America, November 16, 2023. mediamatters.org/twitter/musk-endorses-antisemitic-conspiracy-theory-x-has-been-placing-ads-apple-bravo-ibm-oracle.

425 **By Friday, Apple, IBM, Disney, and more than a hundred**: Mac, Ryan, Brooks Barnes, and Tiffany Hsu. "Advertisers Flee X as Outcry over Musk's Endorsement of Antisemitic Post Grows." *New York Times*, November 17, 2023, sec. Technology. nytimes.com/2023/11/17/technology/elon-musk-twitter-x-advertisers.html.

425 **It would lead to more than $75 million**: Mac, Ryan, and Kate Conger. "X May Lose up to $75 Million in Revenue as More Advertisers Pull Out." *New York Times*, November 24, 2023, sec. Business. nytimes.com/2023/11/24/business/x-elon-musk-advertisers.html.

426 **Texas's attorney general, a Musk ally, also announced an investigation**: Conger, Kate, and Ryan Mac. "X Sues Media Matters over Research on Ads Next to Antisemitic Posts." *New York Times*, November 21, 2023. nytimes.com/2023/11/20/technology/x-sues-media-matters-antisemitic-posts.html.

426 **In a bind, Musk flew to Israel on Monday, November 27**: Toh, Michelle, Lauren Izso, and Alex Stambaugh. "Elon Musk Visits Israel's Destroyed Kibbutz and Meets Netanyahu in Wake of Antisemitic Post." CNN Business, November 27, 2023. cnn.com/2023/11/27/tech/elon-musk-isaac-herzog-israel-meeting-intl-hnk/index.html.

426 **He landed in Austin in the early hours of Tuesday morning**: "ElonJet (@Elonjet.net). "Landed in Austin, Texas, US." 2023. Bluesky Social, November 28, 2023. bsky.app/profile/elonjet.net/post/3kfaifcseto2c.

426 **suggesting that the Pizzagate conspiracy theory "is real."**: Sardarizadeh, Shayan. "Elon Musk has now deleted the tweet that promoted the pizzagate conspiracy theory. Pic.Twitter.Com/Jfwcny7rlr." Twitter, November 28, 2023. twitter.com/Shayan86/status/1729595410220499250.

Image Credits

p. 1, top left: ZUMA Press, Inc. / Alamy Stock Photo

p. 1, top right: Pauline Lubens/MediaNews Group/The Mercury News via Getty Images

p. 2, top: Sipa USA / Alamy Stock Photo

p. 2, middle: Francois Durand / Getty Images for Twitter

p. 2, bottom: UPI / Alamy Stock Photo

p. 3, top: David Paul Morris/Bloomberg via Getty Images

p. 4, top: Kevin Dietsch/Getty Images

p. 4, middle: Abaca Press / Alamy Stock Photo

p. 4, bottom left: David Paul Morris/Bloomberg via Getty Images

p. 4, bottom right: ZUMA Press, Inc. / Alamy Stock Photo

p. 5, top: ZUMA Press, Inc. / Alamy Stock Photo

p. 5, middle left: Patrick T. Fallon/Bloomberg via Getty Images

p. 5, middle right: Amy Osborne/AFP via Getty Images

p. 5, bottom left: David Paul Morris/Bloomberg via Getty Images

p. 7, top: Taylor Hill/Getty Images

p. 7, bottom: DPPI Media / Alamy Stock Photo

p. 8, bottom: Tayfun Coskun/Anadolu Agency via Getty Images

Index

ABC, 156
Abramson, Nelson, 385
Accra, 308-9
acquisitions, 124, 139, 155-56, 160, 219
    hostile takeovers, 122, 155-56, 161
    nondisclosures in, 161
advertising, 105, 325, 388
    on Twitter, 2, 17, 50, 76, 79, 87, 88, 91, 93, 97, 137, 153, 185, 187, 192, 199, 217
        Musk's takeover and, 249-51
        revenue from, 217, 281, 291, 292
    on Twitter under Musk, 271, 281, 291-92, 297, 301, 313, 320, 345, 350, 364, 406-8
    for Apple, 365, 366, 425
        content moderation and, 275, 282, 288, 289-91, 302, 320, 338, 339, 357, 365, 366, 418, 419, 425-29
        imposter accounts and, 338, 339
        meetings with advertisers, 289-92, 294, 301-2
        Musk's political views and, 406, 423-29
        revenue from, 294, 320, 325, 330, 339, 340, 354, 357, 365, 380, 381, 388, 401, 403, 409, 417, 419
        for SpaceX, 340
        Yaccarino and, 417-18
    Ukraine war and, 97, 152, 325
    Upfronts presentation and, 408
Africa, 35-36, 44, 45, 79
Afshar, Omead, 115, 134
Agarwala, Vineeta, 85, 95, 212
Agrawal, Parag, 46, 55, 77-78, 81-87, 89-93, 94-98, 105, 149, 163, 164, 175, 185, 193-94, 212, 219-20, 258-59, 261, 263, 371
    appointed CEO of Twitter, 81-84, 91-92, 94, 95, 106-8, 113, 137, 163, 167, 168, 185, 193, 212
    Beykpour and Falck fired by, 186-88, 191, 209
    Bluesky project of, 87-88, 90-91, 220-21, 433, 435
    bots and spam accounts and, 182-83, 190-92, 225
    in coffee meet-and-greets, 219-20
    content moderation and, 91-93
        Saturn project, 93, 146-48, 172-75, 206, 208, 220-22, 229, 245, 276
    cost-cutting plans of, 97-98, 101, 142-43, 148-51, 153, 167
    Dorsey and, 78, 81-82, 86, 87, 91, 106, 150, 167, 168-69, 193, 212
    education of, 84-85
    firing of, 263-66, 343-44
    Gadde and, 169-72, 191
    Musk and, 85-86, 106-9, 111-12, 114, 116-21, 123-27, 135, 157, 163, 186, 191-92, 193, 212, 215, 251, 263-66, 276, 295, 343-44
        Twitter acquisition and, 128, 129, 138, 141-42, 146-52, 158, 163, 164, 167, 169, 171, 174, 180, 185-88, 195-96, 203, 204, 207, 209, 211-12, 221, 222, 225, 229, 234-35, 239, 250-51, 258-59
    paternity leave taken by, 95-97, 113, 186
    post-Twitter life of, 435
    Zatko and, 224, 225
Alemayehou, Mimi, 78
Alexander, Ali, 44, 391-92
alien life, 199
Allen & Company, 212
Allyn, Bobby, 404
Al-Mahmoud, Mansoor bin Ebrahim, 381
Alphabet, 52
Altman, Sam, 435
Al Waleed bin Talal Al Saud, 233-34
Amazon, 86, 87, 158, 356
    Web Services, 86, 371
American Express, 44, 294
American Girl, 334
American Jewish Committee (AJC), 418
America Online, 139
Andreessen, Marc, 176, 383
Andreessen Horowitz, 176, 179, 212, 241, 284, 382, 383
Android, 280, 318, 346
Anglin, Andrew, 355
Annecy stabbing attack, 410-11

Anti-Defamation League (ADL), 294, 392, 406, 418–19, 421, 424–25, 427
antisemitism, 392, 406, 418, 419, 424
  Musk and, 424–25, 427–29
antitrust law, 139
AOL, 13
Apple, 11, 16, 38, 60, 96, 156, 158, 179
  advertising for, 365, 366, 425
  App store, 317–19, 333, 365–66
  iPhone, 16, 314, 365
  iTunes, 14
Apple Park, 366
Arab Spring, 16
Arizona, 338–39
Armstrong, Brian, 179
artificial intelligence, 35, 92, 100, 333, 395, 435
  deepfakes, 61–62, 64
  OpenAI, 205, 412, 435
  xAI, 435
Associated Press, 32
AT&T, 51
Austin, Sheen, 248–50, 256, 267–68, 273, 274, 350, 384–88

Babylon Bee, 104, 115, 245, 274–76, 340, 354, 375
Baker, Jim, 368–69
Balajadia, Jehn, 239, 241, 279–80, 314–15
banking, 29–31
Bankman-Fried, Sam, 105, 136, 175–76, 179
Bassett, Natasha, 117, 124, 194, 204
BBC, 404
Beckham, David, 381
Beer Hall, 306
Benioff, Marc, 108
Berenson, Alex, 69
Berghain, 109
Berland, Leslie, 44–45, 60, 80, 196–99, 239–43, 280, 289, 295, 298
Beykpour, Kayvon, 46, 55, 72, 76–77, 83, 84, 150, 167
  firing of, 186–88, 191, 209
Bezos, Jeff, 124
Biden, Hunter, 67–68, 170, 367
Biden, Jill, 396
Biden, Joe, 65, 67, 68, 73, 104, 139, 368, 391, 393, 396, 321, 434
Bieber, Justin, 118, 285, 290
Biles, Simone, 46
Binance, 179
Binder, Matt, 379
Birchall, Jared, 27, 28, 37, 102, 106, 110, 111, 121, 124–26, 134–36, 155, 156, 175, 178, 179, 191, 241, 260, 261, 298–300, 341–42, 352, 357, 359, 369, 376, 381
birth rates, falling, 205, 213, 214, 389
Bitcoin, 44, 57, 78, 79, 87, 177, 179, 233, 308, 358, 433
Bjelde, Brian, 299

Black, Julia, 205
Blackbirds, 362
Black Lives Matter, 19, 21, 362
BlackRock, 141, 149
Blind, 348
Block, 233
blockchain technology, 87, 90, 119, 134, 179
Blogger, 13, 15
BloodPop (Michael Tucker), 290, 291
Bloomberg, Michael, 177
Bloomberg Businessweek, 24
Bluesky, 87–88, 90–91, 106, 174, 220–21, 413, 433, 435
Boca Chica Village, TX, 423
Bolsonaro, Jair, 317–18, 390
Boorstin, Julia, 420–22
Boreing, Jeremy, 409
Boring Company, 23, 37, 134, 179, 239, 241, 267, 268, 314
Boucher, Claire Elise (Grimes), 23, 95, 100, 124, 204–6, 377
Box, 382
Brand, Dalana, 80, 271, 272
Brazil, 257, 275, 283, 307, 317, 390
Breton, Thierry, 400, 417
Britton, Sam, 157, 159–60
Brown, A. J., 409
Brown, Michael, Jr., 18–19, 183
Buhari, Muhammadu, 75
Burger King, 290
Burning Man, 335
Business Insider, 200, 205, 285
BuzzFeed, 25, 285, 379

Calacanis, Jason, 134–35, 143, 177–79, 241, 279, 285, 290, 291, 296–98, 313, 317, 335, 378
Caldwell, Nick, 271, 344
Capital Cities, 156
Cardaci, Chris, 374
Carlson, Tucker, 404–6, 424
Catturd2, 390
censorship, *see* content moderation; freedom of speech
Center for Countering Digital Hate, 392
Cernovich, Mike, 104
Chappelle, Dave, 375
Charles Schwab, 284
Chase Center, 171–72, 344, 375
Chen, Jon, 260–63
child sexual exploitation, 16, 20, 92, 146, 372, 392–93
China, 198–99, 244
Christchurch mosque shootings, 42
Christie, Jen, 80
CIA, 27, 173, 183
Clinton, Bill, 223
Clinton, Hillary, 67, 193, 287, 288, 315
CNBC, 406, 408, 420
CNET, 30
CNN, 23, 225, 378

INDEX

Coca-Cola, 294, 311, 363
CODE Conference, 420-22
Cohen, Lara, 309
Cohn, Jesse, 49-57, 76-78, 110
Coinbase, 179
Coleman, Keith, 173-74
Combs, Sean "Diddy," 302
Comer, James, 391
Compaq, 29, 156
Confinity, 30
Conger, Kate, 6
Congress, U.S., 57, 58, 59-60, 62, 88, 147, 225, 264
conspiracy theories, 2, 4, 25, 44, 60, 61, 93, 287, 406, 432
   Great Replacement Theory, 424
   Pizzagate, 25, 104, 390, 426
   QAnon, 25, 391, 392, 433
   about September 11 attacks, 433
Constitution, U.S., 275, 282, 363
content moderation, 4-6, 13, 15, 146, 206, 435
   activist groups and, 294, 320, 392
   Digital Services Act and, 244
   at Facebook, 92
   at Twitter, *see* Twitter, content moderation at; Twitter under Elon Musk, content moderation at
   see also misinformation and disinformation
Conway, Kellyanne, 65-66
Cook, Tim, 366, 420
Costolo, Dick, 11, 17, 20-21, 42, 149, 170
COVID-19, 53-57, 59, 69-70, 113, 148, 173, 174, 180, 218, 325, 326, 414
   misinformation about, 61-64, 68-69, 91, 110, 114, 222, 391
   Musk and, 62-64, 99, 114, 128, 136, 375
   Tesla and, 63, 96, 99
   Twitter and, 53, 56-57, 61-64, 68-69, 77, 83, 90, 91, 97, 98, 185, 186, 255
   vaccines for, 68-69, 391
Craft Ventures, 177
Crawford, Esther, 79, 240, 241, 243, 284-86, 297-98, 306, 310-12, 314-19, 323-24, 328-29, 331, 332, 335, 388, 398-99, 402
   firing of, 398-99
   sleeping photo of, 316
credit card companies, 29-30
crime, 406, 432
Crown Estate, 364
Cruz, Ted, 334, 392
cryptocurrency, 79, 105, 106, 136, 179, 182, 253
   Bitcoin, 44, 57, 78, 79, 87, 177, 179, 233, 308, 358, 433
   Dogecoin, 141, 144, 179
Culver, Leah, 273-74

Daily Mail, 373
Daily Stormer, 355
Daily Wire, 408-9
DARPA, 223, 224

data centers, 87, 349, 385-88, 394
Davis, Dantley, 84, 94
Davis, Steve, 134, 241, 260, 261, 267, 271, 279, 283, 298, 324-25, 344, 358-61, 364, 371, 386, 389, 395, 398, 411-12
DealBook Conference, 426-30
deepfakes, 61-62, 64
Delaware, 210, 351, 434
   Court of Chancery, 210, 215-18, 232
   Division of Corporations, 258
Deloitte, 236
Democrats, 43, 68, 100, 321, 393, 419, 432
Denholm, Robyn, 39
Derella, Matt, 72
DeSantis, Ron, 392
Deutch, Ted, 418
Digital Services Act, 244, 400
disinformation, *see* misinformation and disinformation
Disney, 139, 156, 357, 364, 425, 427-29
diversity, equity, and inclusion, 99-100, 197
Dogecoin, 141, 144, 179
Dominion Voting Systems, 405
Döpfner, Mathias, 104, 115
Dorsey, Jack, 6, 11-17, 22, 84, 102, 114, 150, 153, 163-64, 172, 185, 206, 211, 371, 433
   Africa trip of, 44, 45
   Agrawal and, 78, 81-82, 86, 87, 91, 106, 150, 167, 168-69, 193, 212
   Analyst Day goals and, 76-77
   Berland and, 44-45, 196, 239
   Bitcoin and, 44, 57, 78, 79, 87, 177, 233, 308, 433
   Block company of, 233
   Bluesky project of, 87-88, 90-91, 106, 174, 221, 433, 435
   Cohn and Elliott Management and, 49-57, 77, 78, 94-95, 106, 120, 150
   congressional testimonies of, 57, 58, 59-60
   content moderation and, 15, 43, 57, 59-60, 62, 64-75, 91-92
   child sexual exploitation material and, 372
   early life of, 12
   Ferguson protests and, 18-19, 183, 362
   in Hawaii, 57-58
   health regimen of, 22, 86
   LiveJournal posts of, 13
   Musk and, 50, 53, 78-79, 105-6, 110, 112, 113, 123, 163
      Twitter acquisition and, 123, 142, 143, 163-67, 168-69, 232-33, 308, 432-33
      Twitter employee layoffs and, 308, 309
   NASA and, 40-41
   Nostr and, 382, 433
   at Odeo, 13, 14
   at OneTeam events, 41-42, 44-48, 55, 86
   physical appearance and personal style of, 11-14, 17, 19, 21, 57, 59, 77
   politics and, 43, 321
   in Polynesia, 69-70, 72

Dorsey, Jack (cont.)
  remote work policies and, 53, 57, 336
  at Square, 16, 22, 40, 42, 44, 49, 51, 52, 55, 233
  Trump ban and, 70–75, 245
  as Twitter CEO, 5, 11, 12, 14, 15, 17, 21, 40–42,
    49–58, 60, 77–78, 81, 84, 129, 138, 140, 142,
    149, 165, 166, 177, 194, 308, 432, 433
    resignation as, 79, 80–83, 91, 105, 106, 108,
      163–64, 193
  Twitter created by, 13–14, 16, 194
  Twitter offices toured by, 44
  Twitter's board and, 15–17, 49–51, 53, 73, 77, 78,
    81–82, 105, 112, 142, 164, 194, 433
  on Twitter's rebranding as X, 416
  Zatko and, 223
Dorsey, Marcia, 21, 45
dot-com bubble, 30
DoubleClick, 81
Dow Chemical, 139
Dowell, Christian, 332, 393
D'Souza, Dinesh, 133
DuPont, 139
Durban, Egon, 54–56, 95, 106–7, 127, 138–39,
  143–45, 151, 163, 194–95, 209, 228

eBay, 31, 51
Eberhard, Martin, 31, 32
economy, 152, 158, 188–89, 204, 336,
  364, 388
  financial crisis of 2008 in, 160, 325
  global, 324, 325, 364
  inflation in, 314
Edgett, Sean, 66, 73, 80, 210, 226, 231, 247, 248,
  257, 368
  firing of, 263–65
Edwards, Jon, 201
elections, 317, 342
  in Brazil, 257, 275, 283, 307, 317, 390
  in U.S.
    midterm, 275, 283, 302, 318, 319, 321, 324,
      325, 327, 328, 338–39
    of 2016, 21, 22, 50, 61, 67, 193, 332
    of 2020, 52, 64–65, 67, 68, 70–73, 133, 390,
      405, 431
    of 2024, 355, 431, 433
Eli Lilly, 338
Elkann, John, 177
Elliott Management, 49–58, 77, 78, 83, 90, 91,
  94–95, 106, 111, 120, 122, 123, 139,
  150, 186
Ellison, Larry, 104, 117, 176, 179
Emanuel, Ari, 194, 228, 325, 414
Endeavor, 194
Enjeti, Saagar, 170
Enron Energy Services, 349
Erdoğan, Recep Tayyip, 381, 400
Ethereum, 179
European Union, 103, 244, 330, 400
Excession LLC, 27
extraterrestrial life, 199

Faber, David, 406
Facebook, 1, 16, 17, 50, 60, 72, 92, 108, 177, 187,
  197, 221, 255, 262, 271, 292, 311, 332, 356,
  388, 435
  artificial intelligence and, 35
  IPO of, 155
  revenue of, 21
  satellite launch, 35–36
  Twitter accounts promoting, 382–83
  user base of, 21
  WhatsApp acquired by, 139
Falck, Bruce, 46, 55, 76–77, 84, 150, 167
  firing of, 186–87, 191, 209
Falk, Rebecca, 372–73
Federal Bureau of Investigation (FBI), 67, 315, 318,
  319, 369
Federal Reserve, 152
Federal Trade Commission (FTC), 139, 224–26, 228
  and Twitter under Musk, 244, 257, 274, 323,
    329–30, 336, 337, 340, 342, 367, 374, 387,
    391, 393–94, 414
Ferguson protests, 18–19, 183, 362
Fernandez, Carrie, 256
Fidelity, 434
FIFA World Cup, 179, 350, 380–82
financial crisis of 2008, 160, 325
First Amendment, 393
Floyd, George, 99, 325, 362
Fogarty, Marianne, 264–65, 329–30, 337, 340
Forbes, 34, 139
Fortune 500 companies, 357
For Whom the Bell Tolls (Hemingway), 212
Foster, Norman, 366
Founders Fund, 177
4chan, 287
Fox Corporation, 405
Fox News, 404–5
freedom of speech, 92, 147, 206, 352, 393
  Bluesky and, 90
  freedom of reach versus, 90, 147, 276, 302, 354,
    392, 407, 409, 436
  India and, 208, 209, 244, 400–401
  Musk's commitment to, 100, 101, 103, 201, 234,
    244–45, 250, 275, 276, 282, 294, 320, 354,
    363, 372, 378, 379, 382, 392, 400–401, 409,
    418, 435–36
  see also content moderation
Friedman, Milton, 214
Frohnhoefer, Eric, 346–48, 350
FTX, 105, 136, 175, 176, 179
Fuentes, Nick, 392, 419, 425

Gabbard, Tulsi, 321
Gadde, Vijaya, 19–20, 43, 60–61, 64, 66–69, 71–75,
  79, 80, 83, 141, 146–49, 160, 164, 168–73,
  195, 208–10, 221, 222, 229, 230, 235, 242,
  247, 263, 274, 277, 278, 296, 368, 369,
  373–74, 391, 393, 401, 435
  Agrawal and, 169–72, 191
  firing of, 263–66, 343–44

Musk and, 169–72, 191, 193, 243–46, 257–58, 263–66, 343–44
Game of Thrones, 303
Gamergate, 18, 19, 22
Gates, Bill, 139
Genentech, 160
General Data Protection Regulation, 367
General Motors, 290, 301
Ghana, 308–9
Ginsburg, Ruth Bader, 211
Glass, Noah, 14
Gökçe, Nusret, 381
golden parachutes, 229, 235, 263, 264, 343
Golden State Warriors, 108, 171, 172, 344, 375
Goldman Sachs, 54, 55, 76, 122, 123, 141, 149, 152, 154, 157–59, 162, 179, 203, 207, 258
Google, 13, 49, 51, 52, 81, 86, 87, 108, 115, 139, 156, 158, 173, 197, 255, 280, 292, 395, 435
    Android, 280, 318, 346
    Cloud, 52, 86, 371
    Google+, 311
Gorman, James, 143, 144
government surveillance and overreach, 173, 199, 208, 244
Graber, Jay, 89–91, 220–21, 413, 433, 435
Gracias, Antonio, 103, 177, 207–8, 236, 241, 252–54, 260, 261, 279, 281–82, 286, 294, 299, 341, 434
Graham, Paul, 382
Grant, Jonah, 331, 333, 335
Great Recession of 2008, 160, 325
Great Replacement Theory, 424
Greenblatt, Jonathan, 294, 418–19, 427
Greene, Marjorie Taylor, 287, 391
Griffin, Kathy, 320
Griffin, Ken, 115
Grimes (Claire Elise Boucher), 23, 95, 100, 124, 204–6, 377
Grimes, Michael, 155, 175–76, 180, 184, 185, 266, 401–2
Groypers, 392

hackers and hacking, 67, 68, 223–25, 368
Haile, Tony, 242–43, 298
Hamas attacks, 424, 426
Hansbury, Mary, 277
Harvey, Del, 20, 66, 70, 71, 73–75, 87, 146, 373–74
Harwell, Drew, 379
hate speech, 1, 42, 93, 147, 183, 417, 418, 432, 436
    antisemitism, 392, 406, 418, 419, 424
    Musk and, 424–25, 427–29
Hawaii, 57–58, 97, 113, 117, 124, 128, 135, 173
Hawkins, Tracy, 277, 361
Hayes, Julianna, 236, 254
Heard, Amber, 100
Hemingway, Ernest, 212
Hershey's, 290
Herzog, Isaac, 426
Hewlett Packard, 156
Hitler, Adolf, 392, 425

Hobbs, Katie, 338–39
Hoffman, Reid, 176–77, 232
Hollander, Nicole, 358, 364, 412
Holocaust, 406
Homsany, Ramsey, 171
Horizon Media, 289, 290
Hughes, Tim, 374
Hurricane Harvey, 41

IBM, 425
IBP, Inc., 210
Iger, Bob, 212, 427–29
immigrants, 255, 297, 393, 404, 432
    Great Replacement Theory and, 424
India, 22, 208, 209, 244, 245, 400–401
Indonesia, 164
Information, 402
Infowars, 60, 425, 435
Instagram, 88, 90, 240, 287, 311
    Threads, 413–14, 433
    Twitter accounts promoting, 382–83
internet, 7, 12, 29–31
    dot-com bubble, 30
    satellite launch and, 35–36
    Starlink service, 102, 103, 244–45, 280, 332, 363, 374, 400
    see also social media
Internet Archive, 89
Internet Research Agency, 332
Interpublic Group (IPG), 294
iPhone, 16, 314, 365
Iran, 15–16
Irving, Kyrie, 334
Irwin, Ella, 307, 341, 356, 359–60, 372, 378, 409
Isaacson, Walter, 101, 240–41, 313, 328, 404
Islamic State (IS), 392
Israel
    Musk's trip to, 426–28
    October 7 Hamas attacks in, 424, 426
iTunes, 14

James, LeBron, 403
January 6 Capitol riots, 1, 70–74, 170, 390, 392, 405
January 6th Committee, 245
Jay-Z, 28, 58
Jewish people
    antisemitism and, 392, 406, 418, 419, 424
    Musk and, 424–25, 427–29
    Great Replacement Theory and, 424
Jobs, Steve, 11, 38, 54, 240–41, 310, 427
Johnsen, Bret, 352
Jones, Alex, 60, 425, 435
Jones, Evan, 316
Jordan, Jim, 390–91, 393
JPMorgan, 141, 149, 152, 154, 158, 162
Justice Department, 139, 173

Kaiden, Robert, 236, 247, 253, 286, 291, 293
Kalanick, Travis, 384, 420
Kansas City Chiefs, 396–97

Kardashian, Kim, 70
Kennedy, Robert F., Jr., 433
K5 Global, 136
Khan, Lina, 391
Khashoggi, Jamal, 234
Kieran, Damien, 226, 256, 257, 274, 329–30, 337, 340, 342, 393
King, Gayle, 115, 133–34
King, Martin Luther, Jr., 45
Kingdom Holding, 233–34
Kissner, Lea, 226, 256, 257, 329–30, 337, 340, 342, 393
Kives, Michael, 136, 175
Klein, Alan, 161, 162, 198, 203–4, 207, 209, 211
Klum, Heidi, 292
Koenigsberg, Bill, 290, 291
Kohm, Jim, 394
Kordestani, Omid, 49, 51–52, 55, 56
Korman, Marty, 156, 160–62, 198, 201–2, 203–4, 207, 209, 211, 254
Kraft, Robert, 28
Krishnan, Sriram, 241, 284–85, 312, 341
Ku Klux Klan, 60
Kushner, Jared, 381

Lady Gaga, 117, 290
Lake, Kari, 338–39
Lane Fox, Martha, 80, 106–7, 110, 111, 126, 141, 149, 158, 159, 163, 248
La Russa, Tony, 311
Levchin, Max, 30
Levie, Aaron, 382
Levine, Rachel, 104
LGBTQ people, 103, 146
  transgender, 100, 103, 104, 110, 114, 197, 199, 275, 354, 435
  "What Is a Woman?" and, 408–9
LibsofTikTok, 146
Lindell, Mike, 361
LinkedIn, 156, 176, 298
LiveJournal, 13
Lonsdale, Joe, 115
Lula da Silva, Luiz Inácio, 317, 390
Lutz, Bob, 38

Mac, Ryan, 6, 25n
Magneto, 406
Maheu, Jean-Philippe, 281, 282, 289–92, 294–95, 298, 301
Malone, Post, 296
Mann, Kyle, 115
Mars, 26, 31, 47, 100, 141, 213, 423
Mashable, 379
Mastercard, 301
Mastodon, 320, 378, 382–83, 433
Mattel, 294
Mbappé, Kylian, 380
McCarthy, Kevin, 321, 390, 391
McCormick, Kathaleen, 217–18, 227, 228, 232, 434
McDonald's, 311

McSweeney, Sinéad, 278, 279, 293–94, 329–30, 337, 342, 354
Media Matters for America, 294, 425–26
Mendoza, Pablo, 381
Merrill, Marc, 134
Messi, Lionel, 380
Meta, 158, 388, 413
Michael, Emil, 384
Micheletti, Ed, 216–17
Microsoft, 54, 158, 177, 301
Miller, Stephen, 356
misinformation and disinformation, 1, 193, 276, 288, 307–8, 317–19, 436
  conspiracy theories, 2, 4, 25, 44, 60, 61, 93, 287, 406, 432
  Pizzagate, 25, 104, 390, 426
  QAnon, 25, 391, 392, 433
  about September 11 attacks, 433
  about COVID, 61–64, 68–69, 91, 110, 114, 222, 391
  about elections, 70, 257, 275, 276, 283, 307, 317–19, 338–39, 390, 405
  see also content moderation
Mittal, Lakshmi, 381
Modi, Narendra, 208, 400–401
Mohammed bin Salman, 234
Montano, Mike, 84, 94
Morgan Stanley, 102, 106, 121, 125, 137, 143–45, 152, 154, 155, 176–80, 232, 260, 263, 266, 284, 401, 402
  Tech, Media & Telecom conference of, 401–2
Mudge (Peiter Zatko), 222, 223–27, 228, 343
Murdoch, James, 177, 180
Murdoch, Kathryn, 177
Murdoch, Lachlan, 405
Murdoch, Rupert, 177, 396
Musk, Andrew, 250, 267, 272, 350, 386
Musk, Elon
  Agrawal and, 85–86, 106–9, 111–12, 114, 116–21, 123–27, 135, 157, 163, 186, 191–92, 193, 212, 215, 251, 263–66, 276, 295, 343–44
  Asperger's of, 341
  in Berlin, 109
  bodyguards of, 7
  businesses of
    Boring Company, 23, 37, 134, 179, 239, 241, 267, 268, 314
    Neuralink, 23, 38, 179, 205, 250, 434
    SpaceX, *see* SpaceX
    Tesla, *see* Tesla Motors
    X.com (bank), 29–31, 415
    Zip2, 29, 207
    see also Twitter under Elon Musk; X
  Chappelle show incident, 375
  children of, 95, 100, 124, 197, 204–5, 213, 256, 279, 289, 290, 364, 376–78, 427
  COVID pandemic and, 62–64, 99, 114, 128, 136, 375
  data scientist and, 1–5, 344, 346
  at DealBook Conference, 426–30
  dog of, 144

INDEX > 461

Dorsey and, 50, 53, 78-79, 105-6, 110, 112, 113, 123, 142, 143, 163
drug use of, 39, 99
Durban and, 194-95
education of, 29
ElonJet account and, 79, 378
"episode" of, 383-84, 388, 389
flight attendant's allegations against, 200
free speech as commitment of, 100, 101, 103, 201, 234, 244-45, 250, 275, 276, 282, 294, 320, 354, 363, 372, 378, 379, 382, 392, 400-401, 409, 418, 435-36
Gadde and, 169-72, 191, 193, 243-46, 257-58, 263-66, 343-44
Halloween gala attended by, 292
Israel trip of, 426-28
leadership style of, 115, 279, 323-24
loyal supporters of, 3, 5, 37-38, 128, 134-36, 279, 291, 310, 348, 358, 359, 415
media as viewed by, 32, 37, 102, 198
neck injury of, 414, 416
paranoia of, 3, 7, 32, 34, 256, 257, 286, 287, 297, 344, 412
PayPal and, 30-31
political views of, 99-103, 199, 279, 321, 327, 375, 389-91, 406, 423-29, 432
racial and social justice issues as viewed by, 99-100
relationships of, 23, 30, 32, 34, 95, 100, 103-4, 117, 124, 194, 204-6
Roth tweet of, 373-74, 376
stalker and, 377-78
at TED conference, 124, 127-28, 136-38, 140, 204
Twitter acquisition of, *see* Twitter, Musk's acquisition of
on Twitter board, 50, 105, 107-12, 113-15, 117-21, 136
Twitter competitor threat of, 105, 108, 110, 195, 201, 215
Twitter employees' view of, 114, 116, 128-29, 165
Twitter OneTeam video call of, 47-48, 198
Twitter parody account and, 32-33, 311
Twitter shares of, 102, 103, 105-8, 110-11, 115, 125-26, 129, 136, 138, 175, 217
Twitter under ownership of, *see* Twitter under Elon Musk
Ukraine war and, 102-3, 184, 332
Unsworth accused of pedophilia by, 23-29, 47, 114, 135-36
wealth of, 6, 34, 39, 96, 101, 126, 128, 129, 136, 138-40, 153, 189, 252-53, 376, 394, 425, 434
wokeism and, 99, 100, 103, 362, 375, 406, 435
work ethic of, 1, 29
Zuckerberg and, 35-36, 262
    cage match challenge, 414, 416-17
Musk, Errol, 29
Musk, Exa Dark Sideræl, 205
Musk, James, 250, 267, 272, 347-50, 354, 386, 397
Musk, Justine, 30, 32, 205
Musk, Kimbal, 29, 101, 119, 177
Musk, Maye, 279, 290-92
Musk, Techno Mechanicus, 204
Musk, X Æ A-12, 124, 205, 206, 241, 243, 256, 279, 289, 290, 364, 376-78, 427
Myanmar, 22

Nadella, Satya, 177
Napa Valley, 301
Napster, 177
NASA, 32, 40-41
National Public Radio, 404
Nazism, 184, 282, 355, 392, 425
NBA, 171-72, 380, 403
NBCUniversal, 301, 407, 408, 420
NeighborNest, 255-56
Netanyahu, Benjamin, 426
Netflix, 158, 357
Neuralink, 23, 37, 179, 205, 250, 434
New York Post, 66-68, 170, 171, 204, 222, 367-68
New York Stock Exchange, 17, 102, 164, 263
New York Times, 6, 31, 32, 39, 69, 94, 95, 378, 393, 427
    DealBook Conference of, 426-30
    Roth's op-ed in, 372-73
Nigeria, 75
Night Parrot, 296, 297
Nike, 338
9/11 attacks, 433
Nintendo, 334, 335
Nippon Telegraph and Telephone (NTT), 385
Niwa, Yoshimasa, 333
Nordeen, Ross, 250
Nosek, Luke, 30
Nostr, 382, 433
Notopoulos, Katie, 379
NPR, 404

Obama, Barack, 99, 117, 285
Ocasio-Cortez, Alexandria, 317
October 7 Hamas attacks, 424, 426
Odeo, 13-15
Office Space, 389
Olympics, 380
O'Malley, Pat, 180-81
O'Neal, Shaquille, 280
OnlyFans, 297, 365
OpenAI, 205, 412, 435
Owens, Candace, 44

Pacini, Kathleen, 242, 264-65, 277, 284, 286, 295, 299, 304, 340
Palace Hotel, 401
Pandjaitan, Luhut Binsar, 164
Paris Agreement, 431
Parker, Sean, 177
PayPal, 30-31, 176, 177
Pelosi, Nancy and Paul, 2, 4, 287, 289, 344
Pence, Mike, 70, 71

Penn, Sean, 58
Perelman, Ronald, 155–56
Periscope, 21
Personette, Sarah, 249–50, 271, 289, 301
Perverted Justice Foundation, 20
Peterson, Jordan, 354
Philadelphia Eagles, 396–97
Pichai, Sundar, 52, 59, 280
Pichette, Patrick, 51–52, 54–56, 126, 141, 149, 153, 158, 159, 224, 248
Pixar, 156
Pizzagate, 25, 104, 390, 426
poison pill strategy, 127, 209
　of Twitter, 127, 129, 138, 141, 146, 148, 209, 218
Politico, 170
Pool, Tim, 43
population decline, 205, 213, 214, 389
pornography, 61, 297, 365–66
Posobiec, Jack, 390, 392
Post, 382
Powell Jobs, Laurene, 54
Pravda, 37
presidential elections
　of 2016, 21, 22, 50, 61, 67, 193
　of 2020, 52, 64–65, 67, 68, 70–73, 133, 390, 405, 431
　of 2024, 355, 431, 433
Pride Month, 408
Princess Bride, The, 406
Prism project, 97–98, 101, 142–43, 148–51, 153, 167, 214, 277
Public Investment Fund, 38
Publicis Groupe, 289
Putin, Vladimir, 102, 184

QAnon, 25, 391, 392, 433
Qatar, 380–81
Qatar Investment Authority, 179, 254, 381
Quinn, Zoë, 18
Quinn Emanuel, 28, 343, 363

racial issues, 19, 170, 432
　Black Lives Matter, 19, 21, 362
　diversity and inclusion, 99–100, 197
　Ferguson protests, 18–19, 183, 362
　George Floyd protests, 99, 325, 362
　hate speech on Twitter, 302, 392
　Musk and, 99–100
　Tesla and, 100, 114, 197
　Twitter employees and, 362
Raytheon, 139
Redstone, Shari, 212
Republicans, 43–44, 65, 68, 199, 287, 321, 325, 327, 339, 390–94, 431
Ressi, Adeo, 109, 134
Revlon, 156
Revolt, 302
Rezaei, Behnam, 3
Rice, Kathleen, 59–60
Riley, Talulah, 34, 103–4, 205

Ringler, Mike, 156–57, 160–62, 195, 198, 201, 203, 208–9, 211, 215, 226
Rive, Lyndon, 124
Rive, Peter, 124
Roche, 160
Rock, Chris, 375
Rogan, Joe, 39, 43, 115, 170, 177
Rolling Stone, 367
Ronaldo, Cristiano, 117
Rosenblatt, David, 81, 151, 158
Rossman, Andrew, 216
Roth, Benjamin, 210–11
Roth, Yoel, 61–62, 64–66, 70–74, 146–48, 172–74, 206, 229, 242, 257, 268, 274–76, 301, 302, 356, 368, 391, 393
　graduate school thesis of, 373
　Musk's tweet about, 373–74, 376
　New York Times op-ed of, 372–73
　resignation of, 340–42, 354, 372
　Yaccarino and, 420–21
Rubin, Dave, 390
Rubin, Rick, 78
Russia, 21, 61, 67, 264, 332
　Ukraine invaded by, 97, 102–3, 113, 152, 184, 325, 332
Russia Today (RT), 103, 404

Sacks, David, 30, 31, 177, 241, 243, 260, 261, 285, 287, 298, 300, 312, 313, 362, 378
Sacramento (SMF) data center, 385–88, 394
Salen, Kristina, 180–81
Salesforce, 108, 139
Salt Bae, 381
Samuels, Detavio, 302
Sanders, Bernie, 120, 321
Sandy Hook school shooting, 60
Santa Monica Observer, 287–88
Saturday Night Live, 119, 341
Saturn project, 93, 146–48, 172–75, 206, 208, 220–22, 229, 245, 276
Saudi Arabia, 38, 106, 233–34
Savitt, Bill, 210–11, 215–17, 222, 227, 232, 247–48, 258
Schiller, Phil, 366
Schmidt, Eric, 335
Securities and Exchange Commission (SEC), 38, 39, 49, 102, 110–11, 115, 122, 123, 125, 126, 136, 152, 179, 202, 225, 231, 326
Segal, Ned, 76, 80, 83, 94–96, 137, 143, 148–53, 164, 172, 180–81, 184, 194, 203, 204, 207, 208, 210–14, 235, 237, 247, 251–54, 257–59, 260, 261, 263, 265, 286, 344, 435
　firing of, 263–66, 343–44
September 11 attacks, 433
Sequoia Capital, 179
Sethi, Rinki, 223–24
Shapiro, Ben, 408
Shareworks, 284, 286
Shatner, William, 403
Shotwell, Gwynne, 201

INDEX

Silicon Valley, 8, 12, 41, 54, 55, 67, 95, 108, 128, 138, 156, 329, 361, 366, 382, 384, 435
Silver Lake, 54–56, 91, 95, 106, 138–39, 143
Simon, Luke, 345
Simpson, O. J., 334
Simpson Thacher & Bartlett, 141, 161
Singapore, 364
Singer, Paul, 49
60 Minutes, 13
Skadden, Arps, Slate, Meagher & Flom, 155–56, 162, 190, 201, 216, 254
Skype, 54
Slack, 80, 114, 128, 165, 171, 194, 200, 231, 272–73, 293, 298, 299, 303–6, 307, 308, 310, 319, 336, 345, 346, 348, 349, 351, 367, 368, 389, 398
Smith, Kiko, 349, 387
Snapchat, 42, 50, 158, 199, 240
Snowden, Edward, 382
social media, 1, 3, 57, 119, 151, 175, 413, 433–34, 436
    activist groups and, 294, 392
    Bluesky, 87–88, 90–91, 106, 174, 220–21, 413, 433, 435
    content moderation on, *see* content moderation
    decentralized, 87, 89–90, 221
    Facebook, *see* Facebook
    global regulations for, 323
    Mastodon, 320, 378, 382–83, 433
    Nostr, 382, 433
    Threads, 413–14, 433
    Truth Social, 356, 382–83, 432
    verification on, 311
SoftBank, 177
SolarCity, 124, 211
Solomon, Sasha, 347, 348
Sorkin, Andrew Ross, 427–30
Soros, George, 406, 408
South by Southwest, 15
Soviet Union, 37, 184, 193, 364
SpaceX, 1, 23, 24, 31–36, 47, 96, 101, 115, 137, 165, 167, 175, 181, 208, 241, 250, 256, 290, 291, 299, 316, 317, 320, 322–24, 331, 336, 341, 344, 348, 350, 352, 369, 374, 377, 379, 382, 384, 393, 411, 428, 434, 436
    acquisitions of, 124
    Amos-6 satellite of, 35–36
    board of, 115
    COVID pandemic and, 63
    Musk's loan from, 253
    sexual harassment at, 200–201
    Starbase facility of, 78–79
    Starlink service of, 102, 103, 244–45, 280, 332, 363, 374, 400
    Starship test launch of, 423–25
    tunnel project of, 314
    Twitter advertising of, 340
Spears, Britney, 290
special purpose vehicles (SPVs), 178
Spiegel, Evan, 199

Spiro, Alex, 27–28, 155, 156, 182, 191, 216, 219, 225, 226, 228–30, 247, 271, 277–79, 283, 293, 298–99, 321–23, 337, 341, 342, 343, 348, 359, 363, 367, 369, 374, 414
sports, 380–81
    NBA, 171–72, 380, 403
    Super Bowl, 85, 380, 396–97, 407
    World Cup, 179, 350, 380–82
Squad, 240
Square, 16, 22, 40, 42, 44, 49, 51, 52, 55, 233
Srinivasan, Balaji, 382
SR-71 Blackbird, 364
Stanley, Christopher, 344–45, 354
Starbucks, 314
Starlink, 102, 103, 244–45, 280, 332, 363, 374, 400
Staudinger, Sarah, 194
Steinberg, Marc, 52
Stewart, Patrick, 17
Stone, Biz, 15, 16, 46
Stone, Roger, 287
Stop the Steal, 392
Strine, Leo, 210, 211, 216
Substack, 367, 433
Sullivan, Jay, 92–93, 146, 147, 172–75, 186–88, 213–14, 221, 229, 267, 271
    resignation of, 271, 272
Sun Valley Lodge, 211–13
Super Bowl, 85, 380, 396–97, 407
Swift, Taylor, 118
Swisher, Kara, 420–21

Taibbi, Matt, 367–68
Tang, Yang, 395
Tarpenning, Marc, 31
Tate, Andrew, 355
Taylor, Bret, 82, 106–9, 111, 119–21, 123–26, 138, 141, 142, 144, 149, 153, 157–59, 209, 211–13, 247, 248, 435
Tech, Media & Telecom conference, 401–2
TED Conference, 43, 124, 127–28, 136–38, 140, 204
Teigen, Chrissy, 46
Telegram, 350
Teller, Sam, 293, 298–300
tender offers, 122
    by Musk, 122–23, 128, 140–41, 144, 147–49, 153, 158, 161
terrorism, 392
    September 11 attacks, 433
Tesla Motors, 23, 24, 33–37, 96, 100, 101, 103, 107, 120, 135, 137, 165, 175, 181, 205, 207–8, 241, 244, 248–49, 256, 268, 290, 291, 294, 300, 313, 316, 317, 320, 322–24, 331, 334, 336, 341, 350–52, 366, 369, 370, 376, 377, 379, 382, 383, 384, 388, 402, 411, 420, 428, 430, 434, 436
    acquisitions of, 124
    Afshar at, 115, 134
    autopilot team at, 272
    board of, 31, 38, 39, 140
    COVID pandemic and, 63, 96, 99

Tesla Motors (*cont.*)
　Cyber Rodeo gala of, 115–16
　Cybertruck, 116
　Gigafactory of, 116, 134, 164
　initial public offering of, 32
　investors in, 376
　Model S, 32, 34, 135, 249
　Model X, 85–86
　Musk as CEO of, 32
　Musk as chairman of, 31, 39
　Musk's political views and, 390, 406
　Musk's shares in, 6, 31, 96–97, 101, 140, 152–53, 188–90, 252, 326, 338, 357–58
　Musk's Twitter acquisition and, 189
　privatization of, 38–39, 106, 126, 127, 144, 234
　racial discrimination at, 100, 114, 197
　recession and, 325
　Roadster, 31–32, 207
　SolarCity acquired by, 124, 211
　stock price of, 38, 96–97, 101, 140, 175, 188–91, 216, 358, 376, 383, 434
　Texas factory of, 115–16
Thailand cave rescue, 23–25
Thiel, Peter, 30, 31, 177
Thoma Bravo, 139, 149–50
Thorn, 393
Thorne, Bella, 335
Threads, 413–14, 433
TikTok, 42, 88, 240
Time Warner, 139
To Catch a Predator, 20
Toys "R" Us, 216
transgender people, 100, 103, 104, 110, 114, 197, 199, 275, 354, 435
　"What Is a Woman?" and, 408–9
Tribe, 382
Trump, Donald, 25, 38, 43, 68, 99, 119, 133, 139, 193, 238, 338–39, 381, 391, 407, 431–32
　arrest of, 431
　impeachment of, 43
　January 6 Capitol riot and, 70–74, 170
　in presidential elections
　　of 2016, 21
　　of 2020, 64–65, 67, 68, 70–73, 390, 431
　　of 2024, 355, 431
　Truth Social and, 356, 432
　on Twitter, 21, 42–43, 64–66, 71–74, 148, 245
　　banning, 70–75, 148, 170, 190, 222, 245, 432
　　reinstatement, 290–91, 294, 355–56, 431–32
Trump, Donald, Jr., 287
Truth Social, 356, 382–83, 432
Tucker, Michael (BloodPop), 290, 291
Tundra project, 299
Turkey, 400
21st Century Fox, 139
Twitter
　advertising on, 2, 17, 50, 76, 79, 87, 88, 91, 93, 97, 137, 153, 185, 187, 192, 199, 217
　　Musk's takeover and, 249–51

　political, 321
　revenue from, 217, 281, 291, 292
Agent Tools system of, 268, 386
Agrawal's appointment as CEO of, 81–84, 91–92, 94, 95, 106–8, 113, 137, 163, 167, 168, 185, 193, 212
Analyst Day event of, 76–77
Birdhouse directory of, 303
Bluesky project at, 87–88, 90–91, 106, 174, 220–21, 433, 435
Blue subscription service of, 2, 3, 79, 137, 217, 242–43, 285, 295
board of, 56, 80, 102, 144, 196
　Dorsey and, 15–17, 49–51, 53, 73, 77, 78, 81–82, 105, 112, 142, 164, 194, 433
　Musk on, 50, 105, 107–12, 113–15, 117–21, 136
　Musk's acquisition negotiations with, *see* Twitter, Musk's acquisition of
bots and spam on, 37, 106, 117, 119, 120, 135, 138, 267, 285–86
budget and expenditures of, 325, 344, 370–71
celebrity accounts on, 32, 38, 117–18, 135, 285, 333, 402, 403
character limit on, 14, 33, 42
Chase Center and, 171–72, 344
Cohn and Elliott Management and, 49–58, 77, 78, 83, 90, 91, 94–95, 106, 110, 111, 120, 122, 123, 139, 150, 186
content moderation at, 7, 15, 16, 20, 43, 64–75, 79, 87, 91–93, 111, 138, 146–48, 169–75, 208, 222, 274, 277, 393
　Babylon Bee and, 104, 115, 245, 274–76
　Birdwatch system for, 173–74
　child sexual exploitation material and, 16, 20, 146, 372
　Congress and, 57, 59–60, 62
　Digital Services Act and, 244
　Dorsey and, 15, 43, 57, 59–60, 62, 64–75, 91–92
　free speech issues and accusations of censorship, 16, 65–68, 101, 107
　as international issue, 244–45
　LibsofTikTok and, 146
　of misinformation and disinformation, 21, 61–64, 68–69, 88, 89, 91, 110, 114, 147, 183, 193, 222, 257, 264
　Musk's acquisition and, 104, 115, 125–27, 133, 134, 148, 164, 169–72, 206, 222, 249–50
　New York Post and, 67–68, 170, 171, 222, 367–68
　as politically biased, 5, 43–44, 65–66, 93, 102, 104, 115, 127, 128, 133, 367, 367–69, 391
　Saturn project for, 93, 146–48, 172–75, 206, 208, 220–22, 229, 245, 276
　Trump and, 70–75, 148, 170, 190, 222, 245
　tweet labeling, 62, 64–66, 68, 70, 71, 400–401
　and Twitter's role as public square, 43, 52, 67, 75, 88, 104, 127, 197, 410, 436

INDEX

cost cutting at, 186
Project Prism, 97–98, 101, 142–43, 148–51, 153, 167, 214, 277
Costolo as CEO of, 11, 17, 20–21, 42, 149
COVID-19 pandemic and, 53, 56–57, 61–64, 68–69, 77, 83, 90, 91, 97, 98, 185, 186, 255
creation of, 13–14, 16, 194
data centers of, 87, 349, 385–88, 394
Dorsey as CEO of, 5, 11, 12, 14, 15, 17, 21, 40–42, 49–58, 60, 77–78, 81, 84, 129, 138, 140, 142, 149, 165, 166, 177, 194, 308, 432, 433
  Africa trip and, 44, 45
  office tour and, 44
  resignation, 79, 80–83, 91, 105, 106, 108, 163–64, 193
ElonJet account on, 79, 378
employees of, 272
  annual review and bonus process for, 142
  hiring of, 77, 95, 97, 186
  layoffs of, 21, 97–98, 142–43, 150, 153, 167, 220, 241, 242, 277
  Musk as viewed by, 114, 116, 165
  Musk's acquisition as viewed by, 128–29, 165, 218, 231, 235–37, 239–41, 256
  parental leave policies for, 95–97, 113, 186, 187, 308
  quitting of, 174
  remote work policies for, 53, 57, 91, 97, 108, 185, 197, 199, 336
firehose of, 173, 182–83, 190, 195, 198, 201–2, 209, 215
growth of, 15–16, 42, 151, 186, 244, 308
goals for, 76–77, 94, 95, 142, 147, 186
Halloween celebrations at, 250, 254–56, 260, 264–66, 272
impersonation and parody accounts on, 32–33, 311
initial public offering of, 17
international presence of, 244–45, 278
investors and, 5, 17, 76, 94, 137–38, 142, 149, 157, 185, 186, 195
journalist accounts on, 285, 333, 378, 379, 401–3, 436
Lodge of, 304
log-in page of, 266–67, 271
Market of, 304, 305
meals provided by, 360–61
mismanagement of, 5, 12, 149
Musk's acquisition of, 1–8, 168–71, 174, 175, 179–80, 184–88, 191, 192, 289, 375, 378–79, 401, 402
  advertisers' concerns about, 249–51
  Agrawal and, 128, 129, 138, 141–42, 146–52, 158, 163, 164, 167, 169, 171, 174, 180, 185–88, 195–96, 203, 204, 207, 209, 211–12, 221, 222, 225, 229, 234–35, 239, 250–51, 258–59
  all-hands meeting and, 195–201
  announcement of, 164–67, 176

bot and spam issues and, 161, 164, 181–83, 184–85, 190–92, 193, 199, 201, 203, 204, 207, 213, 215–17, 220, 225, 226, 230, 232
breakup fee and lawsuit clause in, 162, 217
change-of-control terms and exit packages in, 235, 259, 263, 264, 271
completion of, 236–37, 239, 247, 250–54, 256–59, 262–63, 266, 395
content moderation and, 104, 115, 125–27, 133, 134, 148, 164, 169–72, 206, 222, 249–50
Dorsey and, 123, 142, 143, 163–67, 168–69, 232–33, 308, 432–33
due diligence in, 160, 180, 184–85, 190, 216, 227, 292
employees' feelings about, 128–29, 165, 218, 231, 235–37, 239–41, 256
employee stock bonuses and, 236
financial data and, 203–4, 208–9
firehose data and, 182–83, 190, 195, 198, 201–2, 209, 215
funds transfer for, 253–54, 258, 262, 266
idea for, 104–5, 113, 121
Musk's agreement and demand for speed in, 159–62, 168, 202
Musk's attempt to terminate, and Twitter's lawsuit, 208–14, 215–18, 219–22, 226–27, 228–30, 231–32, 242, 243, 253, 343, 363, 364
Musk's financing of, 175–79, 184, 189–90, 207, 217, 229, 232–34, 252–54, 280, 292, 417
Musk's sink-carrying entrance following, 238–39, 363, 434
Musk's stalling of, 185–86, 190–91, 195, 201–2, 203, 206–7
nondisclosure agreement in, 160–61, 180, 185, 225
offers in, 122–29, 133–45, 146–54, 155–67, 203, 204, 227, 228–30, 231–32, 402, 405, 436
outside vendor payment freeze in, 247
price per share in, 7, 126, 128, 140, 149, 152, 153, 157–60, 164, 195, 227, 231, 233, 234, 236, 248, 382
privatization in, 123, 125, 128, 164, 216, 236
resignation of directors following, 247
Turkey and, 400
Twitter operations following, see Twitter under Elon Musk
Twitter's "just say yes" defense in, 155–67, 253–54
Twitter's poison pill strategy in, 127, 129, 138, 141, 146, 148, 209, 218
Wachtell hired in, 209–11
Wachtell's success fee in, 247–48, 258
welcome box for Musk in, 256
Zatko's allegations and, 226–27, 228
Musk's ownership of, see Twitter under Elon Musk
Musk's shares in, 102, 103, 105–8, 110–11, 115, 125–26, 129, 136, 138, 175, 217
Musk's threat to build competitor to, 105, 108, 110, 195, 201, 215

Twitter (*cont.*)
  name of, 14, 415
  NeighborNest and, 255–56
  OneTeam events of, 41–42, 44–48, 53, 55, 86, 89, 97, 198, 344
  outages of, 15, 50, 85
  Periscope acquired by, 21
  political action committee of, 321
  Prism project at, 97–98, 101, 142–43, 148–51, 153, 167, 214, 277
  privacy program and security issues of, 194, 223–26, 244, 257, 274, 296, 336, 337, 343, 367
  Resource Plan for, 77
  revenue of, 5, 17, 21, 42, 57, 76, 77, 88, 137, 142, 143, 150, 152, 153, 181, 187
    from advertising, 217, 281, 291, 292
    from subscriptions, 3, 137, 217, 285, 286
  share buyback program of, 95
  Silver Lake and, 54–56, 91, 95, 106, 138–39, 143
  Slack channels of, 80, 114, 128, 165, 171, 194, 200, 231
  Spaces feature of, 274
  stock price of, 42, 44, 50, 52, 54, 57, 78, 81, 96, 113, 122, 125–27, 142, 149, 157, 162, 163, 191, 195, 227, 231, 234
  succession plan at, 77–78, 81
  Trump's use of, 21, 42–43, 64–66, 71–74, 148, 245
    banning, 70–75, 148, 170, 190, 222, 245, 432
    reinstatement, 290–91, 294, 355–56, 431–32
  user numbers of, 12, 15, 17, 20–21, 42, 76, 94, 137, 142, 152, 161, 173, 181, 184–85, 190, 192, 217, 226
  user verification systems of, 135, 217, 285–86, 297, 311
  violent imagery on, 410–11
  Williams as CEO of, 15–17
  Williams as chairman of, 14
Twitter under Elon Musk
  and accounts promoting competitors, 382–83
  advertising on, 271, 281, 291–92, 297, 301, 313, 320, 345, 350, 364, 406–8
    for Apple, 365, 366, 425
    content moderation and, 275, 282, 288, 289–91, 302, 320, 338, 339, 357, 365, 366, 418, 419
    imposter accounts and, 338, 339
    meetings with advertisers, 289–92, 294, 301–2
    Musk's political views and, 406, 423–29
    revenue from, 294, 320, 325, 330, 339, 340, 354, 357, 365, 380, 381, 388, 401, 403, 409, 417, 419
    for SpaceX, 340
    Yaccarino and, 417–18
  Apple App store and, 317–19, 333, 365–66
  Blue subscription service of, 285, 295, 297, 306, 310–19, 323, 328–35, 338–41, 344, 363–66, 388, 402, 403, 432
  bots and, 285–86, 402
  Caracara conference room of, 2, 328, 330, 335, 363, 370, 376, 388, 415
  code freeze at, 7, 256–57, 345–46
  code reviews at, 272–73
  conservative voices elevated on, 391–93
  content moderation at, 3–4, 275–76, 334, 341, 342, 354
    advertisers and, 275, 282, 288, 289–91, 302, 320, 338, 339, 357, 365, 366, 418, 419, 425–29
    child sexual exploitation material and, 372, 392–93
    elections and, 257, 275, 276, 283, 307, 317, 390
    Europe and, 400, 417
    hate speech and, 1, 417, 418, 424, 436
    racial slurs and, 302, 392
    reduction of teams for, 307–8, 317–18, 331, 341
    stabbing video and, 392
    state-affiliated media label and, 404
    Yaccarino and, 409, 417–19
  debt of, 247, 324, 338, 357–58, 401, 417, 434
  dilution of power of, 433–34
  direct messaging features of, 296–97
  Dorsey's absence from, 433
  employees at, 272, 326, 435
    badge swipes of, 398
    benefits for, 235, 297, 326, 389
    Black, 362
    dissent firings of, 346–50, 354, 398
    engineers, 272–74, 277, 279, 283, 296–97, 307, 339, 345–49, 351, 352, 365, 384, 389, 395–98, 403, 410, 412
    in Europe, 294
    evaluations of, 260, 283, 398
    former, post-Twitter, 434–35
    "ghost," Musk's concern about, 286, 291, 293
    "hardcore" requirements for, 351–53, 360
    laid-off, return of, 345
    layoffs of, 2, 7, 241, 242, 267–68, 271–72, 277–79, 281–82, 283–84, 286, 287, 293–300, 301, 303–6, 307–9, 310, 318, 319, 320–24, 329, 338, 341, 344, 345, 347, 353, 369, 380, 381, 389, 393, 398, 402, 405, 411
    Musk's "fork in the road" plan for, 350–54
    Musk's interactions with, 280, 337–38
    Musk's mistrust of, 7, 344–50, 353, 398
    stock programs for, 236, 263, 277, 284, 293, 402, 411, 434
    support networks created by, 298
    Yaccarino and, 410–12
  executives at, 283
    firing of, 263–68, 271, 273, 294–95
    golden parachutes of, 229, 235, 263, 264, 343
    investigation interviews with, 343–44
    lawsuit by, 435
    resignations of, 271–72, 336, 337, 340–41, 343, 344, 354
  expenditures of, 370–71, 385
    cuts made in, 324–25, 358–62, 385–86, 389, 393, 401, 402, 411

first day of, 238–43
FTC issues of, 244, 257, 274, 323, 329–30, 336, 337, 340, 342, 367, 374, 387, 391, 393–94, 414
imposter and parody accounts on, 311, 317, 318, 320, 330, 331, 334–35, 337–39, 343, 364
infrastructure organization of, 307
investors sought for, 381–83
log-in page and, 266–67, 271
meals at, 361
Musk as CEO, 265
    resignation, 384
    resignation poll, 383–84
    Yaccarino named successor to, 407–8
Musk's account on, 432, 434
    falling engagement with, 394–97
Musk's fears about data stealing and, 412–13
Musk's "goons" at, 250, 256, 260, 261, 266, 268, 274, 293, 295–96, 326, 341, 342, 344, 348, 362
Musk's leadership style and, 279, 323–24
Musk's pet projects for, 296–97, 306
offices of, 326–27, 358–59, 371
    auction of items from, 365
    cleanliness of, 361–62
    decorations in, 362–63
    nonpayment of rent for, 364
    turned into apartments, 359–60
outages of, 381, 387, 394
out-of-network tweets on, 397
payment processing feature of, 363, 401
paywalled video idea for, 297
product development at, 295–96, 302, 310, 315, 316, 319, 323, 411, 421
recommendations on, 365, 392, 393, 397, 402–4
reinstatement and returns of users to, 355, 391–92, 435–36
    Babylon Bee, 274–76, 340, 354
    Peterson, 354
    Trump, 290–91, 294, 355–56, 431–32
revenue of, 313, 357, 388, 400, 402
    from advertising, 294, 320, 325, 330, 339, 340, 354, 357, 365, 380, 381, 388, 401, 403, 409, 417, 419
    from subscriptions, 330, 402, 417
site slowness and, 346–47
Slack channels of, 272–73, 293, 298, 299, 303–6, 307, 308, 310, 319, 336, 345, 346, 348, 349, 351, 367, 368, 389, 398
Spaces feature of, 274, 379
Threads as challenge to, 413–14
transition team in, 236, 242, 250, 260–61, 265, 266, 273, 284, 299, 303, 315, 316
tweet view limits on, 413
Twitter Files and, 367–69
user numbers of, 302
users' sharing of information from, 412
user verification on, 285–86, 297, 306, 310–19, 323, 328–35, 338, 339, 402, 403, 417, 432, 434

valuation of, 233, 248, 402, 417, 434
as X, 8
"everything app" vision for, 8, 230, 262, 401, 415
rebranding, 415–16
Tyson Foods, 210

Uber, 329, 384
Ukraine, 67, 97, 102–3, 113, 152, 184, 325, 332
Ultimate Fighting Championship, 194, 414
underpopulation crisis, 205, 213, 214, 389
United Technologies, 139
University of Pennsylvania, 29, 396
Unsworth, Vernon, Musk's accusations against, 23–29, 47, 114, 135–36
Upfronts, 408

vaccines, 433
    COVID, 68–69, 391
Valor Equity Partners, 207–8
Vanguard Group, 102, 141, 149
VDARE, 424
venture funds, 177
Vy Capital, 179, 381

Wachtell, Lipton, Rosen & Katz, 209–11
    Savitt at, 210–11, 215–17, 222, 227, 232, 247–48
    Strine at, 210, 211, 216
    success fee paid to, 247–48, 258
Wall Street Journal, 234, 253
Washington Post, 20, 124, 225, 234, 241, 378, 379
Washington Square Park, 43
Watkins, Ron, 392
Watt, J. J., 46
Web3 Foundation, 179
WeChat, 198–99
Weiss, Bari, 368
West, Kanye, 33, 355
"What Is a Woman?," 408–9
WhatsApp, 139
Wheeler, Robin, 301–2, 340, 341
Whistleblower Aid, 225
White, Dana, 414
white nationalists, white supremacists, 42, 320–21, 392, 419, 424, 425, 435
Widom, Jennifer, 85
Wikipedia, 173–74
Williams, Ev, 13–16
Wilson, Christine, 394
Wilson, Fred, 133
Wilson, Vivian Jenna, 197, 377
Wilson Sonsini Goodrich & Rosati, 61, 141, 156, 160, 168, 230
Wired, 32
wokeism, 19, 21, 99, 100, 103, 362, 375, 406–8, 435
Wood, L. Lin, 26–28
World Cup, 179, 350, 380–82
World War Z, 330, 334
World Wrestling Entertainment, 180, 194
WPP, 289

X, 8
   "everything app" vision for, 8, 230, 262, 401, 415, 415
   Twitter rebranded as, 415–16
   see also Twitter under Elon Musk
xAI, 435
X.com (bank), 29–31, 177, 240, 415
X Holdings, 415
Xinhua News Agency, 404

Yaccarino, Linda, 301, 410–12, 417–22, 425–26
   appointed CEO, 407–8
   at CODE Conference, 420–22
   at DealBook Conference, 427, 428
   Roth and, 420–21

Y Combinator, 382
YouTube, 52, 60, 156
YubiKeys, 194
Yue, Yao, 349

Zatko, Peiter (Mudge), 222, 223–27, 228, 343
Zhao, Changpeng, 179
Zilis, Shivon, 205–6, 213
Zilis, Strider Dax, 205
Zilis, Valkyrie Alice (Azure), 205–6
Zip2, 29, 207
Zuckerberg, Mark, 59, 87, 140, 199, 413
   Musk and, 35–36, 262
      cage match challenge, 414, 416–17
   satellite launch and, 35–36